Paste Tailings Management

Erol Yilmaz • Mamadou Fall
Editors

Paste Tailings Management

Springer

Editors
Erol Yilmaz
Cayeli Bakir Isletmeleri A.S.
Rize, Turkey

Mamadou Fall
Department of Civil Engineering
University of Ottawa
Ottawa, ON, Canada

ISBN 978-3-319-39680-4 ISBN 978-3-319-39682-8 (eBook)
DOI 10.1007/978-3-319-39682-8

Library of Congress Control Number: 2017930397

Printed on acid-free paper

This Springer imprint is published by Springer Nature
The registered company is Springer International Publishing AG
The registered company address is: Gewerbestrasse 11, 6330 Cham, Switzerland

Contents

About the Editors

Erol Yilmaz is a Mining Engineer who graduated from the University of Istanbul, Turkey. He received his M.Sc. in mining engineering from Karadeniz Technical University in 2003 and Ph.D. in environmental sciences from the University of Quebec at Abitibi-Temiscamingue (UQAT) in 2010. He completed his Post-doctoral Research at UQAT and has since then been conducting research at First Quantum Minerals Ltd. He has more than 15 years of experience in industry, academia and research. He is the author or co-author of many scientific publications and presentations in the areas of mine waste rocks and tailings. His expertise covers paste integrated solutions, mine backfill implementation and sulphidic tailings management.

Dr. Mamadou Fall is a Full Professor in the Department of Civil Engineering at the University of Ottawa (Canada) and the Director of the Ottawa-Carleton Institute for Environmental Engineering. He obtained his Ph.D. in geotechnical engineering from the Freiberg University of Mining and Technology in Germany. He was the Coordinator of the German Research Chair of Environmental Geosciences and Geotechnics. At the University of Ottawa, Dr. Fall has been leading several projects on mine waste management and mine backfill. He has over 15 years' experience in fundamental and applied research on mining geotechnics, mine waste management and mine backfills. Dr. Fall has over 150 publications to his credit. He has been repeatedly invited as keynote speaker or lecturer, and regularly acts as a consultant as well as a reviewer for several scientific committees, peer review journals, and funding agencies. He also serves on the editorial board of international journals.

Chapter 1
Introduction to Paste Tailings Management

Erol Yilmaz and Mamadou Fall

1 Scope of the Book

This book focuses on efficient management of large volumes of sulfide-rich tailings created during the mining and milling operations. These tailings consist of coarse and fine fractions that are either reactive (acid or leachate generating) or nonreactive, and can be risky on the environment if they do not need to be managed appropriately at mine sites. On the whole, the coarse fraction of the tailings is used for underground backfilling while the remaining fines are usually deposited on the surface into tailings dams. However, it is witnessed that these structures are subjected to some risks of failure as a result of leakage, instability, liquefaction, erosion, and poor design. Therefore, the effective, economic, and environmental management of the tailings has become a major issue for all mining operations around the world. There is an apparent necessity for sulfidic tailings management methods that are technically suitable, economically viable, environmentally sustainable, and socially responsible. Paste technology is widely considered as an emerging technique to assess mining wastes in an environmentally sound way. Sulfidic tailings mixed with cements and water are often defined as paste backfill and used to refill mined stopes, resulting in a maximum extraction rate from an underground ore deposit. In addition, surface paste disposal promises an attractive technology and viable solutions for tailings treatment. The water coming from sulfidic tailings is recovered and can be used repeatedly while the requirement of constructing the big tailings dams at surface can be minimized using paste technology.

E. Yilmaz (✉)
Cayeli Bakir Isletmeleri A.S., P.O. Box 42, Madenli Beldesi, Cayeli, Rize TR 53200, Turkey
e-mail: yilmazer@fqml.com

M. Fall
Department of Civil Engineering, University of Ottawa, 161 Colonel By, Ottawa, ON, Canada, K1N 6N5

E. Yilmaz, M. Fall (eds.), *Paste Tailings Management*,
DOI 10.1007/978-3-319-39682-8_1

This book endeavors to collect an operational knowledge on management and treatment of sulfidic paste tailings generated at most modern underground mine sites. In reality, there are several types of wastes including sulfidic tailings, waste rocks, effluents, sludges, leachates, slags, smelter wastes, and chimney dusts. Most of these different wastes are generated from the other sources instead of producing at mine sites. Thus, this book essentially addresses on sulfidic tailings which are inert and generated after processing of pyrite and pyritic ores. This book deals only with paste-integrated solutions being used right now at modern mines including cemented paste backfill (CPB) and surface paste disposal (SPD). These new techniques have been explained in detail from the industrial application point of view. CPB is an engineered mix that consists of total tailings without removing the fines, or desliming (70–85 wt% solids), and single, binary, or ternary binders (3–7 wt%) in order to meet stability requirements and combined with mix water (20–25 wt%). CPB provides an efficient and environmentally nondestructive tailings management technique due to its technical, operational, and environmental advantages over other types of backfilling such as hydraulic and rock fills. The most attractive features of CPB are that it allows mine operators to use the total tailings without desliming for paste production; it significantly reduces surface tailings disposal by placing them safely in underground stopes, eliminating the damages due to environmentally harmful sulfidic tailings in the form of acidic mine water or leachates; and it reduces post-mining rehabilitation costs. SPD is a new technique used by the mining industry for safe disposal of sulfidic paste tailings on the surface, since it allows for a smaller tailings dam area, increased water management capabilities, and less environmental impact. SPD consists in the deposition of tailings at a solid concentration of 70 wt% in relatively thin layers that are allowed to drain and dry before the next layer is put in place. The term "paste" typically signifies that the tailings do not result in segregation of fine and coarse particles during the transport and exhibit minimal bleeding after the final placement of sulfidic paste tailings. When comparing with conventional tailings disposal, SPD could offer operational and environmental advantages, such as a better water management, no need for complex retaining tailings dams, a reduced footprint of the tailings disposal area, and the possibility to use progressive reclamation.

This book has been organized into 12 chapters which document the different aspects of paste tailings management, based on underground paste backfill and surface paste disposal applications. Chapter 2 presents practical importance of tailings characterization for cemented paste backfill. Chapter 3 addresses the rheological properties and natural dewatering aspects of paste tailings for managing successfully their improved solids concentrations. Chapter 4 describes the physical, mechanical, hydraulic, thermal, chemical, and microstructural aspects and characteristics of fresh or cemented paste tailings backfill materials. Chapter 5 provides issues related to the design and characterization of the paste backfill used for underground mines while Chap. 6 provides field properties and performance of paste tailings for surface disposal. Chapter 7 presents aspects of instrumentation of paste backfill or tailings disposal covering from preparation to application. Chapter 8 focuses on internal pressure monitoring and field testing of cemented paste backfill placed within an underground stope. Chapter 9 presents the overall risk assessment of underground mine backfill and surface paste disposal techniques. Chapter 10 presents a closer look at paste backfill system and operations of the Cayeli mine. Chapter 11 outlines Boulby mine backfill system while Chap. 12 presents case studies related to paste backfill plants.

- Chapter 1 focuses on the problem statement about tailings management and also gives important definitions of two tailings management techniques: cemented paste backfill for underground mines and surface paste tailings disposal.
- Chapter 2 presents the fundamental ingredients (tailings, binder, and water) required for the preparation of paste backfill mixtures and their practical importance on quality and performance assessment. This chapter also deals with the intrinsic and extrinsic factors which may greatly affect the strength and stability of cemented paste backfills.
- Chapter 3 provides the fundamental aspects of the rheology which are an essential tool in the overall management of both surface and underground paste tailings systems. It also gives the effects of major variables such as flow rate, shear history, particle size, shape, and concentration for the success of a paste tailings system.
- Chapter 4 outlines the physical, mechanical, hydraulic, thermal, chemical, and microstructural properties of cemented paste backfills which have a profound effect on their ultimate strength and durability. The chemical interactions and microstructural properties are closely related to the long-term backfill performance since the backfill with loose pore structure will not function properly its ground support role in comparison with the backfill with dense structure. All these aspects are described meticulously in this chapter.
- Chapter 5 describes the quality and behavior of cemented paste backfill considered for refilling underground mined stopes. This chapter deals with an overview of underground paste backfill, including paste ingredients, mix preparation, delivery, placement and curing conditions, barricade construction and monitoring, and quality control testing.
- Chapter 6 summarizes the fundamental aspects of surface paste disposal method. It also examines the behavior of paste material placed in field conditions in many ways and other geochemical (acidic mine water risk, heavy metal mobilization) and geotechnical (hydraulic conductivity, consolidation, crack propagation, and solidification behavior) properties in all kind of weather conditions.
- Chapter 7 outlines the instrumentation of underground backfill and surface disposal from controlling plant operational performance to the monitoring of backfill pipeline pressure. This chapter also presents the paste production process that meets the target design specifications for further improving operational efficiencies. The backfill may be optimized with the instrumentally collected key parameters such as fill sequences, stoping sequences, and binder usage in paste backfill.

- Chapter 8 covers a brief statement of the pressure monitoring and field testing of cemented paste backfill used for underground mines. The field performance of mine backfill is of great importance in designing an efficient and more completive cemented paste backfill material. Most often, paste backfill design is done on laboratory results but always underestimates the events that happen at underground paste-filled stopes. This section gives more details on field behavior and characteristic of paste backfill.
- Chapter 9 presents the overall risk assessment of both underground mine backfill and surface tailings disposal. It also addresses the application of various risk management techniques in order to take preventive and corrective actions in a timely manner and reduce or remove possible negative consequences on the operation and environment.
- Chapter 10 presents an overview of the Cayeli paste backfill system and operations. It also addresses the application of paste backfill method to better enhance underground mining operations and reduce or remove possible negative consequences on surface-related tailings management. All aspects including mining methods, paste fill plant, reticulation system, and backfill strategy and applications are given in this section.
- Chapter 11 deals with the operation of the Boulby mine backfill system. It also introduces certain constraints (backfilling for environmental reasons) as well as certain design freedoms that allow for unusual design features. The backfill has provided the potash industry with a proven alternative method of tailings disposal to the conventional options of disposal at sea or placement in a conventional surface tailings depository.
- Chapter 12 presents the two case studies (Barrick Goldstrike Mines, Inc.—BGMI; AuRico Gold, Inc., Young-Davidson Mine—AuRico) based on the distinct approaches to paste backfill plant design. It also includes the inherent characteristics of the mill tailings and backfill strength requirements. This chapter also deals with different equipment selection, sizing, and process design for each paste plant.

There have been a number of books, reviews, papers, and proceedings which demonstrate all the technical aspects of both underground backfill and surface disposal. However, no complete publication exists on their industrial aspects, like the one in the present document. Thus, the absence of such a document in the literature has triggered the realization of this work.

This book not only provides an enhanced understanding of the quality and behavior of both underground paste backfill and surface paste disposal, but also presents a valuable reference for mining practitioners, students, consultants, researchers, and interested individuals. It makes an original contribution to the efficient mine tailings management and treatment techniques. This book has contributed to a better understanding of the industrial perspective of cemented paste backfill and surface paste disposal, including all operational and industrial points from the very beginning to application. Moreover, it reflects a solid reference book for those who want to acquire further information on paste technology for sustainable mining. Finally, it is expected that this book can help backfill researchers and practitioners better understand the quality and behavior of paste backfill and disposal techniques. Due to space limitations, the scope of this book does not allow for the comprehensive study of each individual chapter although such investigations are site specific and needed to better evaluate the mines studied.

Chapter 2
Practical Importance of Tailings for Cemented Paste Backfill

Bayram Ercikdi, Ferdi Cihangir, Ayhan Kesimal, and Haci Deveci

1 Introduction

Mineral processing methods, such as flotation, are used to treat low-grade ores from mining operations; however, a significant amount of tailings are then generated (typically 95–98% of the feed ore). Therefore, mineral processing must also be essentially regarded as a "waste/tailings management" project (Ercikdi et al. 2012). The disposal, stability and safety of tailings, and their effects on water and soil, are important technical and environmental problems. For example, sulphide (i.e. pyritic) tailings can lead to the generation of acid mine drainage (AMD) when they are disposed under atmospheric conditions. This results in the release/mobility of heavy metals, such as arsenic (As), copper (Cu) and zinc (Zn) with the concomitant pollution of water resources and soil (Fig. 2.1).

The increasingly stringent environmental and regulatory constraints today require the management of tailings in a safe and secure manner. Today, tailings generated during ore processing are managed via (a) disposal into tailings dams, (b) discharging into available deep sea zones or (c) backfilling into underground mine openings.

In practice, deep-sea discharging is infrequently used since the mine site is required to be very close to the sea to provide a suitable environment for the disposal of tailings (Cetiner et al. 2006). Instead, tailings dams, which are well constructed, supervised and controlled, are widely used for tailings management/disposal around the world. Chambers and Higman (2011) reported that more than 3500 tailings dams are found worldwide. However, a number of tailings dam accidents have occurred, especially between 1960–1980 and 1980–2011, which is an average of 2–5 incidents per year, respectively. These accidents have caused loss of human lives, structural

B. Ercikdi (✉) • F. Cihangir • A. Kesimal • H. Deveci
Mining Engineering Department, Karadeniz Technical University, Trabzon 61080, Turkey
e-mail: bercikdi@ktu.edu.tr

E. Yilmaz, M. Fall (eds.), *Paste Tailings Management*,
DOI 10.1007/978-3-319-39682-8_2

Fig. 2.1 The formation of acid mine drainage as a result of mineral processing

damage, devastation of agricultural and forestry lands and adverse environmental impacts (such as water pollution) (Rico et al. 2008). A total of 198 tailings dam accidents took place prior to 2000, and 22 after 2000 worldwide (Azam and Li 2010; WISE 2016). These incidents are a result of (a) inappropriate designs, construction, operation and management of the tailings dams; (b) adverse climatic conditions (such as heavy rainfall); (c) inadequate height of the body of the dam and excessive disposal; (d) soil conditions, liquefaction, slope instability and displacement; and (e) the drainage conditions, leakage and pore water pressure (Azam and Li 2010). Paste backfill technology, which was first successfully used in the Bad Grund mine in Germany in 1980, is a safe way of storing tailings into underground mined-out voids (Landriault 2006). However, the success of a tailings backfill design and operation ultimately depends on the characteristics and flowability properties of the tailings material, and environmental conditions of the underground voids.

2 Paste Backfill Technology

Cemented paste backfilling is regarded as a proper waste management method for mill tailings since it leads to the disposal of the tailings back into underground mines. Compared with rock and hydraulic fill, cemented paste backfill (CPB) offers significant technical, economic and environment benefits as follows.

- Cemented paste backfilling permits the placement of a large portion of the tailings back into underground openings and thereby space for tailings disposal and rehabilitation costs are substantially reduced.
- There is no segregation of the paste backfill during its transportation to the mined-out underground voids. Therefore, the use of CPB allows the entire production opening to be filled, thus preventing the collapse of the hanging and side walls and providing a safe work environment.
- About 90% of the processed water can be recovered via thickening and filtration processes prior to the cement backfilling of the tailings. Additionally, this prevents the leakage of pore water outside and the passing of underground water through the paste backfill material owing to its low permeability and high degree of saturation.
- CPB has a low permeability and hence acts as a barrier to prevent underground water seepage. It also reduces the formation of acid mine drainage (AMD) by inhibiting the diffusion of oxygen.
- The transportation of CPB through a pipeline system reduces problems (damage to support systems, traffic problems, etc.) caused by conventional transport systems (conveyors, mobile equipment, etc.).
- Paste backfilling is a relatively fast operation in which the CPB shows a rapid gain of strength for a given binder content. This shortens the mining cycle period compared to hydraulic filling.
- Cement added into CPB provides strength gain, and reduces permeability and the formation of AMD by increasing acid neutralisation potential.
- Cemented paste backfilling increases the volume of ore extraction via substitution and recovery of the ore from pillars, and allows safe working platforms.

However, there are some drawbacks to paste backfilling, which are as follows.

- The dewatering of tailings and pumping operations are capital-intensive and expensive operations.
- Paste backfilling requires a high pumping pressure to transfer high-density backfill material into the underground voids. This increases the pumping maintenance and energy costs.
- The oxidation of sulphide minerals in the presence of oxygen and water may cause long-term instability problems in the CPB of sulphide-rich tailings.
- Transportation problems may occur with changes in the particle size, density and specific gravity of the tailings, and amount of water.
- Paste backfilling requires qualified and accurate engineering work.

In underground mining, the use of CPB to fill production voids or provide support is common, particularly in Canada and Australia. Paste backfill technology has made significant progress over the last 30 years. Over 100 paste backfill plants with capacities that range from 12 to 200 m^3/h are active throughout the world. Additionally, around 30 plants are known to be at the design and installation stage (Fig. 2.2) (Yumlu 2010).

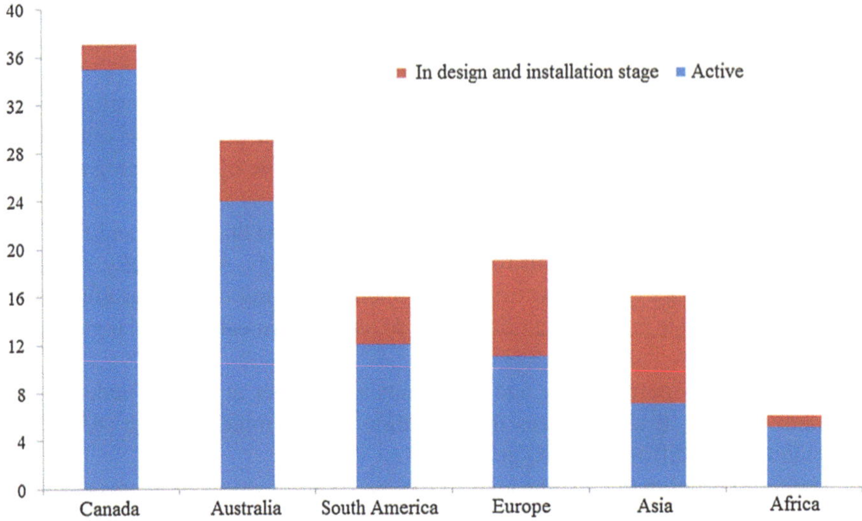

Fig. 2.2 Use of paste backfill technology worldwide (data from Yumlu 2010)

3 Operational Aspects of Cemented Paste Backfilling

Cemented paste backfill (CPB) is an engineered mixture of fine process tailings (75–85% solids by weight), a hydraulic binder (3–9% by total dry paste weight) and mixing water for a solid density of 70–80% by weight (Fig. 2.3). The addition of a binder is essential for the strength and stability of CPB. CPB has to contain sufficient water content to achieve the desired consistency for its transport from the paste plant to the underground openings. In general, a granular material must have at least 15 wt.% finer material than 20 μm to retain sufficient colloidal water so as to form paste with the desired flow properties for its transport through a borehole or pipeline. The index testing consists of a series of water separation and standard slump cone tests that are designed for the assessment of the colloidal properties of an uncemented material (Brackebusch 1994; Kesimal et al. 2003; Ercikdi et al. 2013). The physical, chemical and mineralogical characteristics of the components of CPB, that is, the tailings, binder and mixing water, have a significant role in its short- and long-term performances (i.e. strength and stability), transportation and placement into underground openings.

3.1 *Strength and Stability*

Many factors such as the preparation method of the paste backfill mixture, transportation of the paste backfill to the underground openings, barricade design, strength and stability of the paste backfill, flowability properties of the paste backfill, effects on water quality and cost of backfilling all influence the design of a paste backfill system (Belem and Benzaazoua 2008). Problems during the design stage may result in failures, which will cause production and labour loss, subsidence and environmental

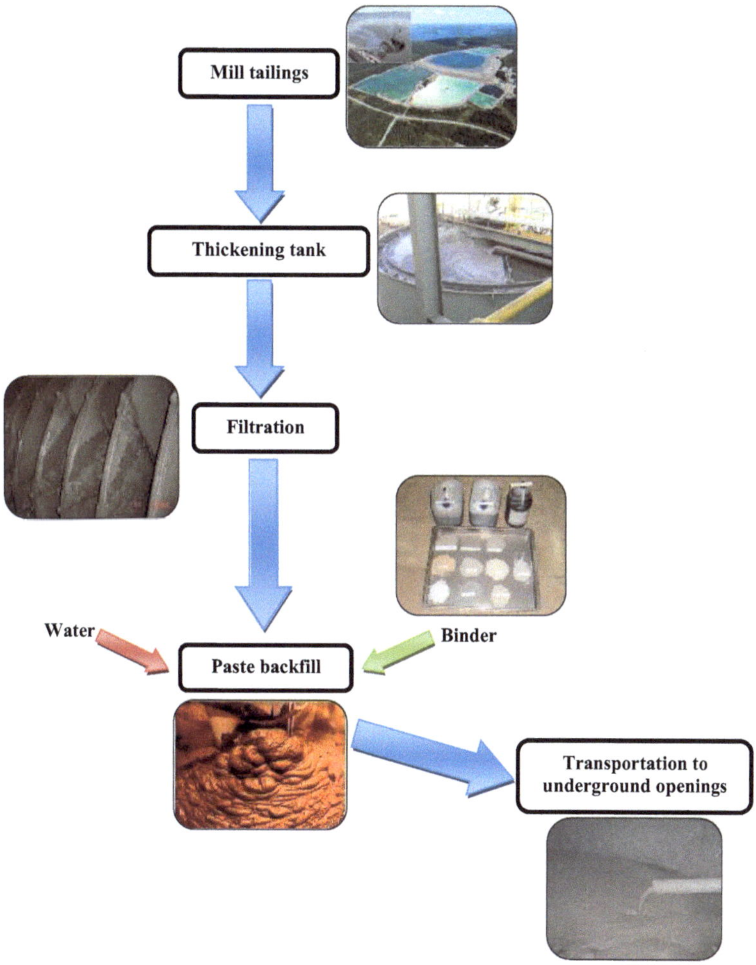

Fig. 2.3 Paste backfill production process

problems. During the paste backfill plant installation, the effect of the paste backfill on groundwater quality, its flowability properties and its short- and long-term strength and stability properties should be examined in detail and optimised. In practice, the strength of paste backfill is determined by unconfined compressive strength testing owing to its simplicity and ease of application as well as low cost (Fig. 2.4).

The required function of CPB in underground stopes determines the designed value of its unconfined strength. For example, to prevent liquefaction risk and barricade collapse, a minimum of 0.15 MPa is required at the early curing stages (Beenet al. 2002; Roux et al. 2004). In mines where cut-and-fill and sublevel mining methods have been used, paste backfill material placed into underground voids provides stability during the mining of the adjacent stopes (Fig. 2.5a). For this purpose, CPB is suggested to have a minimum strength of 0.7 MPa after 28 days of curing (Brackebusch 1994; Landriault 1995). In addition, CPB, which serves as a

Fig. 2.4 Laboratory-scale tests of paste backfill materials: (**a**) preparation of the mixture, (**b**) tamping, (**c**) curing and(**d**) uniaxial compressive and deformation testing

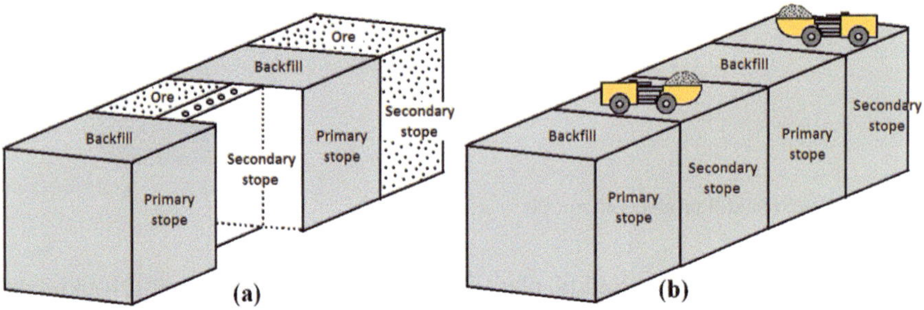

Fig. 2.5 Function of paste backfill in underground mining voids (modified after Hassani and Bois 1992; Belem and Benzaazoua 2008)

working platform for equipment and workers (Fig. 2.5b), is said to have high strength gain at the early ages of curing (Belem and Benzaazoua 2008). Therefore, for roof support, CPB is required to have a uniaxial compressive strength (UCS) value of ≥4 MPa (Grice 1998). The rate in the gain of the desired strength has practical importance for reducing the waiting period for the mining of the adjacent stopes. An appropriate engineering design is therefore required for CPB to achieve the desired strength.

Studies have shown that the CPB design used for a specific type of mill tailings cannot be generalised for use with other mine tailings. Therefore, each type of tailings requires an appropriate and separate mix design with an optimal binder type and mix proportion (Kesimal et al. 2004; Tariq and Nehdi 2007; Ercikdi et al. 2009a, b, 2010a, b; Nasir and Fall 2010; Cihangir et al. 2012).

3.2 Rheology

Rheology is among the most important properties of CPB material which will determine its transportability. CPB mixture should have an appropriate consistency during its transportation through a pipeline system into mined-out underground openings (Simon 2005). An efficient and economical means of transporting paste backfill is only possible when the volume of solid material placed into underground production voids is maximised with minimum energy used. However, an increase in the solids content reduces the fluidity of the mixture, causes friction loss in the pipelines which impedes the rate of material transfer and reduces the efficiency of backfilling (Huynh et al. 2006). Therefore, another important factor that should be considered in the CPB mix design is the flowability properties of the mixture. The physical, chemical and mineralogical characteristics of the paste backfill components significantly affect the paste material consistency.

The main factors that affect the fluidity of paste backfill are the solid ratio, water-to-cement ratio, binder type and proportion, mineral and chemical additives, particle size distribution of the binder and tailings material, shape of the particles, density and surface area of the tailings, surface properties (e.g. zeta potential, water retention), chemical and mineralogical compositions of the tailings, and chemical properties of the mixing water (ion concentration and pH) (Nguyen and Boger 1998; Clayton et al. 2003; Henderson et al. 2005; Huynh et al. 2006).

The flowability properties of paste backfill are usually assessed by slump testing due to its simplicity in practice. For CPB applications, the slump value ranges from 6 to 10 in. A standard slump cone, that is, a right circular cone, is used in slump testing. The right circular cone is 12 in. in height, 8 in. at the base and 4 in. in diameter at the top (Landriault et al. 1997; Cooke 2007). The slump of a mixture is defined as the difference in the level of the original height of the mixture poured into a slump cone (Fig. 2.6a) and the collapse of the material after removal from the slump cone (Fig. 2.6b).

When the water content of a CPB mixture is increased, the slump value also increases (Fig. 2.7) and the mixture can be more easily transported to underground voids through pipelines (Brackebusch 1994; Grabinsky et al. 2002). However, excess water prolongs the curing time of a mixture and also reduces its strength and durability. Chemical plasticisers enhance the flowability properties and thus CPB can be transported at a lower water/cement ratio into the underground voids (Huynh et al. 2006). A reduced water/cement ratio improves the strength and durability of a CPB mixture because the microstructures are refined (i.e. there is a reduction in the

Fig. 2.6 Schematic of the cone slump test

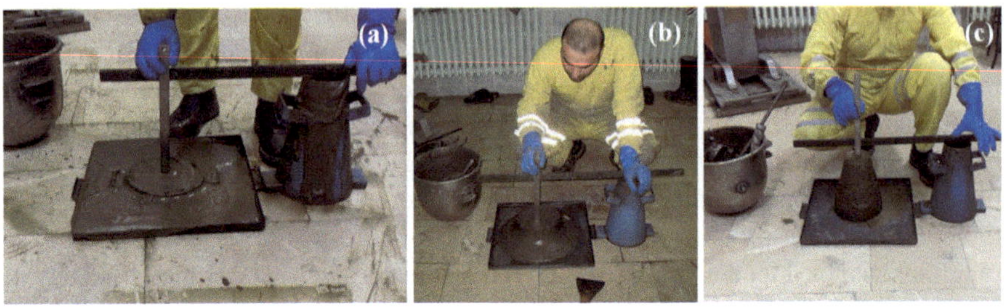

Fig. 2.7 Slump with (**a**) a high value, (**b**) a normal value and (**c**) and a low value

porosity). A reduced water/cement ratio also minimises liquefaction risk in the early curing ages (Kesimal et al. 2005; Ercikdi et al. 2008). An appropriate slump value for CPB mixtures not only prevents **pipeline blockage, but also ensures** safe and efficient transportation of the mixture (Clark et al. 1995). Some researchers have even suggested yield testing and viscosity measurements of CPB mixtures to understand their flowability properties and determine the risk of pipeline blockage in the transportation of these mixtures (Clayton et al. 2003; Fourie and Dunn 2007; Moghaddam and Hassani 2007).

Yield stress is the pressure required for overcoming the static friction of the fluid materials. The viscosity of a fluid is a measure of its resistance to gradual deformation by shear stress or tensile stress. Clayton et al. (2003) stated that two mixtures with the same slump value might have different yield stresses. Mixtures with an overly low yield stress can produce low strength values in a given binder dosage. Table 2.1 presents the different yield stresses for materials with the same slump value, but different density. Therefore, in addition to the slump, the yield stress should be determined for tailings with different particle sizes and densities and mineralogical and chemical compositions since it has practical importance for the successful transportation of paste backfill.

3.3 *Cost*

One of the most important factors that should be taken into consideration in paste backfill design is the cost of the CPB. Since a typical paste backfill plant would have dewatering equipment, concrete pumps, cement-mixer plants, pipeline transportation

Table 2.1 Slump and yield stress values of different mixtures (Clayton et al. 2003)

	Coal tailings	Gold tailings	Lead-zinc tailings
Specific gravity (kg/m^3)	1450	2800	4100
Solids concentration (%)	36	75	75
Slurry density (kg/m^3)	1120	1930	2310
Slump height (mm)	203	203	203
Yield stress (Pa)	160	275	330

Table 2.2 Initial cost of a sample of paste backfill plants (Golder Associates 2012)

Company name/location	Installation	Project capital cost (USD)
Greens Creek silver mine/Juneau, Alaska, US	Capacity: 84 tons/per hour Pipeline system: 6–8 in.	6 million
Zinkgruvan zinc and copper mine/Zinkgruvan, Sweden	Capacity: 90 tons/per hour and pipeline system	6 million
Red Lake Gold Mine/Ontario, Canada	Capacity: 50 tons/per hour and pipeline system	6 million
Bulyanhulu Gold Mine/Tanzania, Africa	Capacity: 180 tons/per hour and pipeline system	6 million
Campbell Gold Mine/Red Lake, Ontario, Canada	Capacity: 60 tons/per hour and pipeline system	6 million
Cayeli Copper Mine/Rize City, Turkey	Capacity: 90 tons/per hour and pipeline system	6 million
Cannington Silver and Lead Mine/Queensland, Australia	Capacity: 175 tons/per hour and pipeline system	6 million

system and computer control systems, the cost of the investment is high. However, in recent years, there is a trend in which the cost of dewatering equipment and concrete pumps has declined due to advances in technology. As seen in Table 2.2, the cost of investing in a paste backfill plant could range between \$5 and \$7 million USD (http://www.golder.com/modules.php?name=Services&sp_id=1221 Golder Associates 2012) depending on the capacity and the pipeline transport system. Paste backfill operation costs make up 10–20% of the total cost of mining operations (Grice 1998). The binder consumption can represent up to 75% of the cost of paste backfill.

Naylor et al. (1997) stated that as a general rule, the cost of a paste backfill mix with a binder of 1 wt.% is \$1/ton. De Souza et al. (2003) indicated that the cost of a paste backfill mix with 3 wt.% of cement constitutes 42% of the total paste backfill operation cost. On the other hand, Fall and Benzaazoua (2003) stated that a paste backfill design with a binder of 5–9 wt.% comprises approximately 50–75% of the total paste backfill operation cost. It is evident that a higher binder content results in a higher total operation cost. Therefore, it is practically important to select an optimum binder type and dosage to provide the desired strength and stability so as to reduce operation costs. For this purpose, the utilisation of chemical agents (i.e. plasticisers, aqueous sodium silicate, sodium hydroxide) or pozzolanic minerals (i.e. blast furnace slag, silica fume, fly ash, pumice) as additives to ordinary Portland cement (OPC) has been shown to mitigate the binder cost as well as improve the stability performance of CPB (Benzaazoua et al. 2004; Klein and Simon 2006; Tariq and Nehdi 2007; Ercikdi et al. 2009c, 2010a, b, 2015; Fall et al. 2010; Cihangir 2011; Cihangir et al. 2011, 2012, 2015).

4 Factors That Affect Strength and Stability of Paste Backfill

Many factors affect the short- and long-term strength and stability of paste backfill. They can be mainly classified as intrinsic and extrinsic factors (Table 2.3). Intrinsic factors include all of the parameters related to the physical, chemical and mineralogical properties of the three main components of paste backfill (the tailings, binder and mixing water) as well as its mixing properties (e.g. water-to-cement ratio) (Brackebusch 1994; Landriault 1995; Amaratunga and Yaschyshyn 1997; Ouellet et al. 1998; Bernier et al. 1999; Belem et al. 2000; Benzaazoua et al. 2002, 2004; Cihangir 2011). Extrinsic factors are related to all phenomena that occur on a stope filled with paste backfill and its interaction with the adjacent rock: for example, the effect of the interface of the paste fill–rock wall, consolidation due to pressure changes, ground vibration due to production blasting that could generate crack formation in the fill mass, drainage and cracks in the rock which may have an effect on the amount of water within the paste fill. Another extrinsic factor that could affect CPB properties is the in situ curing temperature, which can vary greatly with depth, the type of rock that surrounds the backfill mass, geographical location of the mine, mine ventilation, blasting operations, etc. (Benzaazoua et al. 1999; Aubertin et al. 2003; Henderson et al. 2005; Fall and Samb 2009; Yilmaz et al. 2006, 2008a, b, 2009; Ercikdi et al. 2008, 2009a).

Table 2.3 Factors that affect strength and stability of paste backfill

Intrinsic factors	Extrinsic factors
Tailings properties	*In situ conditions*
• Particle size distribution	• Curing conditions (temperature)
• Particle shape	• Self-weight consolidation
• Specific gravity	• Drainage conditions
• Mineralogy	• Stope size/geometry
Binder properties	• Groundwater conditions
• Binder type and dosage	• Paste fill–rock wall interface
• Mineral admixture usage	• Blast-induced ground vibrations
• Chemical admixture usage	• Backfill placement type (continuous or gradual)
Mixing water properties	• Stability of backfill barricade
• SO_4^{2-} content (mg/lt)	• Arching effect and fill mass shrinkage
• pH	• Confining pressure
Mixing properties	
• Water/cement ratio	
• Solid ratio (%)	

4.1 *Intrinsic Factors*

4.1.1 Particle Size Distribution

One of the most important characteristics of a fill material is the particle size distribution. In general, as the fines content increases in a fill, it becomes more difficult for water to flow through the fill. The amount of fines present is very important in the transportation of slurry fills through pipes. The fines help to float the coarse grains in the slurry and provide a non-settling slurry flow within the network of pipes. This is especially important in paste fill applications; there is a certain amount of fines that should be present for the cohesion of the paste fill during transportation through the pipelines (Kuganathan 2005). Generally, the tailings material used in a paste backfill mixture must contain 15 wt.% of particles with size finer than 20 μm in order to retain sufficient water and hence form a paste (Landriault 1995). Mill tailings used in paste backfill mixtures are categorised as coarse, medium and fine depending on the amount of −20 μm fractions in the tailings (Landriault 2001) (Table 2.4).

Modification of the particle size distribution of the tailings material will improve the performance of the paste backfill. For many years, hydrocyclones have been extensively utilised to produce several grain size classes that correspond to fine, medium and coarse tailings by selecting the proper size of hydrocyclones and level of operating variables. However, there is no consensus in the current literature on the optimum size distribution requirements or measurements for backfill materials. This is because different mineralogical contents and variability of tailings on the cement quality need to be taken into consideration as well as variations in delivery systems (e.g. unlined boreholes that allow groundwater to flow into the paste mixture). As shown in Fig. 2.8, a fill material is well graded if its constituent particles demonstrate a wide range of sizes, and poorly graded or uniform if there is a narrow range of sizes. Landriault (2001) suggested that fine particles in a well-graded backfill may fill the voids between larger particles. This reduces the volume occupied by the cement gel (binder) and possibly leads to the formation of stronger bonds as illustrated in Fig. 2.9.

Indices such as the "coefficient of uniformity (C_u)" and "coefficient of curvature (C_c)" are generally used to characterise and quantify the particle size distribution of backfill materials (Eqs. (2.1) and (2.2)). It has been reported that a high proportion of fines leads to a more uniform particle size distribution of the tailings (Kesimal et al. 2003; Fall et al. 2005, Kuganathan 2005; Ercikdi et al. 2013). Kesimal et al.

Table 2.4 Size distribution categories of paste backfill (Landriault 2001)

Tailings type	Finer than 20 μm content (wt.%)	7 in. slump solid content (wt.%)	Explanation (depending on water-to-cement ratio
Coarse	15–35	78–85	High strength acquisition
Medium	35–60	70–78	Lower strength acquisition
Fine	60–90	55–70	Poor strength acquisition

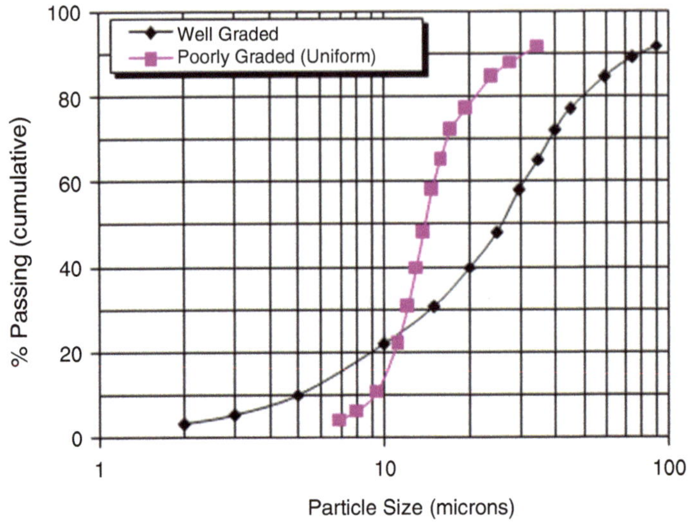

Fig. 2.8 Typical particle size distribution curves for poorly and well-graded backfill materials

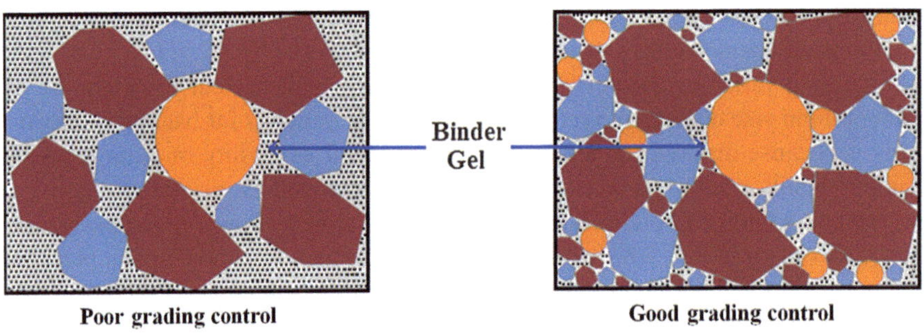

Fig. 2.9 A model of the benefits of fine particles in fill material (modified after Landriault 2001)

(2010) investigated the development of the strength of paste backfill as a function of C_u (D_{60}/D_{10}) for three types of tailings. They noted that paste backfill samples have the highest strength with a C_u of approximately between 4 and 6 (Table 2.5):

$$(\text{Coefficient of Uniformity})\quad C_U = \frac{D_{60}}{D_{10}}, \tag{2.1}$$

$$(\text{Coefficient of Curvature})\quad C_C = \frac{(D_{30})^2}{(D_{10})\times(D_{60})} \tag{2.2}$$

where D_{10} = grain size at 10% passing, D_{30} = grain size at 30% passing and D_{60} = grain size at 60% passing.

Table 2.5 Evaluation of the mechanical strength (UCS) of paste backfill as function of the coefficient of uniformity C_u (D_{60}/D_{10}) (Kesimal et al. 2010)

Tailings type	Coefficient of uniformity (Cu)	UCS (kPa)		Binder type	Binder dosage (wt.%)	Slump inch
		14 day	28 day			
Tailings T1	13.57	886	954	CEM II 42.5 R	6	7.5
	4.72	1117	1321			
Tailings T2	13.33	377	408			
	3.62	780	817			
Tailings T3	11.36	781	815			
	5.56	1030	1258			

The removal of slimes (i.e. desliming) induces changes in the physical, chemical, mineralogical and flowability properties of tailings (Ercikdi et al. 2013). Figure 2.10 shows a typical particle size distribution of sulphide-rich reference (as-received) and deslimed mill tailings. The details on the physical and chemical characterisation of these tailings are also provided in Table 2.6. The reference tailings samples, SP and BN, which have a similar particle size distribution, contain 49.7 and 51.0 wt.% fines (−20 μm), respectively (Fig. 2.10). After desliming, the fines (−20 μm) content of the SP and BN tailings materials appears to decrease to 27.7 and 16.0 wt.%. These values suggest that the as-received and deslimed tailings can be classified as a medium and coarse size tailings material in accordance with the work by Landriault (2001), respectively. The C_u values for SP and BN are 13.57 and 13.33, respectively, with corresponding values of 4.72 and 3.62 for the deslimed tailings (Table 2.6). Similarly, the C_c is 1.12 and 0.96 for SP and BN, and 1.22 and 1.08 for the deslimed SP and BN tailings, respectively.

The specific gravity of the tailings, which is determined by using a pycnometer, tends to increase with reduced fines content as the sulphide content increases after the removal of the slimes (Table 2.6). However, a reduction in the fines content also reduces the specific surface area of the tailings. The finer particles increase the specific surface of tailings which is the surface area that must be cemented and wetted.

There are many studies in the literature on the effect of the particle size distribution of tailings on the strength and stability of paste backfill. Kesimal et al. (2003) investigated the effect of the fines (−20 μm) content of mill tailings on the short-term strength of paste backfill. They showed that reducing the fines content via desliming has a positive effect on the strength of their paste backfill samples with the highest strength obtained in the tailings that contain 25% fines (Fig. 2.11a). Kesimal et al. (2002) also reported that deslimed tailings are 12–52% higher in strength than as-received tailings. Fall et al. (2005) investigated the effect of tailings fineness on the short-term strength development of paste backfill produced from the tailings of a gold mining plant. They observed that the strength of the paste backfill samples peaks with a fines content of 25–30 wt.% which was interpreted as the optimum fines content (−20 μm) for the design of paste backfill (Fig. 2.11b). Similarly, Ercikdi et al. (2013) indicated that paste backfill samples prepared from

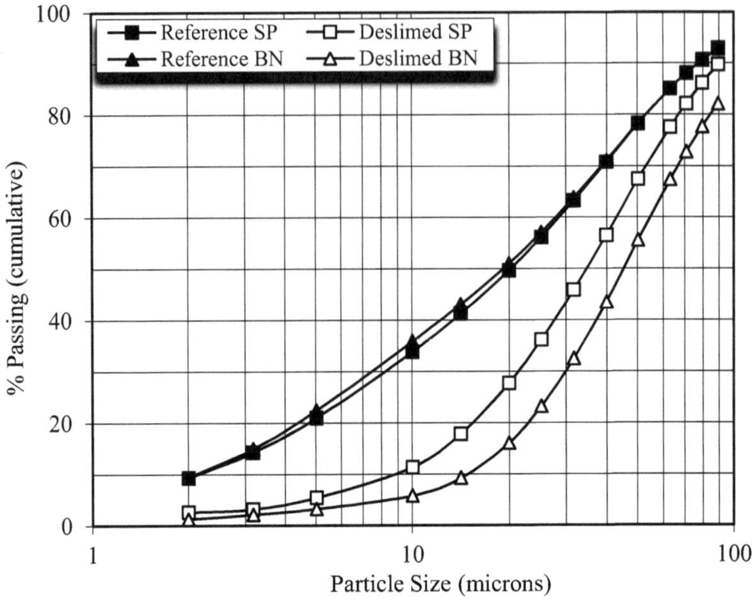

Fig. 2.10 Typical particle size distribution of reference and deslimed tailings

coarse tailings (16–27.7 wt.% finer than −20 μm) result in remarkably higher long-term strength (1.4–4.3-fold) than those prepared from medium tailings (49.7–51 wt.% finer than −20 μm) (Fig. 2.12). They also stated that desliming could allow for a reduction of 13.4–23.1% in binder consumption depending on the inherent characteristics of the tailings. These studies confirm that desliming can be used to improve the strength and stability of paste backfill and reduce binder consumption.

Changes in the tailings properties (e.g. particle size distribution or sulphide content) induced by desliming can affect the flowability properties of paste backfill. It is well known that fine tailings require more water than coarse tailings to reach the targeted consistency, which in turn yield a higher moisture level and lower solids content (Kesimal et al. 2003; Fall et al. 2005). On the other hand, paste backfill samples produced with coarse tailings release more water (by drainage) than those produced with medium or fine tailings. The loss of water through drainage may lead to the settling of the solids (increase in the packing density) and the consequent reduction of the total porosity and void ratio of the backfill material. Ercikdi et al. (2013) performed water separation tests to assess the flowability properties of uncemented as-received and deslimed sulphide-rich mill tailings. They observed that the water retention capacity of the as-received tailings is 7–12 times higher than that of the deslimed tailings over 6 h (Fig. 2.13). In general, as the fines content is reduced in a fill, the settling rate of the tailings increases. The relatively high amount of water separation from the deslimed tailings could lead to settling or segregation problems in a paste plant mixer or pipeline during the transporting of the CPB mixture into underground openings.

Table 2.6 Example of physical and chemical properties of sulphide-rich reference and deslimed mill tailings (Ercikdi et al. 2013)

Characteristic	SP tailings		BN tailings	
	Reference (%)	Deslimed (%)	Reference (%)	Deslimed (%)
Chemical composition				
SiO_2	13.96	11.12	13.16	10.43
Al_2O_3	3.85	2.22	4.81	2.46
Fe_2O_3	46.83	49.47	48.41	53.93
MgO	2.31	1.90	1.13	0.91
CaO	2.05	1.58	1.83	1.28
Na_2O	0.22	0.18	0.19	0.18
K_2O	0.14	0.08	0.64	0.23
TiO_2	0.07	0.06	0.08	0.07
P_2O_5	0.02	0.02	0.02	0.02
MnO	0.08	0.07	0.06	0.05
Cr_2O_3	0.011	0.006	0.012	<0.002
$BaSO_4$	0.84	1.16	2.54	2.02
Loss on ignition (LOI)	27.4	27.6	26.9	28.2
Total	97.78	95.47	99.78	99.78
Sulphide content (S^{-2}) (%)	34.7	39.2	37.4	42.2
Pyrite content (FeS_2) (%)	65.0	73.5	70.1	79.2
Physical property				
Specific gravity	4.12	4.16	4.09	4.32
Specific surface area (cm^2/g)	3594	2432	3662	1956
Coefficient of curvature ($C_c=(D_{30})^2/(D_{10} \times D_{60})$)	1.12	1.22	0.96	1.08
Coefficient of uniformity ($C_u=(D_{60}/D_{10})$)	13.57	4.72	13.33	3.62

The particle size distribution of tailings is also critically important to the microstructure (e.g. total porosity, void ratio) of paste backfill. It has been reported that the overall porosity of backfill tends to decrease with increases in the fines content of the tailings (Kesimal et al. 2003; Fall et al. 2004; Ercikdi et al. 2013; Cihangir et al. 2014). The microstructure of the paste material is strongly influenced by the drainage ability of the fresh backfill. The drained paste backfill samples show both less porosity and smaller void ratios. Fall et al. (2005) reported that the packing density of the tailings decreases when the proportion of fine tailings is less than 45%. A lower packing density is linked to a higher volume of void spaces between the tailings particles that are available for the development of cement hydration products. Indeed, coarse tailings particles mean less particle-to-particle contact per unit volume, and thus, there are larger pores between the particles. Particle size distribution also determines the permeability of paste fill. A higher fines content means lower permeability. As can be observed in Fig. 2.14, the paste backfill samples made of coarse tailings have a higher permeability for a given consistency (Fall et al. 2009).

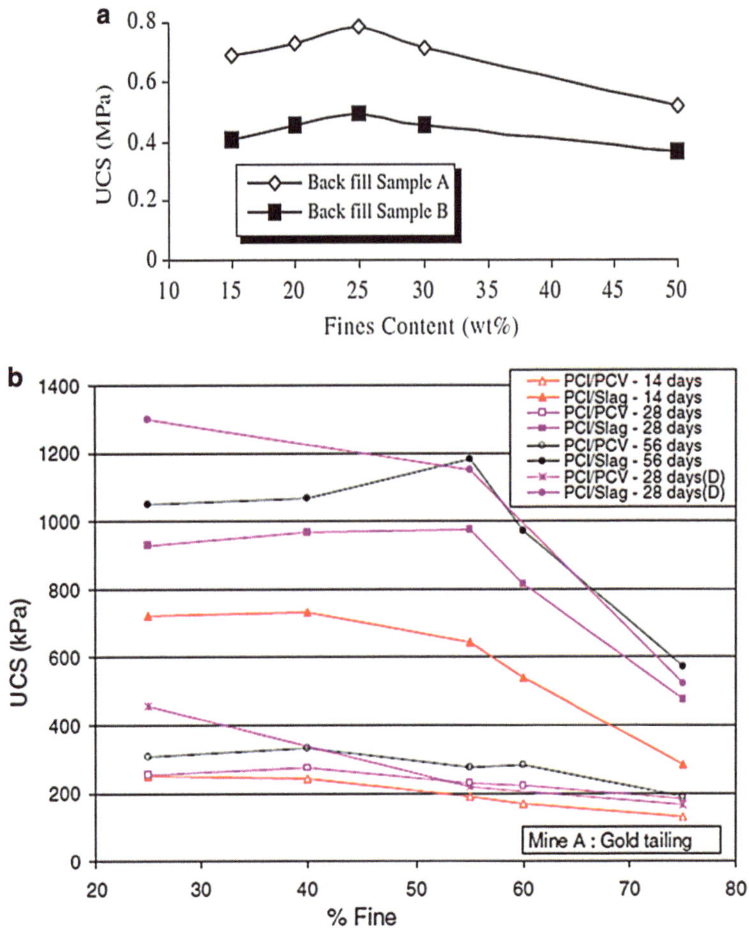

Fig. 2.11 Effect of fines content on strength development of paste backfill: (**a**) Kesimal et al. 2003; (**b**) Fall et al. 2005

4.1.2 Particle Shape and Specific Gravity

The particle shape and specific gravity of tailings material can affect the paste backfill performance. Due to the blasting, crushing and grinding process, most tailings particles are very angular in shape and rough in texture. Henderson and Revell (2005) indicated that mineral particles that are flat in shape will generally settle more slowly than particles that are round with equal specific gravity, thus affecting the thickening and consolidation, and drainage time in the case of fill. For example, mica minerals are characterised by their platy geometry and smooth surface, both of which reduce the strength of fill when cement is used. In addition, the smooth surface of mica makes it difficult for cement to develop high strength aggregation of particles (Revell 2004). Particle shape can also affect the size of voids and path connections to hold and transport fluids (Fig. 2.15).

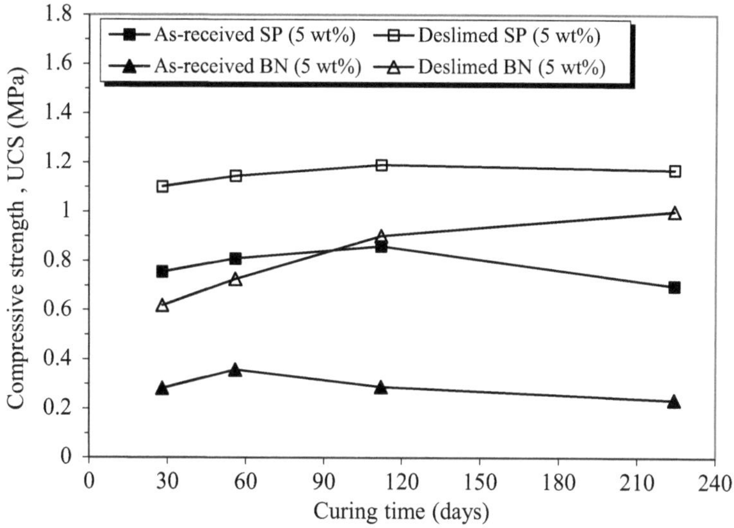

Fig. 2.12 Long-term strength development of paste backfill samples produced from reference and deslimed tailings (Ercikdi et al. 2013)

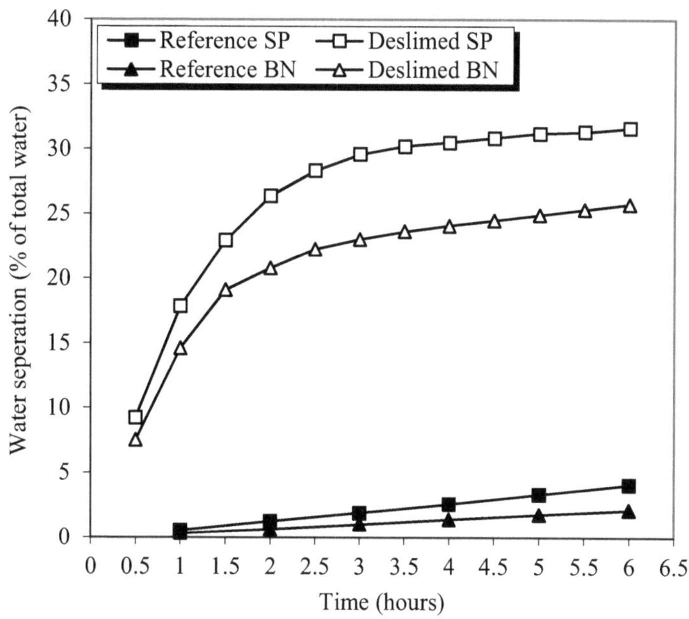

Fig. 2.13 Time-dependent separation of water from reference and deslimed tailings (Ercikdi et al. 2013)

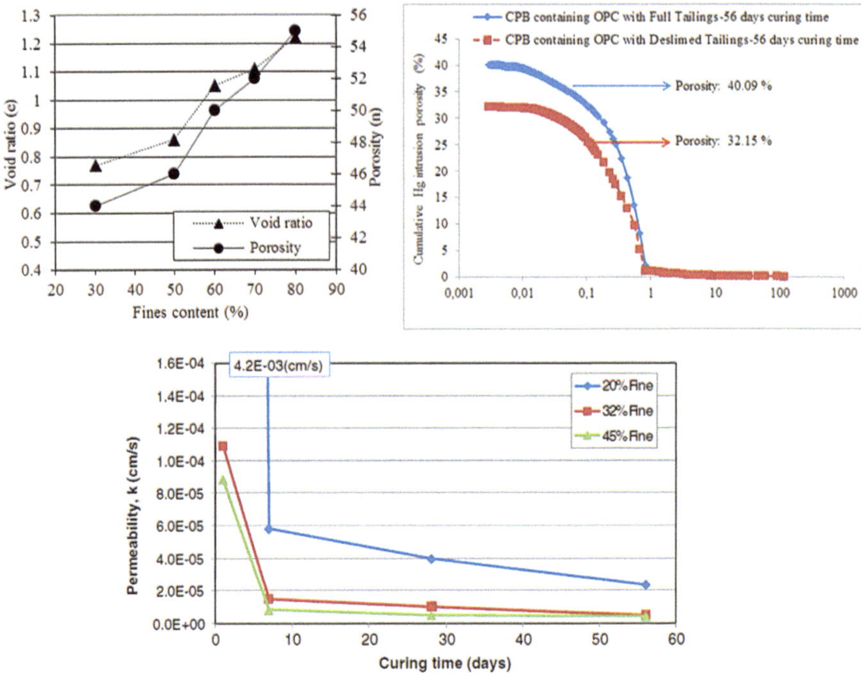

Fig. 2.14 Effect of fines content on microstructure and permeability of paste backfill (Fall et al. 2004, 2009; Kesimal et al. 2015)

Fig. 2.15 Effect of particle shape on interconnection of tailings in backfill (modified after Hassani and Archibald 1998)

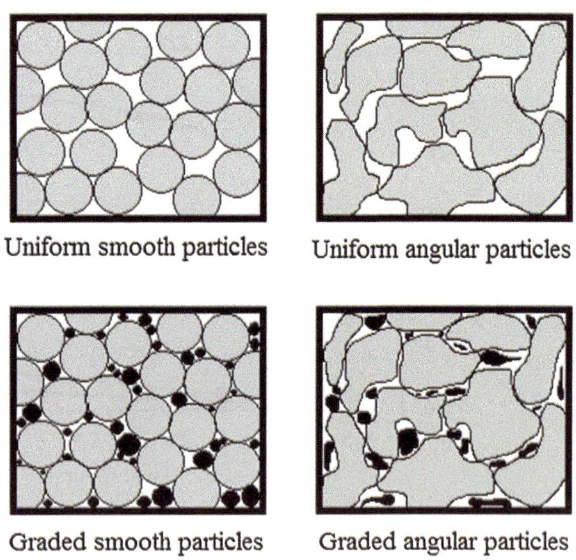

There is a linear relationship between paste backfill strength and the specific gravity of tailings material. Binder is generally added into a paste fill mixture based on the solids content (tailings plus binder on a dry basis). Therefore, tailings with a high specific gravity require more binder (in volume). However, an increased binder content will add to the operating cost of using paste backfill.

4.1.3 Mineralogy

The mineralogy of tailings influences a number of other paste backfill characteristics, such as water retention, strength, settling characteristics and abrasive action. Clay minerals, mica and sericite have been identified to contribute to water retention in paste fill (Henderson and Revell 2005; Kesimal et al. 2005; Ercikdi et al. 2013). Silicate minerals (particularly quartz) can be very abrasive and result in a high wear rate of a pipeline system. Similarly, sulphide minerals (e.g. pyrite) found in tailings may adversely affect the strength and stability of paste backfill. Sulphide reactivity is dependent on the sulphide type. Hassani and Archibald (1998) listed the order of sulphide mineral reactivity as pyrrhotite > arsenopyrite > pyrite > chalcopyrite > sphalerite > galena > chalcocite. Increased specific surface area, oxygen availability and temperature increase the rate of sulphide oxidation (Hassani and Archibald 1998). Pyrite is prone to oxidation when cured in the presence of air and moisture with the concomitant formation of acid and sulphate (Eq. (2.3)):

$$4FeS_2 + 15O_2 + 8H_2O \rightarrow 2Fe_2O_3 + 8SO_4^{-2} + 16H^+ \tag{2.3}$$

The acid generated by the oxidation of sulphide tailings can attack and destroy the structure of the calcium silicate hydrate (C–S–H) bonds. Hence, there is a reduction in the binding properties with the eventual loss of the paste backfill stability as none of the hydration products are stable at pH < 9 (Benzaazoua et al. 1999; Hassani et al. 2001; Tariq and Nehdi 2007; Ercikdi et al. 2009b, c; Cihangir et al. 2011, 2012). Cihangir et al. (2012) monitored the generation of acidity (i.e. decrease in the pH) in CPB samples prepared with different binder dosages (5–7 wt.%) and found that an increase in the binder dosage mitigates the generation of free acidity. The benefits of increasing the binder dosage are an increase in the quantity of hydration products (calcium hydroxide (CH) and C–S–H) with a resultant increase in the binding component and buffering capacity of CPB samples towards acid attack. Moreover, an increase in the binder dosage could also increase the surface of the tailings which are fully covered by the cement hydration products with a resultant reduction in the tailings surface available for reaction/oxidation.

There are four primary internal sources of sulphate in paste backfill systems (Orejarena and Fall 2010). The most significant source of sulphate originates from the tailings (Kesimal et al. 2005) and, depending on the type of mineral extraction, it is common to find pyrite (FeS_2) as an important constituent of these tailings. Ercikdi et al. (2013) reported up to 80% pyrite in the tailings in their study, with concentrations of up to 24,000 ppm of sulphate in the processing waters. Fall and Benzaazoua (2005) reported that the presence of sulphate in the paste can be due to the presence of pre-oxidised tailings. Once the tailings are mixed with a binder, tailings oxidation is considered to be negligible due to the high degree of saturation of the paste backfill, which impedes oxygen diffusion through the paste. Another source of sulphate in paste backfill, apart from the sulphide-rich tailings, is the sulphur dioxide/air used for the destroying of cyanide in gold mining (Akcil 2003).

Sulphate can also be found in paste backfill mass based on the type of cement used for the mixture. It is known that gypsum ($CaSO_4{\cdot}2H_2O$) or anhydrite ($CaSO_4$) is often added in the clinker to control the setting of cement and therefore small amounts of sulphate are introduced into the mixture. Finally, the water added to the mixture for hydration may contain free sulphate ions, either from the pre-oxidised tailings or processing water (Orejarena and Fall 2010). Sulphate that is present in the sulphate-rich water of tailings and that produced by the oxidation of pyrite (FeS_2) can react with free calcium ions produced by the dissolution of unstable portlandite ($Ca(OH)_2$) and calcium aluminate (C_3A), thus giving rise to the precipitation of secondary expansive gypsum ($CaSO_4.2H_2O$) and highly expansive ettringite ($3CaO.Al_2O_3.3CaSO_4.32H_2O$) (Eqs. (2.4) and (2.5)). Gypsum and ettringite can expand 2.2 and 2.8 times in volume, respectively, and generate 70–200 MPa of internal stress due to the crystallisation pressure. The stresses generated by the expansion can produce cracking and loss of cohesion between components (Fig. 2.16), which could then culminate in reduced backfill strength and potential collapse of CPB (Ouellet et al. 1998; Benzaazoua et al. 1999; Hassani et al. 2001; Benzaazoua et. 2002; Kesimal et al. 2004, 2005; Fall and Benzaazoua 2005; Tariq and Nehdi 2007; Ercikdi et al. 2009b, c, 2013; Cihangir et al. 2012):

$$Ca(OH)_2 + SO_4^{-2} + 2\mathrm{H_2O} \rightarrow \mathrm{CaSO_4} \cdot 2\mathrm{H_2O} + 2OH \tag{2.4}$$

$$3CaO \cdot Al_2\mathrm{O}_3 + 3\mathrm{CaSO_4} \cdot 2\mathrm{H_2O} + 30\mathrm{H_2O} \rightarrow 3CaO \cdot Al_2\mathrm{O}_3 \cdot 3\mathrm{CaSO_4} \cdot 32\mathrm{H_2O} \tag{2.5}$$

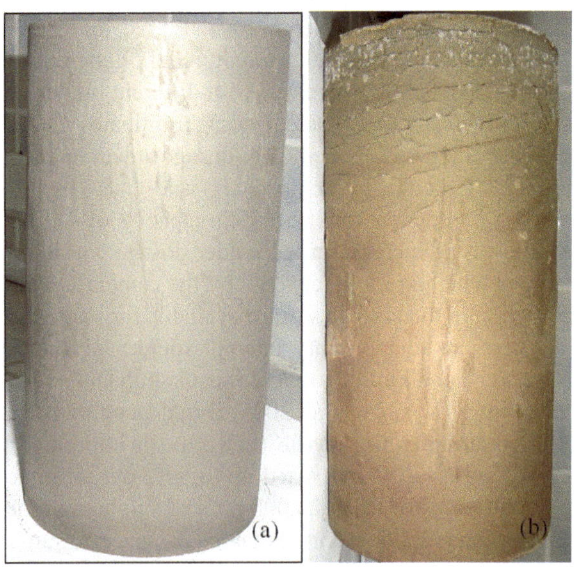

Fig. 2.16 Visual appearance of 28- (**a**) and 360-day (**b**) cured CPB samples. Crack formation due to long-term acid and sulphate attacks

The effect of sulphate on the strength of CPB depends on the sulphate concentration, curing time and cement composition and content (Fall and Benzaazoua 2005). Some studies have demonstrated that there is a reduction in the strength of CPB with time due to internal sulphate attack, which results from the chemical interactions of the sulphate ions with Portland cement hydration products (Hassani et al. 2001, Benzaazoua et al. 2002; Kesimal et al. 2004; Fall and Benzaazoua 2005; Tarig and Nehdi 2007; Cihangir et al. 2012, Ercikdi et al. 2013). These reactions typically form secondary ettringite, gypsum and monosulphoaluminate, which are highly expansive products that generate high internal pressure, thereby deteriorating the strength of the CPB. Fall and Benzaazoua (2005) also found that the early strength of CPB (at less than 28 days) is enhanced with sulphate concentrations less than 2000 ppm due to the precipitation of secondary hydration products (e.g. gypsum, ettringite and brucite). These precipitates fill the void spaces within the CPB, thus contributing to an increase in strength.

4.1.4 New Developments for Characterisation of Paste Backfill Performance

The strength of paste backfill at any given time is one of the most important parameters since the paste backfill structure must remain stable during the extraction of adjacent stopes to ensure the safety of mine workers and avoid ore dilution. UCS testing is often used in practice to evaluate the paste backfill quality since this test is relatively inexpensive and can be incorporated into routine quality control programmes at a mine site. Recently, ultrasonic pulse velocity (UPV) testing has been applied as a non-destructive, low-cost, less time-consuming and easy method in both the field and laboratory, and has been suggested for use in the assessment of the strength properties of paste backfill instead of the conventional compressive strength testing (Yilmaz 2013; Ercikdi et al. 2014; Yilmaz et al. 2014; Yilmaz and Ercikdi 2016). UPV testing applies the principle of measuring the travel velocity of ultrasonic pulses through a material medium. UPV is measured on paste backfill samples by using a portable ultrasonic non-destructive digital indicating tester (PUNDIT) that measures the time of propagation of ultrasound pulses with a precision of 0.1 μs. The 54 kHz transducers of PUNDIT are 42 mm in diameter (Fig. 2.17). The length of the measuring base is determined within an accuracy of 0.1 mm. The end surfaces of CPB samples are polished to provide a good coupling between the transducer face and the sample surface to maximise the accuracy of the transit time measurement.

A thin film of Vaseline® is applied to the surface of the transducers (transmitter and receiver) in order to ensure full contact and eliminate the air pocket between the transducers and the test medium. The direct transmission technique, which is the most satisfactory and reliable method, is used in the testing in which the transmitter and receiver are positioned onto opposite end surfaces of the specimens tested. Repeated readings at a particular location are taken and the minimum transit time is taken as the experimental result. After the measurements, the velocity of the *P*-wave,

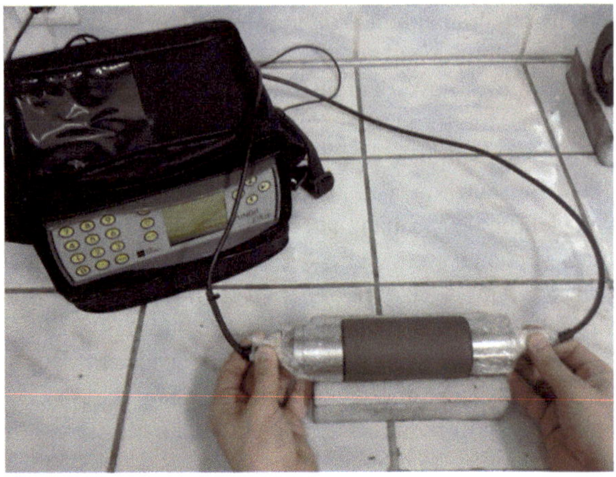

Fig. 2.17 Ultrasonic pulse velocity testing on paste backfill samples

Table 2.7 Relationship between UCS and UPV of paste backfill samples (Ercikdi et al. 2014)

Strength (MPa) (10 × 20)	UPV (m/s) (10 × 20–5 × 10)	
	Tailings T1	Tailings T2
UCS < 0.5	<1450	<1370
0.5 < UCS < 1.0	1450–1690	1370–1600
UCS > 1.0	>1690	>1600

i.e. the UPV, is calculated from the measured travel time and the distance between the transmitter and receiver, as follows (Eq. (2.6)):

$$UPV\left(x,t\right) = x / t \tag{2.6}$$

where UPV (x,t) is the velocity of the *P*-wave in CPB, x is the distance between the transmitter and receiver and t is the travel time.

Ercikdi et al. (2014) conducted UPV tests on paste backfill samples with a diameter x height of 5 × 10 cm and 10 × 20 cm and found a linear relation with a high correlation coefficient between the UCS and UPV of the samples. They also indicated that the strength of large paste backfill samples (10 × 20) can be determined by measuring the UPV in large or small samples over curing time, which can considerably reduce the number of paste backfill samples required and tailings material used (Table 2.7). Furthermore, UPV testing allows the strength of paste backfill in underground environments to be quickly determined.

References

Akcil A (2003) Destruction of cyanide in gold mill effluents: biological versus chemical treatments. Biotechnol Adv 21:501–511

Amaratunga LM, Yaschyshyn DN (1997) Development of a high modulus paste fill using fine gold mill tailings. Geotech Geol Eng 15(3):205–219

Aubertin M, Li L, Arnoldi S, Belem T, Bussiere B, Benzaazoua M, Simon R (2003) Interaction between backfill and rock mass in narrow stopes. In: Soil and rock America mechanics symposium, Essen, Germany, p 1157–1164

Azam S, Li Q (2010) Tailings dam failures: a review of the last one hundred years. Waste Geotech 28:50–53

Been K, Brown ET, Hepworth N (2002) Liquefaction potential of paste fill at Neves Corvo Mine, Portugal. IMM Transact Sect A 111(1):47–58

Belem T, Benzaazoua M (2008) Design and application of underground mine paste backfill technology. Geotech Geol Eng 26:147–174

Belem T, Benzaazoua M, Bussiere B (2000) Mechanical behaviour of cemented paste backfill. In: Proceedings of 53th Canadian geotechnical conference, Montreal, p 373–380

Benzaazoua M, Ouellet J, Servant S, Newman P, Verburg R (1999) Cementitious backfill with high sulfur content: physical, chemical, and mineralogical characterization. Cem Concr Res 29(5):719–725

Benzaazoua M, Belem T, Bussiere B (2002) Chemical factors that influence the performance of mine sulphidic paste backfill. Cem Concr Res 32(7):1133–1144

Benzaazoua M, Fall M, Belem T (2004) A contributing to understanding the hardening process of cemented pastefill. Miner Eng 17(2):141–152

Bernier RL, Lee MG, Moerman A (1999) Effects of tailings and binder geochemistry on the physical strength of paste backfill. In: Proceedings of Sudbury '99, Mining and the environment II, Sudbury, Canada, p 1113–1122

Brackebusch FW (1994) Basics of paste backfill systems. Miner Eng 46(10):1175–1178

Cetiner EG, Unver B, Hindistan MA (2006) Regulations related with mining wastes: European community and turkey. J Mining 45(1):23–34 (in Turkish)

Chambers DM, Higman B (2011) Long term risks of tailings dam failure. Center for Science in Public Participation. Report, 34p

Cihangir F (2011) Investigation of utilisation of alkali activated blast furnace slag as binder in paste backfill. Ph.D. Thesis, Karadeniz Technical University, The Graduate School of Natural and Applied Sciences, Trabzon, Turkey, 207p (in Turkish)

Cihangir F, Ercikdi B, Turan A, Kesimal A, Deveci H, Yazıcı M, Karaoğlu K (2011) Utilisation of sodium silicate activated blast furnace slag as an alternative binder in paste backfill of high-sulphide mill tailings. In: Proceedings of the 14th international seminar on paste and thickened tailings, Perth, Australia, p 465–475

Cihangir F, Ercikdi B, Kesimal A, Turan A, Deveci H (2012) Utilisation of alkali-activated blast furnace slag in paste backfill of high-sulphide mill tailings: effect of binder type and dosage. Miner Eng 30:33–43

Cihangir F, Ercikdi B, Kesimal A, Deveci H, Akyol Y, Ocak S, Kurtuluş M (2014) The effect of binder type on porous structure development of paste backfill. In: Proceedings of the XIth regional rock mechanics symposium, Afyon, Turkey, p 205–212 (in Turkish)

Cihangir F, Ercikdi B, Kesimal A, Deveci H, Erdemir F (2015) Paste backfill of high-sulphide mill tailings using alkali-activated blast furnace slag: effect of activator nature, concentration and slag properties. Miner Eng 83(117–127):2015

Clark CC, Vickery JD, Backer RR (1995) Transport of total tailings paste backfill: results of full-scale pipe test loop pumping tests, USBM, RI 9573, USA

Clayton S, Grice T, Boger DV (2003) Analysis of the slump test for on-site yield stress measurement of mineral suspensions. Int J Miner Process 70(1–4):3–21

Cooke R (2007) Backfill pipeline distribution systems-design methodology review. In: 9th international symposium on mining with backfill, Minefill 2007, Montreal, Canada, 9p

De Souza E, Archibald JF, Dirige APE (2003) Economics and perspectives of underground backfill practices in Canadian mining. In: 105th annual general meeting of the Canadian institute of mining, Metallurgy and Petroleum, Montreal, Canada, 15p

Ercikdi B, Cihangir F, Kesimal A, Deveci H, Alp I (2008) Effect of drainage conditions on the strength of paste backfill. J Mining 47(2):15–24 (in Turkish)

Ercikdi B, Kesimal A, Cihangir F, Deveci H, Alp İ, Yazıcı M, Şahin B (2009a) An environmentally friendly technology: paste backfill—a case study in Cayeli copper mine 3. In: Mining and environment symposium, Ankara, p 139–152 (in Turkish)

Ercikdi B, Cihangir F, Kesimal A, Deveci H, Alp İ (2009b) Utilization of industrial waste products as pozzolanic material in cemented paste backfill of high sulphide mill tailings. J Hazard Mater 168(2–3):848–856
Ercikdi B, Kesimal A, Cihangir F, Deveci H, Alp İ (2009c) Cemented paste backfill of sulphide-rich tailings: importance of binder type and dosage. Cement Concrete Comp 31(4):268–274
Ercikdi B, Cihangir F, Kesimal A, Deveci H, Alp İ (2010a) Effect of natural pozzolans as mineral admixture on the performance of cemented paste backfill of sulphide-rich tailings. Waste Manag Res 28:430–435
Ercikdi B, Cihangir F, Kesimal A, Deveci H, Alp İ (2010b) Utilization of water-reducing admixtures in cemented paste backfill of sulphide-rich mill tailings. J Hazard Mater 179:940–946
Ercikdi B, Cihangir F, Kesimal A, Deveci H (2012) Waste management method for mill tailings: paste backfill technology. Mining Turkey 24:70–75 (in Turkish)
Ercikdi B, Baki H, İzki M (2013) Effect of desliming of sulphide-rich mill tailings on the long-term strength of cemented paste backfill. J Environ Manage 115:5–13
Ercikdi B, Yılmaz T, Külekçi G (2014) Strength and ultrasonic properties of cemented paste backfill. Ultrasonics 54:195–204
Ercikdi B, Külekci G, Yılmaz T (2015) Utilization of granulated marble wastes and waste bricks as mineral admixture in cemented paste backfill of sulphide-rich tailings. Construct Build Mater 93:573–583
Fall M, Benzaazoua M (2003) Advances in predicting performance properties and cost of paste backfill. In: Proceedings on tailings and mine waste'03, Vail, USA, p 73–85
Fall M, Benzaazoua M (2005) Modelling the effect of sulphate on strength development of paste backfill and binder mixture optimization. Cem Concr Res 35(2):301–314
Fall M, Samb SS (2009) Effect of high temperature on strength and microstructural properties of cemented paste backfill. Fire Safety J 44:642–651
Fall M, Benzaazoua M, Ouellet S (2004) Effect of tailings properties on paste backfill performance. In: The 8th international symposium on mining with backfill, Beijing, China, p 193–202
Fall M, Benzaazoua M, Ouellet S (2005) Experimental characterization of the effect of tailings fineness and density on the quality of cemented paste backfill. Miner Eng 18(1):41–44
Fall M, Adrien D, Jelestin JC, Pokharel M, Toure M (2009) Saturated hydraulic conductivity of cemented paste backfill. Miner Eng 22(15):1307–1317
Fall M, Celestin JC, Pokharel M, Toure M (2010) A contribution to understanding the effects of curing temperature on the mechanical properties of mine cemented tailings backfill. Eng Geol 114:397–413
Fourie A, Dunn F (2007) Limitations to the use of the modified slump test for yield stress determination. In: Proceedings of the tenth international seminar on paste and thickened tailings, Paste 2007, Perth, Australia, p 219–228
Golder Associates (2012) Golder associates paste engineering and design services. http://www.golder.com/modules.php?name=Services&sp_id=1221 (23 May 2012)
Grabinsky MW, Theriault J, Welch D (2002) An overview of paste and thickened tailings disposal on surface. In: Proceedings of the symposium sur l'environnement et les mines: Defis et perspectives. Rouyn-Noranda, Canada, 8 p
Grice T (1998) Underground mining with backfill. In: The second annual summit on mine tailings disposal systems, Brisbane, Australia, p 5–15
Hassani F, Archibald J (1998) Mine backfill. In: Canadian Institute of Mine, Metallurgy and Petroleum, Published on CD-ROM Proceedings, Canada, 263p
Hassani F, Bois D (1992) Economic and technical feasibility for backfill design in Quebec underground mines. Final report 1/2, Canada-Quebec Mineral Development Agreement, Research & Development in Quebec Mines. Contract no. EADM 1989–1992, File no. 71226002
Hassani FP, Ouellet J, Hossein M (2001) Strength development in underground high sulphate paste backfill operation. CIM Bullet 94(1050):57–62
Henderson A, Revell MB (2005) Basic mine fill materials. In: Handbook on mine fill, Australian Centre for Geomechanics, 179p

Henderson A, Revell MB, Landriault D, Coxon J (2005) Paste fill. In: Handbook on mine fill. Australian Centre for Geomechanics, 179p

Huynh L, Beattie DA, Fornasiero D, Ralston J (2006) Effect of polyphosphate and naphthalene sulfonate formaldehyde condensate on the rheological properties of dewatered tailings and cemented paste backfill. Miner Eng 19:28–36

Kesimal A, Alp I, Yilmaz E, Ercikdi B (2002) Optimisation of test results obtained from different size slumps with varying cement contents for Cayeli Mine's clastic and spec ore tailings. Karadeniz Technical University, Revolving Fund Project, Turkey

Kesimal A, Ercikdi B, Yılmaz E (2003) The effect of desliming by sedimentation on paste backfill performance. Miner Eng 16(10):1009–1011

Kesimal A, Yılmaz E, Ercikdi B (2004) Evaluation of paste backfill test results obtained from different size slumps with varying cement contents for sulphure rich mill tailings. Cem Concr Res 34:1817–1822

Kesimal A, Yılmaz E, Erçıkdı B, Deveci H, Alp İ (2005) Effect of properties of tailings and binder on the short- and long-term strength and stability of cemented paste backfill. Mater Lett 59(28):3703–3709

Kesimal A, Cihangir F, Ercikdi B, Deveci H, Alp I (2010) Optimization of paste backfill performance for different ore types in cayeli copper mine. Karadeniz Technical University, Revolving Fond Project, Turkey

Kesimal A, Cihangir F, Ercikdi B, Deveci H, Ocak S, Akyol Y (2015) Utilization of alkali activated blast furnace slag cement as an alternative binder to cement in paste backfill of sulphide-rich mine tailings having different fineness and investigation of the performance properties. The Scientific and Technological Research Council of Turkey (TUBITAK Project No: 112 M378), 95p, Turkey (in Turkish)

Klein K, Simon D (2006) Effect of specimen composition on the strength development in cemented paste backfill. Canad Geotech J 43:310–324

Kuganathan K (2005) Rock fill in mine fill. In: Handbook on Mine Fill, Australian Centre for Geomechanics, 179p

Landriault D (1995) Paste backfill mix design for Canadian underground hard rock mining. In: Proceedings of the 97th annual general meeting of the CIM rock mechanics and strata control session, Nova Scotia, Canada, p 652–663

Landriault D (2001) Backfill in underground mining. In: Hustrulid WA (ed) Underground mining methods engineering fundamentals and international case studies. SME, New York, pp 608–609

Landriault D (2006) They said "It will never work"—25 years of paste backfill 1981–2006. In: Proceedings of 9th international seminar on paste and thickened tailings, Limerick, Ireland, p 277–292

Landriault D, Verburg R, Cincilla W, Welch D (1997) Paste technology for underground backfill and surface tailings disposal applications. CIM Bullet 54:112–120

Moghaddam AS, Hassani FP (2007) Yield stress measurement of cemented paste backfill with the vane method and slump tests. In: 9th international symposium on mining with backfill, Minefill 2007, Montreal, Canada, 8p

Nasir O, Fall M (2010) Coupling binder hydration, temperature and compressive strength development of underground cemented paste backfill at early ages. Tunnell Underground Spaces 25:9–20

Naylor J, Farmery RA, Tenbergen RA (1997) Paste backfill at the Macassa mine with flash paste production in a paste production and storage mechanism. In: Proceedings of the 29th annual meeting of the Canadian mineral processors, Canada, p 408–420

Nguyen QD, Boger DV (1998) Application of rheology to solving tailings disposal problems. Int J Miner Process 54:217–233

Orejarena L, Fall M (2010) The use of artificial neural networks to predict the effect of sulphate attack on the strength of cemented paste backfill. Bullet Eng Geol Environ 69:659–670

Ouellet J, Benzaazoua M, Servant S (1998) Mechanical, mineralogical and chemical characterization of a paste backfill. In: Proceedings of the 4th international conference on tailings and mine waste, Colorado, p 139–146

Revell M (2004) Paste-how strong is it? In: Proceedings of the 8th international symposium on backfill, Beijing, China, p 286–294

Rico M, Benito G, Salgueiro AR, Herrero AD, Pereira HG (2008) Reported tailings dam failures: a review of the European incidents in the worldwide context. J Hazard Mater 152:846–852

Roux LK, Bawden WF, Grabinsky MW (2004) Liquefaction analysis of early age cemented paste backfill. In: 8th international symposia on mining with backfill, Beijing, China, p 233–241

Simon D (2005) Microscale analysis of cemented paste backfill. Ph.D. Thesis. Graduate Department of Civil Engineering, University of Toronto, 208p

Tariq A, Nehdi M (2007) Developing durable paste backfill from sulphidic tailings. Waste Res Manage 160(4):155–166

WISE Uranium Project (2016) Chronology of Major Tailings Dam Failures. http://www.wise-uranium.org/mdaf.html (03 March 2012)

Yilmaz E, Belem T, Bussiere B, Benzaazoua M (2008a) Consolidation characteristics of early age cemented paste backfill. In: Proceedings of the 61st Canadian geotechnical conference and the 9th joint CGS/IAH-CNC groundwater conference, Edmonton, Canada, p 797–804

Yilmaz T (2013) Effect of sample size on the strength and ultrasonic pulse velocity of paste backfill, M.Sc. Thesis. Karadeniz Technical University, The Graduate School of Natural and Applied Sciences, 93p (in Turkish)

Yilmaz T, Ercikdi B (2016) Predicting the uniaxial compressive strength of cemented paste backfill from ultrasonic pulse velocity test. Nondestructive Testing and Evaluation, doi:10.1080/10589759.2015.1111891

Yilmaz E, El Aatar O, Belem T, Benzaazoua M, Bussiere B (2006) Effect of consolidation on the performance of cemented paste backfill, Underground Support'06, Quebec, Canada, p 1–14

Yilmaz E, Belem T, Benzaazoua M, Bussiere B (2008b) Experimental characterization of the influence of curing under stress on the hydromechanical and geotechnical properties of cemented paste backfill. In: Proceedings of the 15th international conference on tailings and mine waste, Colorado, p 139–152

Yilmaz E, Benzaazoua M, Belem T, Bussiere B (2009) Effect of curing under pressure on compressive strength development of cemented paste backfill. Miner Eng 22(9–10):772–785

Yilmaz T, Ercikdi B, Karaman K, Külekçi G (2014) Assessment of strength properties of cemented paste backfill by ultrasonic pulse velocity test. Ultrasonics 54:1386–1394

Yumlu M (2010) Mining with paste fill. AusIMM Cobar Mining Seminar, 26p

Chapter 3
Rheological Properties of Fresh Cemented Paste Tailings

Fiona Sofra

1 Introduction

Shifting from disposal of dilute tailings into either large tailings dams or underground as hydraulic backfill to disposal of tailings as dewatered paste tailings involves the implementation of an additional engineering step. This additional step is the specification and design of the paste production unit operation, or paste plant.

Paste tailings technology is employed both above ground and in underground mining for the storage of mineral production waste. The produced paste is most often combined with cement in the underground backfill case. The production and utilisation of paste tailings technology have increased in recent years due to environmental, economic and social pressures to:

- Lower water consumption
- Lower reagent consumption
- Reduce water storage and tailings footprint
- Provide structural integrity and support for overlying and adjacent mining areas
- Increase the structural stability and rehabilitation potential of tailings storage sites

The increased adoption of paste tailings has been facilitated by the improved ability to produce and handle high solids concentration slurry/paste systems. Technological advancements in thickening, filtration, centrifugation, mixing and pumping ability are all dependent on the increased understanding of paste flow behaviour (rheology) and strength characteristics. A combination of these technological developments and an understanding of the material behaviour *throughout*

F. Sofra (✉)
Department of Chemical and Biomolecular Engineering, Rheological Consulting Services Pty Ltd, The University of Melbourne, Parkville, VIC 3010, Australia
e-mail: Fsofra@rheological-consulting.com

E. Yilmaz, M. Fall (eds.), *Paste Tailings Management*,
DOI 10.1007/978-3-319-39682-8_3

the paste production, transport and deposition process allow for effective design, engineering and operation of successful paste systems.

Cemented paste tailings systems require production of paste material that can be delivered to the deposition site at the required solids concentration and cement content. This needs to be achieved with minimal pumping energy, at the flow velocity required and with the yield stress and viscous characteristics for the desired deposition behaviour, either on surface or underground. These flow characteristics must also be reconciled with the strength characteristics required to provide adequate stability for both short- and long-term structural support for surrounding operations.

Understanding the variation in flow characteristics as material moves from pre-thickening operations through to the end-point deposition is where the complexity of paste system design lies. Any physical or chemical change at any point in the process will alter how the paste behaves both at that point and also downstream. Quantification of the effects of processing and operating variables on the flow properties of a given paste is achieved through rheological characterisation.

Rheology is the 'science of deformation and flow of matter' (Oxford). A comprehensive understanding of the rheological characteristics of cemented paste tailings throughout the paste operation is prerequisite to achieving successful paste surface deposition, and/or the desired characteristics for paste backfill. One of the challenges with carrying out accurate rheological testing programmes is that one can generally get results that appear sensible using readily available techniques. The complexity and relevance of the characterisation lie in knowing if that result is correctly interpreted, meaningful and valid to working plant situations. The purpose of this chapter is to elucidate the significance of material preparation and rheological characterisation for the planning, design, operation and optimisation of paste tailings systems.

2 Rheological Aspects of Paste System Design

2.1 Relevance of Rheology to the Overall System Design

Due to legislative, environmental, water availability and operational pressures (Robinsky 1975), the minerals industry is progressing from simply dealing with tailings 'as is' at the end of processing to engineering the tailings to suit the preferred or required disposal scheme. In many cases, this disposal scheme will involve the use of highly dewatered tailings, or pastes, often with the addition of cement. Figure 3.1 illustrates a suggested approach for determination of the tailings disposal system (or paste plant) requirements (Sofra 2001). It is important to note that the design sequence begins at the disposal point and works upstream (back solving) to the dewatering stage.

The complexity of paste system design arises largely due to the changing nature of the thickened material throughout the paste plant. There is no 'straight-line' solution back from deposition, through transport to dewatering, as the flow characteristics are continually changing from one part of the process to another.

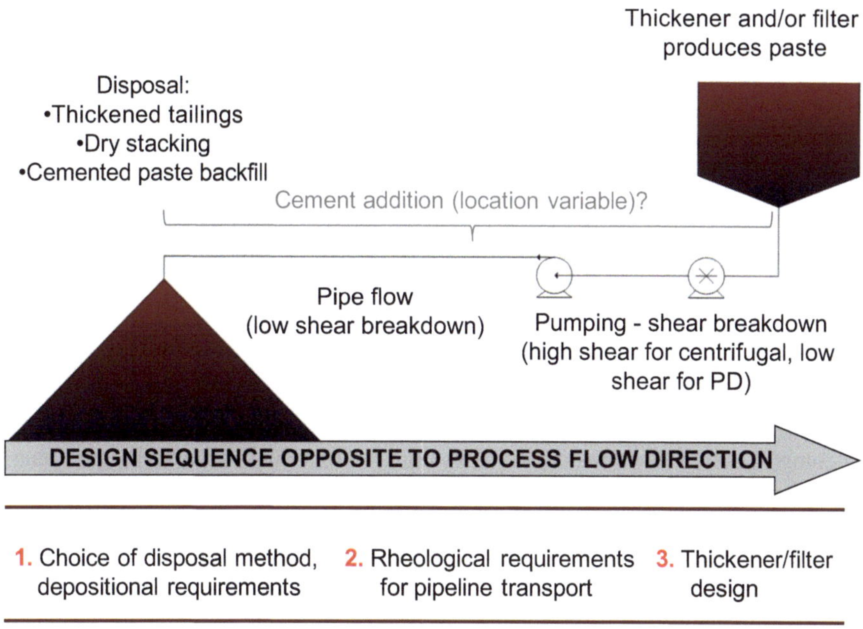

Fig. 3.1 Suggested approach for determination of a tailings management system

A common historical practice in industry has been to use a clarifier or thickener to produce a clear overflow for recycle back to the process, with the waste stream produced being a low-concentration Newtonian fluid suspension which is pumped to a large tailings dam. Alternatively, Fig. 3.1 illustrates the case where the disposal area is receiving thickened tailings or paste. Once the tailings are thickened, and particularly when cement is also added, flow characteristics become non-linear and non-Newtonian characteristics prevail. It is integral to the success of the paste plant to understand this changing behaviour. For a paste tailings system, it is suggested that the engineer designs the tailings management system upstream from the preferred (or stipulated) method of disposal, through transport requirements, ending with the thickening, filtering or other dewatering unit operation.

By considering tailings rheology at all points during the design phase, implementation strategies can be developed for new operations or existing systems optimised to reduce capital and operating costs and/or the impact on the surrounding environment, particularly in regard to recovery of process water. Shear rheological properties, in this instance both viscous flow and yield stress data, are required as a first step for prediction of deposition behaviour, pipeline transport and pump design and the dewatering operation design.

Figure 3.1 briefly summarises the steps involved in determining a tailings management strategy. Point 1 in the design sequence represents the choice of the disposal method, and requires the determination of the rheological properties of the tailings and cement required to achieve the desired footprint and shape of the final deposit, or underground spreading properties (depositional requirements). Point 2 refers to the pumping and pipeline conditions needed for optimal transport whilst ensuring that the tailings reach the disposal site with the rheological properties necessary for the chosen disposal method (pipeline requirements). It must be noted that rheological properties can change in the pipeline itself due to shear intensity, time of shear or both. Rheological testwork must be carried out with this in mind, and shear effects should be quantified where necessary. As shown in Sect. 5, the addition of cement changes may change the rheological properties and this effect must also be taken into account and quantified.

Once the depositional and pipeline requirements have been quantified and optimised on paper, it is possible to begin considering the dewatering operations represented by point 3 in Fig. 3.1. The dewatering stage will need to fulfil two simultaneous requirements: to produce tailings at the solids concentration required by the mass balance, and to generate tailings with the desired flow and spreading characteristics (rheology), taking into consideration any changes that may occur during pumping, pipeline transport and deposition. The dewatering stage should be designed and operated to produce the required tailings' properties via control of size, geometry, residence time, and control of flocculation and pretreatment requirements.

For clarity, the design sequence is depicted in Fig. 3.1 as a linear sequence. However in reality the disposal, transport and dewatering systems are designed using an iterative approach in order to optimise the entire disposal operation. Understanding the impacts of rheology at each of points 1, 2 and 3, and in turn how the design of these systems may alter the rheology and therefore the final tailings characteristics upon disposal, is central to successful paste tailings management. However, obtaining meaningful rheological information at each of points 1, 2 and 3 and therefore gaining an overall understanding of the rheological behaviour are totally dependent on prudent sample selection and preparation.

2.2 *Designing a Rheology Study: The Importance of Sample Selection and Preparation*

2.2.1 Sample Selection

Prerequisite to obtaining meaningful rheological information for design and/or operation of any stage within a paste system is an understanding of the particular design parameters relevant to each stage and reconciling these with appropriate sample selection and treatment. Specific parameters may be, for example, determination of thickener rake torque requirements, pump sizing, calculation of pipeline pressure drop or quantifying end-of-pipe rheology for modelling depositional behaviour. Once the objective is identified, sample selection, preparation and testing conditions must be determined to ensure that information gained through testing will be applicable for design. With the rheology of paste systems being particularly sensitive to numerous physical and chemical/mineralogical factors, this is not always a trivial exercise. For example, the rheology of thickener feed, thickener underflow and end-of-pipe material may vary dramatically—even if all streams were to be compared at the same solids concentration.

Variables to be considered during sample selection include the mineralogy, which can vary significantly with both location and depth within a particular ore body, state of weathering of the sample and water quality. Additionally, wherever possible, samples undergoing rheological characterisation should have similar processing history including beneficiation (particle size distribution is a critical factor), shear history and chemical treatment, including any flocculants used, and process water chemistry.

Obviously, it is not always feasible to match the processing history of samples sent for rheological testing with plant conditions. The numerous combinations and permutations of sample treatment cannot always be replicated in order to generate ideal samples due to time, cost and ore sample availability. For example, a tailings filter cake or thickener underflow will have a different history to that same material which has then been mixed with cement (and possibly water), thus changing the mineralogy, particle size distribution (PSD) and shear history—all of which may alter the rheology. If this material is then pumped along a pipeline to a disposal site (more shear, time and possibly further change in PSD due to attrition) and deposited in a stope or on surface, the rheology can change yet again. In many cases, on-site testing at various locations (sampling point permitting) may be required to get a true indication of the range of rheological characteristics throughout the plant.

2.2.2 Test Sample Preparation

Despite all attempts to obtain materials representative of those encountered under plant conditions, additional laboratory-based treatment may be required prior to testing. Every reasonable attempt should be made to match the anticipated process conditions including PSD, shear history, chemical treatment history and process water quality for the design or operation objective.

The PSD may need to be altered by desliming or addition of coarse material (often sand) to simulate the plant material at various stages of the paste production process. It must also be noted that the addition of cement may alter rheology, in part by altering the PSD of the mixture and this too needs to be considered.

In preparing samples for testing it is important to use site or process water whenever possible as water quality can significantly affect the rheological characteristics of the tailings, particularly if clays are present. Water quality can also alter the cement or binder activity, rheology and interaction with the tailings. In some cases where process water is not available but is known to contain high levels of salts or is particularly acidic or alkaline, it may be necessary to prepare synthetic site water in the lab for use during testing. Air entrainment during sample preparation should be avoided and any air incorporated should be knocked or vibrated out prior to testing.

Shear history is often the most difficult parameter to approximate and simulate and is of particular importance in cases where flocculants are used to enhance dewatering. Flocculants are used to bind fine particles together into agglomerates (flocs) and increase the rate of dewatering via gravitational settling. Flocculation will typically increase the yield stress and viscosity of the dewatered material over and above the increases due to densification alone. However the flocs formed can be delicate and susceptible to breakdown upon the application of shear during mixing, pumping and transport, leading to a progressive reduction in the rheology due to both the rate and duration of shear. The most appropriate method of approximating shear effects is to match the intensity and duration of shear at various points throughout paste production by matching energy input prior to testing prepared samples. However, in a greenfield project, where the paste production route may not be fully defined, it may be necessary to assess the rheology over a wide range of shear histories to provide the operating envelope and indicate the range of behaviour that may be encountered for various paste production options.

Having prepared samples for determination of the rheological properties at particular stages of production, transport and deposition, particular care must be taken during testing not to introduce excessive additional air, shear, resting time or recovery periods during the actual rheology testing. Though specific descriptions of testwork protocols are not within the scope of this chapter, some general information on this aspect is provided in Sect. 4.

3 Key Rheological Concepts

3.1 Non-Newtonian Flow Behaviour in Paste Tailings

Rheology is the study of the deformation and flow of matter. In terms of fluid flow, materials may be classified as either Newtonian or non-Newtonian fluids. Generally speaking, as the solids concentration of a slurry is increased, transitioning from low-solids-concentration 'dirty' water to a high-solids-concentration paste, the rheology transitions from Newtonian behaviour, where the viscosity is constant with flow (or shear) rate, to non-Newtonian, where the viscosity is a function of the shear rate and a yield stress (effectively the pressure required to initiate flow), may be present.

The viscosity (η) of a fluid is defined as the ratio of the shear stress (τ) to the shear rate $\left(\dot{\gamma}\right)$, as shown in Eq. (3.1). In many flows, the shear rate is equivalent to the gradient in velocity:

$$\eta = \frac{\tau}{\dot{\gamma}} \tag{3.1}$$

Inelastic Newtonian fluids exhibit a linear relationship between the applied shear stress and the shear rate, as shown in curve **a** in Fig. 3.2. Flow is initiated as soon as a shear stress is applied. The linear relationship between the shear stress and the shear rate indicates a constant viscosity.

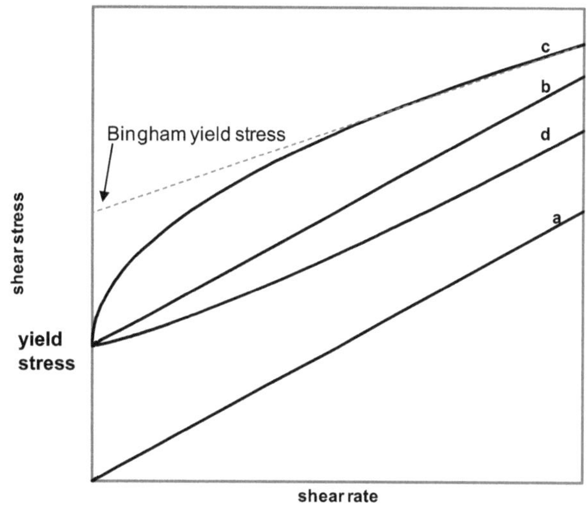

Fig. 3.2 Typical flow curves. (**a**) Newtonian; (**b**) Bingham (yield-constant plastic viscosity); (**c**) yield-pseudoplastic (shear thinning); (**d**) yield-dilatant (shear thickening)

Concentrated mineral tailings often display non-Newtonian flow behaviour in that they possess a yield stress. The yield stress profile (the yield stress as a function of solids concentration and cement content) is the single most important data set for design and operation of paste systems.

The *yield stress* (τ_y) is the critical shear stress that must be exceeded before irreversible deformation and flow can occur. For applied stresses below the yield stress, the particle network of the suspension deforms elastically, with complete strain recovery upon removal of the stress. Once the yield stress is exceeded, the suspension exhibits viscous liquid behaviour—the material will flow. The yield stress is arguably the most important rheological parameter for design and operation of a paste system.

Curve **b** in Fig. 3.2 shows a yield stress followed by a linear shear stress—shear rate relationship, commonly known as Bingham behaviour. Although not a true viscosity according to Eq. (3.1), the gradient of this line is referred to as the Bingham plastic viscosity. In the absence of sufficient paste characterisation testwork, flow data required for rheological analysis is often not available at low shear rates. In these cases, a common practice is to extrapolate the linear portion, say of curve **c** in Fig. 3.2, to the *y*-axis and refer to this as the yield stress. This is an approximated Bingham yield stress which may bear no relationship whatsoever to the true yield stress of the paste at a given solids concentration and cement content. A yield stress determined in this way is far from adequate and can result in gross overdesign for the design of paste tailings dewatering systems, tailings/cement mixing requirements or pumping and pipeline design (Sofra and Boger 2011); more accurate yield stress measurements can be and should be made (see Sect. 4).

Table 3.1 lists typical yield stress values of some familiar items for reference. The values listed for thickened tailings disposal ranging from 30 to 100 Pa are indicative of those used for surface disposal in the minerals industry today. There are other thickened tailings strategies where the material is discharged and flows like a

Table 3.1 Typical yield stress values

Tomato sauce	15 Pa
Yoghurt	80 Pa
Toothpaste	110 Pa
Peanut butter	1500–1900 Pa
Thickened tailing disposal	30–100 Pa
Mine stope fill	250–800 Pa

river delta, where the yield stress would be greater than 0 but below 30 Pa. Paste backfill is a true paste material where the yield stress, in our experience, can reach as high as 800 Pa, but is more commonly in the range of 250 Pa.

In addition to yield stress behaviour, as slurries are thickened toward the paste end of the spectrum, the viscosity of the material often shows *shear rate dependence*. As the shear rate is increased, pseudoplastic or shear thinning materials exhibit a decrease in the viscosity (curve **c**, Fig. 3.2). Dilatant or shear thickening materials exhibit an increase in viscosity with increasing shear rate (curve **d**, Fig. 3.2). Dilatant behaviour, although not common, *is* sometimes observed in mineral suspensions.

The shear thinning nature of many industrial tailings can be attributed to lubrication, alignment of particles or flocs in the flow field direction and/or the breakdown of flocs (Barnes, 1997). An increase in the shear rate from rest can result in alignment of particles in the direction of shear or a reduction in the floc network structure, therefore providing a lower resistance to flow. As such, the suspension will show a decreasing viscosity with increasing shear rate. Shear thickening can be caused by particle jamming and cluster formation both with and without particle contact (Hoffman 1998).

The most complicated non-Newtonian behaviour, which can be observed in cemented pastes even before hydration and curing commence, is a *time-dependent behaviour*. Here the shear stress is a function of both shear rate *and* time of shear. For such materials the viscosity and yield stress depend both on shear rate and time of shear or resting time. When such a material is encountered it is necessary to see how the yield stress varies with time of shear and to measure the shear stress as a function of time at a particular shear rate.

Thixotropic behaviour is the most common form of time dependence observed in pastes, where the viscosity decreases with both shear rate and time of shear. Rheopectic behaviour, the time-dependent dilatant behaviour analogue, is observed less often but when it does occur it can lead to disastrous consequences. Rheopectic materials show an increase in viscosity with shear rate and time of shear. Such behaviour is quite perplexing for the engineering community where the tendency is, when in doubt, to add energy to the system. For example, in a mixing vessel, if the viscosity increases with shear rate and time then any effort to enhance mixing by increasing the impeller speed can result in catastrophic consequences. We have seen this happening on a number of occasions in the past with mineral suspensions, highlighting the importance of understanding and quantifying the rheological phenomena likely to be encountered.

3.2 Flow Models

For the different fluid categories, various empirical flow models are used to describe the flow behaviour in the laminar flow regime. The most commonly used equations are the Ostwald-De Waele model for shear thickening or shear thinning materials, the Bingham model for yield stress materials and the Herschel–Bulkley model for yield stress, shear thinning or yield stress, shear thickening materials:

$$\text{Ostwald-De Waele power law model;} \quad \tau = K\dot{\gamma}^{n} \tag{3.2}$$

$$\text{Bingham Model;} \quad \tau = \tau_{\mathrm{B}} + \eta_{\mathrm{B}}\dot{\gamma} \tag{3.3}$$

$$\text{Herschel – Bulkley Model;} \quad \tau = \tau_{HB} + K\dot{\gamma}^{n} \tag{3.4}$$

In these equations, K and n are experimentally determined constants.

4 Measurement Methods and Rheological Properties of Pastes

The following section gives a brief description of the methods found to be most appropriate for the measurement of paste rheology. The descriptions are not intended to be exhaustive, but rather provide an overview of rheological techniques widely used in industry.

The most common geometry used for measurement of shear stress and shear rate has traditionally been Couette flow using a cup-and-bob rheometer. However, this geometry also has a distinct disadvantage for many pastes in that the gap between the bob and the cup has to be large enough to ensure that the particles themselves do not interfere with the measurement (Nguyen and Boger 1983). When the gap is large, the analysis of data becomes complex and often is not correctly understood, especially when yield stress behaviour is evident. Wall slip is a common problem and shear-induced sedimentation and particle migration away from the solid bob can also occur. Similar problems are encountered using capillary tube rheometry, with the added issue of obtaining low-shear-rate data. Yield stress measurements made with conventional rheometric equipment can be equally problematic.

4.1 Yield Stress Measurement

The shear yield stress is arguably the most fundamental and important rheological parameter for design and operation of a paste system. When measurements are performed correctly, the yield stress gives the best indication of the magnitude of the effects of physical, chemical, mineralogical, shear and time variables on the overall rheological properties of the paste, and should be the starting point for any rheological testing programme.

A graph of the yield stress as a function of solids concentration is referred to as a yield stress profile. As yield stress profiles generally display exponential trends, this profile should be determined over a wide range of solids concentrations encompassing, and extending well beyond the expected operational range. A theoretical design operating point near the 'elbow' of the curve, where the yield stress increases dramatically with very minor variations in solids concentration, would be problematic, with an unfeasibly narrow operating window available. It is difficult to operate dewatering operations to the very narrow solids concentration range that would be required as often <1 wt% control would be required in this 'steep' region for successful operation.

Despite the majority of mineral tailings suspensions exhibiting an exponential yield stress profile, the steepness of this curve and the actual solids concentrations required to produce specific consistencies or yield stresses vary very dramatically, as shown in Fig. 3.3, where the yield stress profiles for a number of paste studies conducted by the author are shown. If a typical target paste plant yield stress of 250 Pa is considered, Fig. 3.3 shows that the target concentration at the deposition point could range from approximately 22 wt% to 80 wt%, depending on the tailings used, the mineral processing plant conditions and the paste production conditions. The yield stress profiles highlight the importance of reliably measuring and understanding the characteristics of the paste yield stress profile from the outset of the design process. All measurements shown in Fig. 3.3 were conducted using a rotational rheometer with a vane geometry.

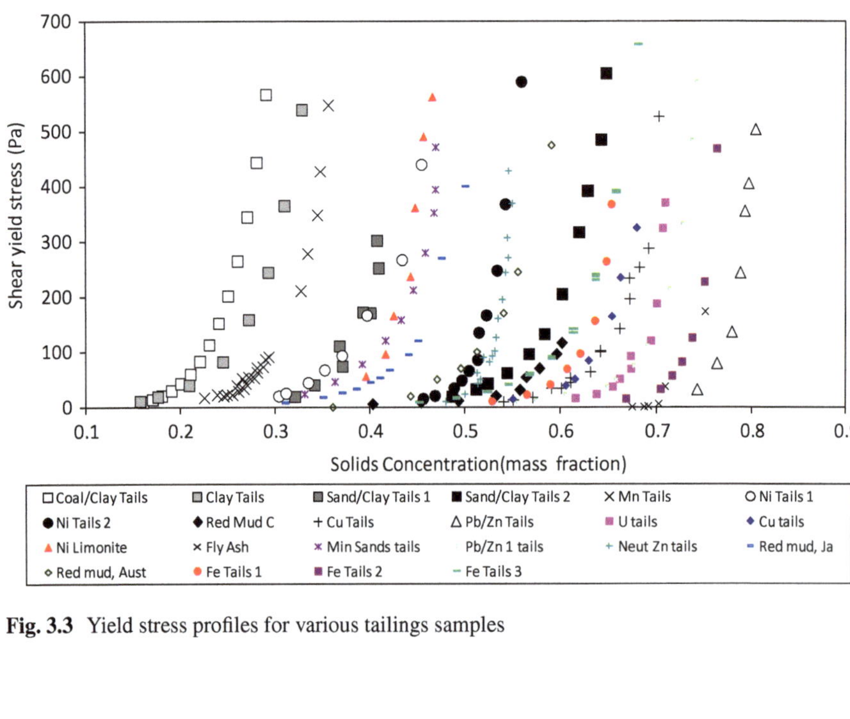

Fig. 3.3 Yield stress profiles for various tailings samples

Figure 3.4 illustrates typical shear stress–shear rate (flow curve) behaviour that is observed for a cemented paste. There are significant misconceptions associated with the measurement and reporting of paste yield stress based on extrapolation of flow curves to define what has been called the Bingham yield stress, τ_B. The Bingham yield stress is a model fitting parameter only and may have no meaning whatsoever in terms of the true yielding of the material. Frequently the extrapolated Bingham yield stress has been used as a basis for design, yet the Bingham yield stress can grossly overestimate the true yield stress, leading to costly overdesign.

The yield stress is the magnitude of the shear stress at which the material initially flows, and in principle can be determined if the measurements of the shear stress as a function of shear rate are made at low enough shear rates. However for high-concentration slurries, slip regularly occurs in concentric cylinder and capillary geometry at low shear rates. Thus, extrapolated values of the shear stress–shear rate data are not easy to obtain and often involve significant error.

Figure 3.5 further illustrates potential errors associated with extrapolation, dependent upon the shear rate region in which data are obtainable. The results in Fig. 3.5 were obtained with a capillary rheometer, a conventional concentric cylinder or bob-in-cup rheometer, and a vane-in-cup device (see Sect. 4) used to eliminate slip. Data obtained with the capillary and the cup-and-bob rheometers agree at the higher shear rates, but at the lower shear rates both deviate from the vane-and-cup data which extrapolates to approximately the correct yield stress. At shear rates less than about 300 s^{-1} (which is significantly greater than the shear rate in most paste systems), both the capillary and Couette data illustrate the dramatic influence

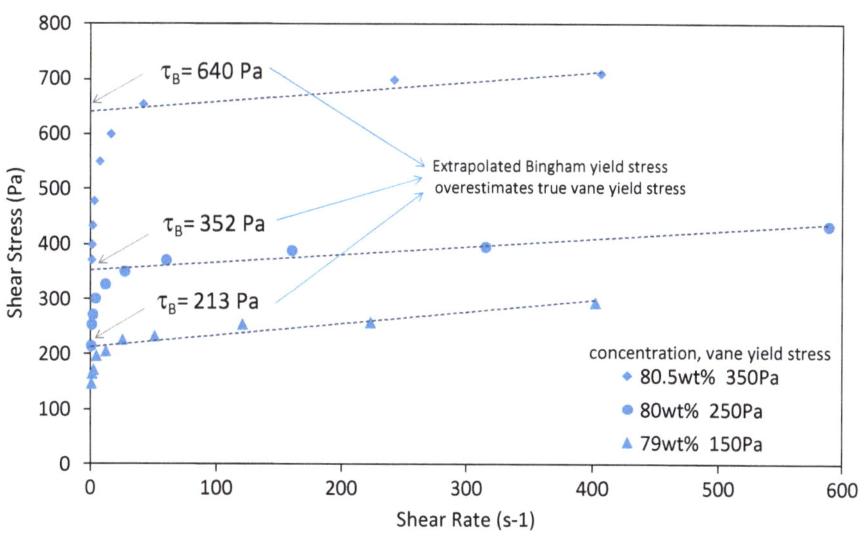

Fig. 3.4 Shear stress–shear rate curves illustrating the overestimation of the Bingham yield stress compared with the true yield stress

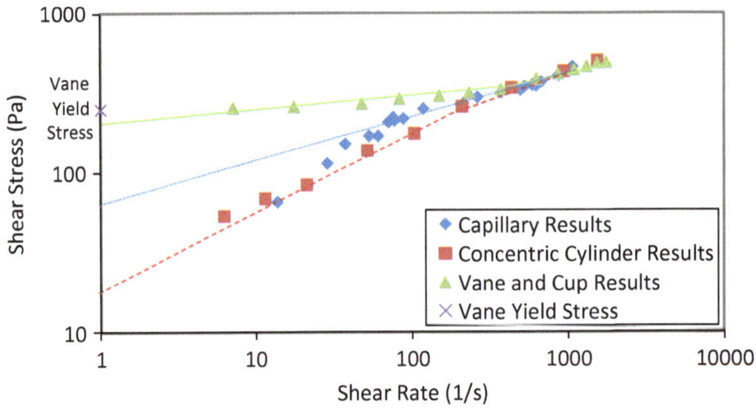

Fig. 3.5 Flow curves for a paste sample (vane τ_y = 250 Pa). Data are obtained using various geometries

of slip flow at a solid surface. At a shear rate of 10 s^{-1}, there is more than a threefold difference in the measured shear stress.

4.1.1 Yield Stress Measurement Using the Vane Method

The true yield stress for the tailings shown in Fig. 3.5 above was obtained using the vane method (Nguyen and Boger 1983, 1985). The extrapolation of the high-shear-rate data obtained with a capillary and concentric cylinder device gives a yield stress of 65 Pa, while the extrapolated value obtained from the lower shear rate data is in the order of 18 Pa. The true yield stress was 250 Pa. The errors involved are significant and this is typically the case for pastes. Thus, if the true yield stress is needed, which is the case for rake design in thickeners, for pump and pipeline restart requirements, pipeline design and deposition flow behaviour, then conventional rheometry geometries should generally not be used and a low-slip geometry like the vane method is far more appropriate.

A summary of the vane method is as follows: the vane is inserted into the paste suspension and rotated at a very low speed where the torque is observed as a function of time. The torque increases until reaching a maximum value, T_m, when the material yields. The maximum torque is related to the yield stress by the vane geometry, provided that end effects are minimised by using an appropriately sized vane as compared to the sample container dimensions. The great advantage of the vane is the material yields on itself; slip is hence generally not an issue and the vane can be inserted into the paste sample in different regions to negate any thixotropic effects. This technique is now used worldwide for measurement of yielding in all matter of materials and was motivated by the need to understand the true yielding behaviour under plant/site conditions.

4.1.2 Yield Stress Measurement Using the Slump Test

The need for a simple method of measuring the yield stress on site, not requiring a power supply or any sophisticated instrumentation, led to the adaptation of the cone slump test used for measuring the consistency of wet cement (Pashias and Boger 1996). The adapted method involves filling an open-ended cylinder with slurry or paste, lifting the cylinder and measuring the distance of the slurry slumps relative to the original container height. The slump is then related to the yield stress by Eq. (3.5) (approximate solution) or Eq. (3.6) (exact solution):

$$\tau_y' = \frac{1}{2} + \frac{1}{2}\sqrt{s'} \tag{3.5}$$

$$s' = 1 - 2\tau_y'\left(1 - \ln\left(2\tau_y'\right)\right) \tag{3.6}$$

τ_y' is a dimensionless yield stress $\left(\tau_y' = \tau_y / \rho g H\right)$, and s' is the dimensionless slump which is the actual slump divided by the height of the cylindrical slump vessel, *H*.

The slump test is an inexpensive, convenient and fast method of determining the yield stress and is ideal for quality control or rapid on-site measurements. However, the slump height is best used for relative measurements as the density of each particular paste must be known in order to calculate the true yield stress from the slump height.

4.2 Shear Stress: Shear Rate Measurements (Viscosity Measurements)

The Couette viscometer is most commonly used for obtaining shear stress–shear rate data for various materials across many industries. The torque on the bob is observed as a function of its rotational speed and these measurements are used to calculate the shear stress and the shear rate. The viscosity is the ratio of the shear stress to the shear rate at the shear rate of interest. However, calculation of the shear rate depends on knowing a functional form between the shear stress and shear rate, i.e. substituting a fluid model. Such a model will not be available a priori; hence approximate techniques are needed to evaluate the shear rate in the cup-and-bob rheometer, particularly when the gap is wide, which is necessary for paste or suspension-like materials that are of interest here. Often a narrow gap assumption is made in order to define the shear rate. This definition, often embedded in instrument software, is not valid for the wide gap required when dealing with paste-like systems, leading to erroneous results.

4.2.1 Viscosity of Pastes Using the 'Bucket Rheometer' (Vane Rheometer) (Fisher et al. 2007)

The shear rate *can* be calculated from first principles if the outer radius of the cup becomes very large (approaches ∞ (Kreiger and Maron 1954)), or is large enough so that the stress decays to zero prior to the outer radius of the cup. The shear stress and the shear rate on the inner bob surface rotating in an infinite medium can then be directly defined and are not dependent on any particular model assumption.

Barnes and Carnalli (1990) first published the idea of using a vane in a cup to minimise slip and a combination of this idea and the infinite medium analysis of Kreiger and Maron (1954) has resulted in the 'bucket rheometer' (Fisher et al. 2007; Sofra et al. 2007a, b), which has many advantages for measuring paste rheology.

Major advantages of using the vane in an infinite medium include the following:

- Insertion of a vane causes less sample disturbance than a cylindrical bob, minimising thixotropic breakdown.
- The vane in cup is less susceptible to errors arising from large particle sizes.
- Yielding occurs between layers of fluid, minimising the effects of wall slip.
- The device is easily portable.
- Shear stress and shear rate are easily determined.
- The analysis is valid for yield stress fluids.

Figure 3.6 shows flow curve data obtained for thickened nickel tailings. The figure indicates the regions in which data would be required for the design and

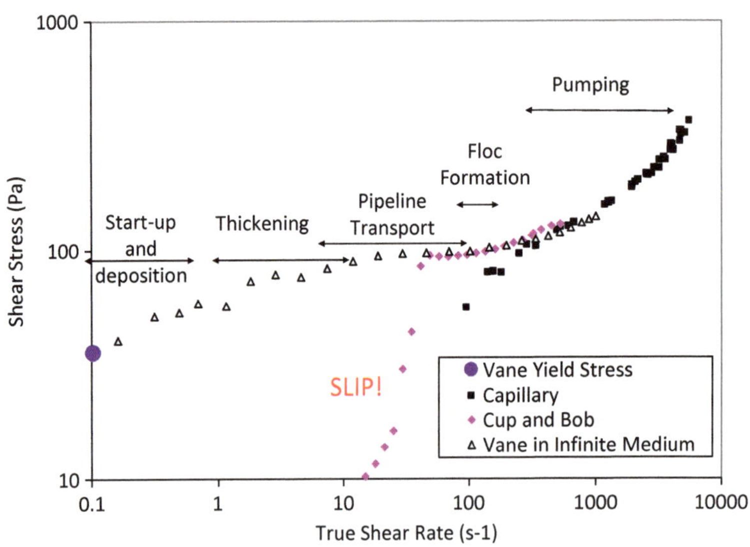

Fig. 3.6 Shear stress—shear rate data for thickened nickel tailings showing the benefits of using the bucket rheometer with vane geometry

operation of various unit operations and also illustrates clearly the slip problems associated with cup and bob and the capillary instruments. As is required for low-shear-rate measurements to be valid, the vane flow curve data extrapolates to the vane yield stress measurement, indicating its suitability for paste viscosity measurement.

5 Effect of Physical and Chemical Factors on Paste Rheological Properties

With the ability to reliably measure various rheological phenomena comes the ability to assess the effect of various physical and mineralogical/chemistry factors on rheology. This information can then be used to either exploit existing conditions or manipulate conditions for optimal paste production and disposal.

5.1 Physical Factors

5.1.1 Solids Concentration

The response of slurry rheology to an increase in solids concentration is unique for each system as exemplified in Fig. 3.3, and often varies significantly even within one ore deposit depending on location, depth and mineralogical or clay inclusions. If one desires to employ thickened tailings or paste system, it is essential that this unique relationship is well understood. Attempting to operate near the 'elbow' of the curve, where the yield stress can vary by up to 100% over a solids concentration of only ±1 wt%, can be highly problematic due to the inevitable process instabilities that must be considered.

The most common effect of solids concentration on the viscosity observed in the minerals industry is a shift from Newtonian behaviour at low solids concentrations, through Bingham behaviour at intermediate concentrations to yield stress shear thinning behaviour at higher concentrations as shown in Fig. 3.7 for a copper tailings paste. As with the yield stress, the specific concentration range at which these transitions occur is material specific and depends on mineralogy and particle size distribution among other factors. Figure 3.7 also shows the effect of cement addition, which, in this instance, is to increase viscosity of the paste at a given solids concentration.

For thixotropic materials, shear history must be taken into account when determining the yield stress–solids concentration relationship. During the design phase, it is unlikely that the extent of structural breakdown through various unit operations will be known. As such, the yield stress–concentration relationship must be determined for structural states ranging from the (relatively) unsheared state to the equilibrium state. From this data, the operating window can be determined.

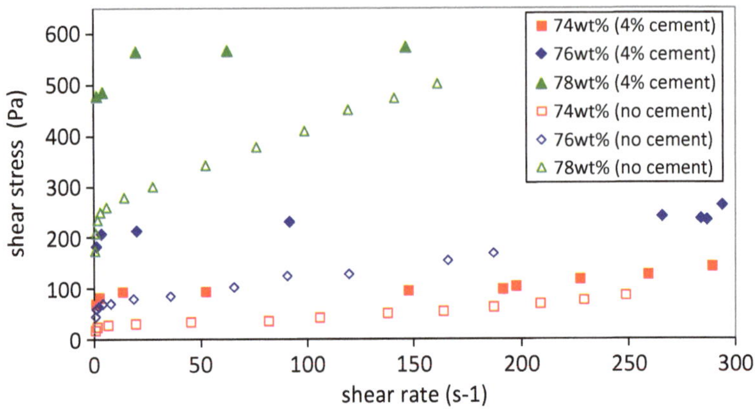

Fig. 3.7 Effect of solids concentration and cement addition on copper tailings paste flow curves

5.1.2 Particle Size

In general, if all else is equal, a finer grind material will have a higher yield stress and viscosity at a given solids concentration due to the finer slurry having a greater particle-specific surface area and therefore a greater area available for interparticle interactions. Finer slurries generally display a more gradual increase in yield stress with increasing solids concentration, whereas coarse materials show a relatively sharp transition from liquid-like to solid-like behaviour as shown in Fig. 3.8.

Knowledge of the effect of particle size and particle size distribution on the rheology is increasingly being exploited to optimise paste backfill operations, where sand is often added to reduce the paste yield stress.

The effect of desliming and sand addition is shown for an industrial lead/zinc/silver tailings paste sample in Fig. 3.9, showing that the deslimed tails generally showed a lower yield stress for a given concentration and that for the blends tested, the variation between the deslimed and final tails is greatest at ~87 wt% tailings blended with 13 wt% sand.

5.1.3 Cement/Binder Addition

The effects of cement (and non-cementitious binders) addition are generally attributable to the fine particle size distribution of the cement as opposed to any short-term hydration and curing effects, provided that testing is completed soon after cement/binder addition—generally within 2 h but preferably within 1 h.

The yield stress versus solids mass fraction curves for cement added to lead/zinc tailings at various addition rates are presented in Fig. 3.10a, along with the tailings and cement PSDs in Fig. 3.10b. The pure tailings sample displayed a lower yield stress at any given solids concentration than tailings blended with

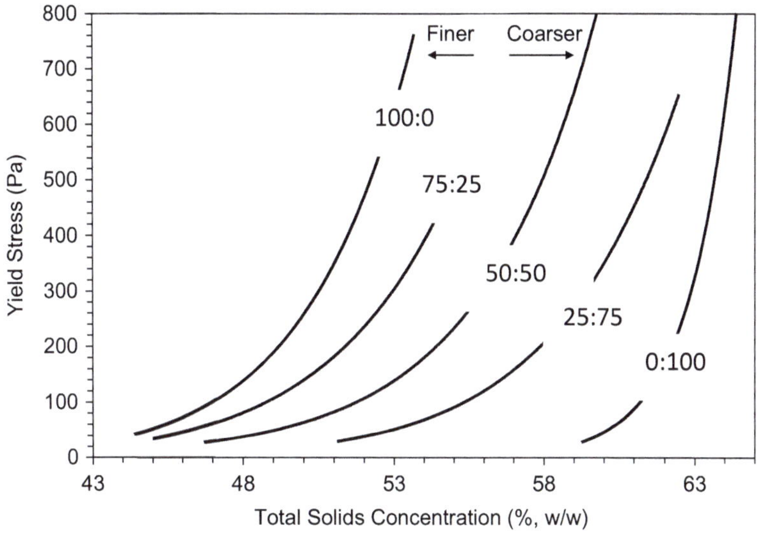

Fig. 3.8 Yield stress profiles for various slime-sand blends (numbers indicate %slimes:%sand)

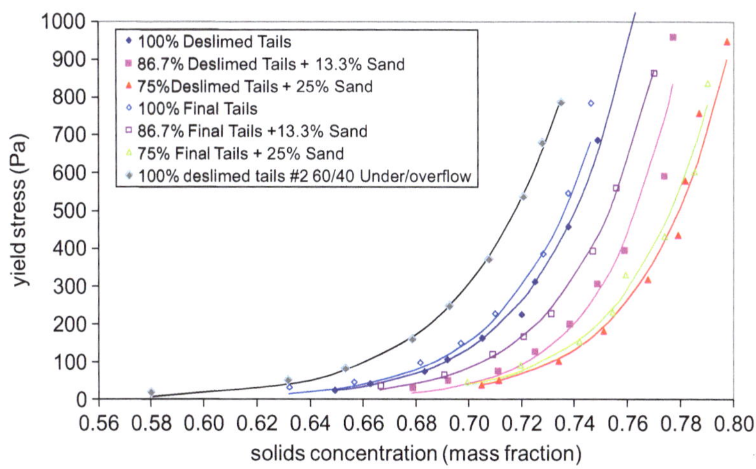

Fig. 3.9 Effect of desliming and sand addition on zinc/lead/silver paste tailings

cement samples within the solids concentration range tested. The higher yield stresses shown by the cement-blended samples are most likely attributable to the finer particle size of the blended cement influencing the tailings' rheology; however the magnitude of this effect will be dependent on the initial PSD of the tailings and the cement/binder type used. As mentioned in Sect. 5.1.2, it is generally the fine fraction which determines the rheological characteristics of cemented pastes.

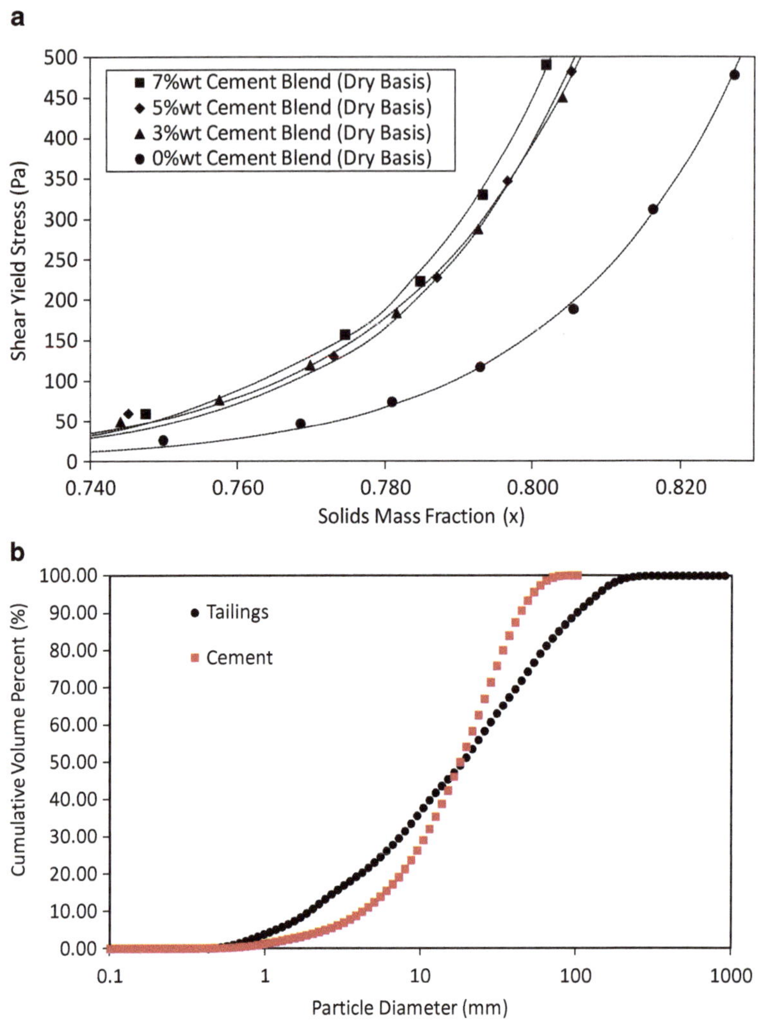

Fig. 3.10 (**a**) Effect of cement addition on lead/zinc tailings. (**b**) Lead/zinc and OP cement tailings PSD

While the 7 wt% cement blend showed higher yield stresses than the 5 wt% and 3 wt% cement blends the differences were minimal compared to the difference between no cement addition and cement addition. Each of the cement-blended samples sat within a solids concentration range of 0.5% of the others for a given yield stress within the range tested. This general effect is typical; when cement addition does have an effect, the relative effect diminishes as the cement percentage is increased beyond the initial addition.

The effect of shear history on the rheology shown in Fig. 3.10a was negligible over the time frame of testing. Reproducible results were obtained when duplicate measurements were taken with extensive shearing between the measurements.

5.1.4 Particle Shape

Particle morphology or shape can have a dramatic effect on rheology. This can be particularly noticeable when plate or needle-like particles with high aspect ratios are present. High-aspect-ratio particles may show yield stress behaviour and a high shear stress (high viscosity) at low flow rates as the particles are forced to move from random orientation to align in the flow direction. Following this a sudden drop in the shear stress (often to below the yield stress) is observed as the particles become aligned at higher flow rates. This type of behaviour, shown in Fig. 3.11 for high-goethite-content nickel tailings, is particularly difficult to analyse, as the resulting flow curve would suggest a negative viscosity due to the decrease in shear stress, which is nonsensical. Nevertheless, it is important to be aware of this behaviour in mixing and agitation applications as it highlights that stagnant zones must be avoided to ensure that localised regions of very high yield stress and viscosity do not develop, which would be difficult to remobilise.

5.1.5 Shear Rate and Shear History

The effects of shear rate and shear history have been briefly discussed earlier in this chapter, but it is worth reiterating that the rheology of pastes is very often dependent on the shear rate under consideration and as such if a viscosity is quoted then the relevant shear rate must also be quoted. In addition, the shear history of a material should be stated (even if this is only possible qualitatively) for all rheological testwork conducted. Manipulation of shear rates and shear history can provide valuable opportunities for optimising rheological characteristics for a particular application.

Figure 3.12 shows the change in the yield stress profiles as a function of shear history, from a relatively unsheared state to fully sheared, where the rheology is no

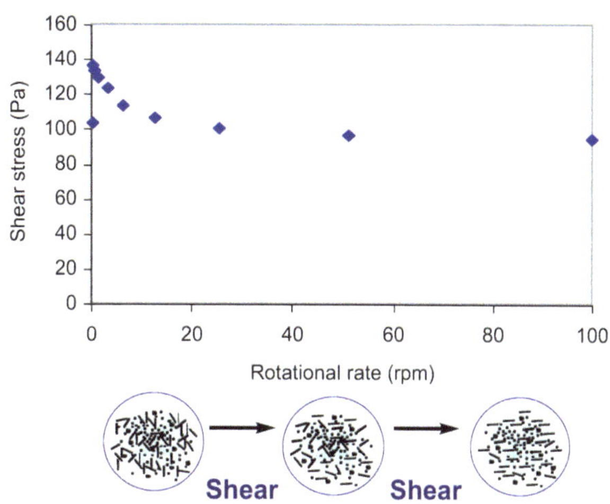

Fig. 3.11 Effect of particle alignment on flow behaviour of high-goethite-content paste

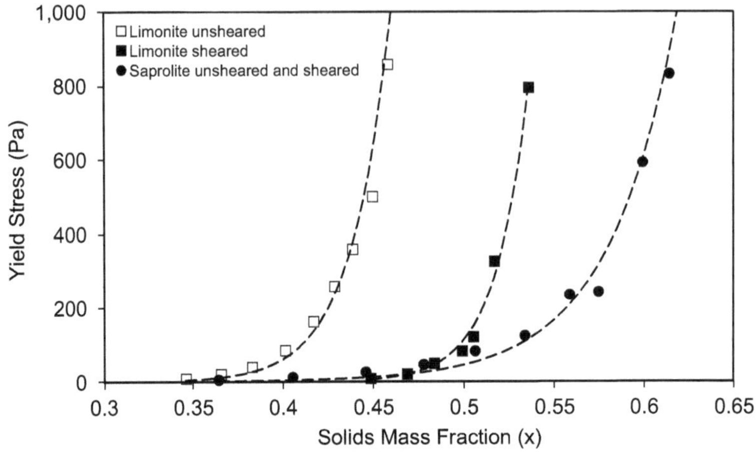

Fig. 3.12 Effect of shear history on the yield stress profiles of lateritic and saprolitic samples from the one same body

longer changing with shear imparted via mixing, pumping or pipe flow (Sofra 2005). Two materials consisting of predominantly laterites and predominantly saprolites from the same ore body are shown. The lateritic sample shows a significant effect of shear history, whereas the saprolitic sample shows no effect of shear history. Figure 3.12 gives important risk mitigation information regarding in-built flexibility that must be designed into the paste plant that will encounter these two materials.

Shear history can have a dramatic effect on pastes where flocculants are used to enhance dewatering via thickening, filtration or centrifugation. It is well understood that flocculants increase the settling and dewatering rate of a material to practical levels. However, this is often at the expense of the extent of dewatering. Flocculants act by binding fine particles together to form more dense aggregates which settle and dewater faster than the individual particles, but water is entrapped within the floc, which in turn creates a three-dimensional network with adjacent flocs which also retains water This water is not fully expelled when a compression force is applied (in the bed of the thickener, filter cake, etc.) and the networked structure that is formed in the bed can have a high yield stress.

Rheological techniques may be used to determine the effect of shear on flocculated structures and this work is being used to identify and develop flocculants that produce desirable dewatering rates, and create flocs that are robust enough to remain intact during the dewatering phase but are then easily sheared and/or compressed to facilitate the escape of inter- and intrafloc water and a reduction of the yield stress (Franks 2005; McFarlane et al. 2006).

It should be noted that not all pastes are shear history sensitive, nor do they all show a reduction in yield stress and viscosity with time or intensity of shear. Some materials display shear-induced aggregation which results in an increase in the structural strength with shear, while other friable materials show an increase in the rheology with time of shear as shown in Fig. 3.13. The increase in rheology shown

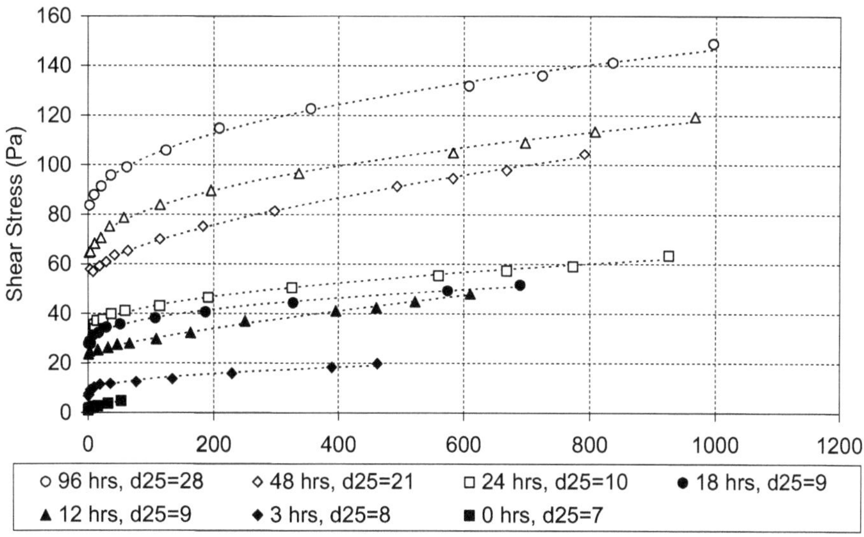

Fig. 3.13 Shear stress versus shear rate as a function of shearing time (pre-cement addition)—increase due to particle attrition during long-distance pipeline transport

in Fig. 3.13 for this material was found to be due to a reduction in the particle size distribution, which can become particularly important during long-distance pipeline transport.

5.2 *Mineralogy/Chemical Factors*

5.2.1 Clays

Many mineral deposits contain clays in the form of kaolinite, illite, montmorillonite and mixed-layer illite/montmorillonite. These clays generally report to the tailings stream for dewatering and storage, are notoriously difficult to dewater and have problematic rheology (De Kretser et al. 1997; Omotoso and Melanson 2014; Helinski and Revell 2014). Upon addition to water (generally in the initial milling process), some clays hydrate and swell and in many cases this leads to complete separation or breakup of the layers composing the clay.

The dispersion of swelling clays can be controlled by suppressing the swelling via increasing the ionic strength of the aqueous medium to which the clays are added (Van Olphen 1977; Callaghan and Ottewill 1974; de Kretser 1997). As the ionic strength is increased, the electrical double layer is compressed, thereby reducing platelet repulsion and suppressing swelling. If a sufficiently high calcium ion concentration is maintained (lime and gypsum are frequently used), calcium ion exchange will occur with the exchangeable interlayer cations. As calcium is a divalent ion, the neutralising effects will be increased and platelet coagulation will occur

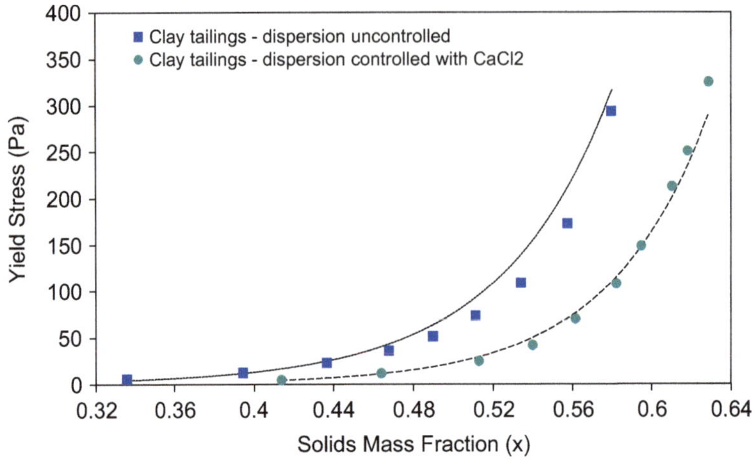

Fig. 3.14 Yield stress profiles for high-clay coal tailings with and without dispersion control

at lower concentrations in addition to the promotion of space-efficient face-face aggregation. The resulting suspension will have a reduced effective particle surface area, fewer interparticle interactions and a weaker network structure, all of which improve the rheology. Figure 3.14 highlights the reduction in the yield stress that can be achieved by controlling the dispersion of high-clay-content coal tailings using calcium ions.

The concept of controlled dispersion is not new; however, implementation for tailings disposal is not widespread due to the belief that both clay swelling and complete platelet separation must be avoided, necessitating high calcium concentrations. The high calcium concentrations required for complete controlled dispersion are often not practical due to economic factors and the limited solubility of many available calcium sources leading to excessive scaling issues. However, complete suppression of clay swelling and breakup is not essential. For improved dewatering performance, suppression of breakup alone can lead to dramatically improved clay dewatering (Sofra et al. 2007a).

5.2.2 Water Chemistry: pH and Ionic Strength

The trend toward processing ever-finer grind slurries has its disadvantages in that the fine particles tend to dominate the rheology, making thickening and pumping more difficult. However, the difficulties with fine suspensions stem from the fact that the finer the particles, the more influential surface chemistry becomes and this can provide opportunities to manipulate the rheology through manipulation of the water chemistry.

In many cases it is considered impractical and uneconomical to alter the process water chemistry. However, as resource, environmental, legislative and other pressures increase, so too does the incentive to invest in advancements in tailings management, including assessing the effect of water quality on the dewatering, flow, spreading and eventual strength properties of tailings. For fine particle systems such as pastes, the effects of flocculants, cement and other binders, viscosity modifiers and other additives are all largely influenced by both particle mineralogy and water chemistry—which is effectively to say by surface chemistry.

Understanding the effect of surface chemistry is achieved by understanding particle surface charge, the effect of surface charge on rheology and how the surface charge can be modified. The surface chemistry can be evaluated by first measuring the zeta potential, which is an indication of the surface charge of the particle as a function of pH. The isoelectric point (IEP) can be determined which is the pH of net zero surface charge and maximum van der Waals attraction. At the IEP, the particles are attracted and coagulated and this pH coincides with a maximum in the yield stress as shown in Fig. 3.15. Away from the IEP, where the magnitude of the zeta potential is high, particles are electrostatically repulsed and the system is dispersed, with a low yield stress. It can be seen in Fig. 3.15 that increasing the pH by one unit can nearly halve the yield stress at high solids concentrations.

Surface interactions and hence the rheology can also be modified by changing the ionic strength of the process water via salt addition. A high ionic strength compresses the electrostatic double layer of the particle, reducing interparticle repulsion and thus flattens the zeta potential curve as shown in Fig. 3.16. As such, the yield stress becomes less sensitive to changes in pH than if the ionic strength was low. This is yet another example of how rheology can be utilised to measure physical parameters that describe the link between complex interparticle interactions and the material behaviour encountered in real processing situations.

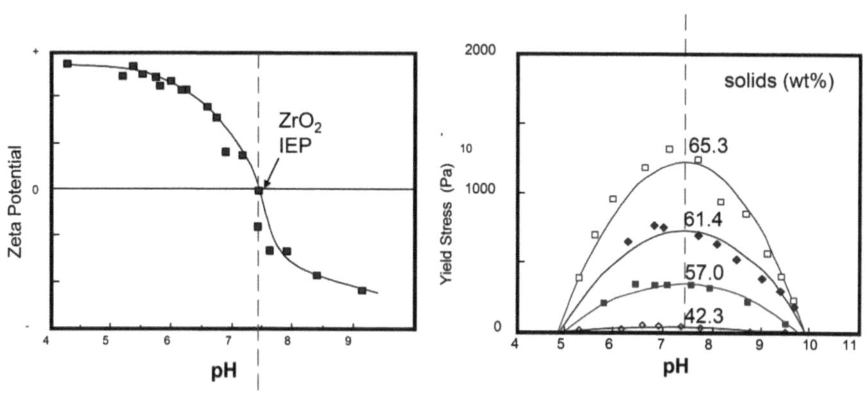

Fig. 3.15 Zeta potential and yield stress curves as a function of pH for various zirconia suspensions

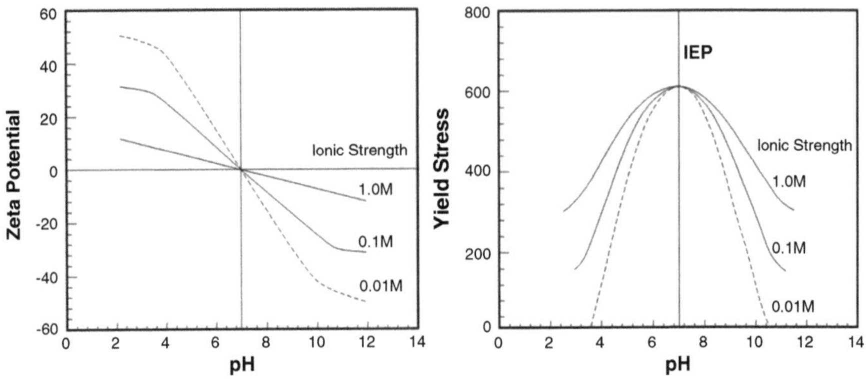

Fig. 3.16 The influence of ionic strength on surface charge and yield stress

6 Conclusion

The increased adoption of paste tailings has been facilitated by the improved ability to produce and handle high-solids-concentration slurry/paste systems. Technological advancements in thickening, filtration, centrifugation, mixing and pumping ability are all dependent on the increased understanding of paste flow behaviour (rheology) and strength characteristics. A combination of these technological developments and an understanding of the material behaviour *throughout* the paste production, transport and deposition process allow for effective design, engineering and operation of successful paste systems.

Understanding the variation in flow characteristics as material moves from pre-thickening operations through to the end-point deposition is where the complexity of paste system design lies. Any physical or chemical change at any point in the process will alter how the paste behaves both at that point and also downstream. Quantification of the effects of processing and operating variables on the flow properties of a given paste is achieved through rheological characterisation.

The science of rheology is an essential tool in the overall management of both surface and underground paste tailings systems. Basic design data can be measured with techniques specifically designed for the purpose and the influence of major variables such as flow rate, shear history, particle size, shape and concentration and the influence of surface chemistry can be evaluated and exploited for the success of a paste tailings system.

References

Barnes HA, Carnali JO (1990) The vane-in-cup as a novel rheometer geometry for shear thinning and thixotropic materials. J Rheol 34(6):841

Barnes HA (1997) Thixotropy a review. J Non-Newtonian Fluid Mech 70:3

Callaghan IC, Ottewill J (1974) Interparticle forces in montmorillonite gels. J Chem Soc Faraday Discuss 57:110

De Kretser R, Scales PJ, Boger DV (1997) Improving clay-based tailings disposal: case study on coal tailings. AIChE J 43(7):1894–1903

Fisher DT, Scales PJ, Boger DV (2007) The bucket rheometer for the viscosity characterization of yield stress suspensions. J Rheol 51(5):82

Franks GV (2005) Improved solid/liquid separation using stimulant sensitive flocculation and consolidation. J Colloid Interface Sci 292:598–603

Helinski M, Revell MB (2014) Fill design and implementation with challenging material: Wambo fill project—case study. In: Potvin Y, Grice T (eds) Proceedings of the 11th international symposium on mining with backfill (Minefill 2014), 20–22 May 2014, Australian Centre for Geomechanics, Perth, p 421

Hoffman RL (1998) Explanations for the cause of shear thickening in concentrated colloidal suspensions. J Rheol 42:111

Kreiger IM, Maron SH (1954) Direct determination of the flow curves of non-Newtonian fluids, III. Standardised treatment of viscometric data. J Appl Phys 25(1):72–75

McFarlane A, Bremmell K, Addai-Mensai J (2006) Improved dewatering behaviour of clay mineral dispersions via interfacial chemistry and particle interactions optimization. J Colloid Interface Sci 293:116–127

Nguyen QD, Boger DV (1983) Yield stress measurement for concentrated suspensions. J Rheol 27:321

Nguyen QD, Boger DV (1985) Direct yield stress measurement with the vane method. J Rheol 29:335

Omotoso O, Melanson A (2014) Influence of clay minerals on the storage and treatment of oil sands tailings. In: Fourie AB, Jewell R (eds) Proceedings of seventeenth international seminar on paste and thickened tailings (Paste 14), June 8–12 2014, Vancouver Canada 2014, Australian Centre for Geomechanics, Perth, p 269

Pashias N, Boger DV (1996) A fifty-cent rheometer for yield stress measurement. J Rheol 40(6):1179

Robinsky EI (1975) Thickened discharge—a new approach to tailings disposal. Canad Mining Metallurg Bulletin 68:47–53

Sofra F (2001) Minimisation of bauxite tailings using dry disposal techniques, PhD thesis. The University of Melbourne

Sofra F (2005) The importance of understanding feed rheology in nickel laterite processing. In: Proceedings of processing nickel ores and concentrates 05, 16–17 November 2005, Cape Town South Africa, MEI, CD-rom only

Sofra F, Scales PJ, Kilcullen A (2007a) Dewatering and clays—the importance of controlling dispersion 'up-front'. In: Fourie AB, Jewell R (eds) Proceedings of tenth international seminar on paste and thickened tailings (Paste07). 13–15 March 2007, Australian Centre for Geomechanics, Perth, p 229

Sofra F, Fisher DT, Boger DV (2007b) The bucket rheometer for thickened tailings and paste flow curve determination. In: Fourie AB, Jewell R (eds) Proceedings of tenth international seminar on paste and thickened tailings (Paste07). 13–15 March 2007. Australian Centre for Geomechanics, Perth, p 249

Sofra F, Boger DV (2011) Rheology for thickened tailings and paste: history, state-of-the-art and future directions. In: Fourie AB, Jewell R (eds) Proceedings of fourteenth international seminar on paste and thickened tailings (Paste 11). 5–7 April 2011. Australian Centre for Geomechanics, Perth, p 121

Van Olphen H (1977) Introduction to clay colloid chemistry, 2nd edn. Wiley, New York

Chapter 4
Properties of Cemented Paste Backfill

A. Ghirian and M. Fall

1 Introduction

Knowledge on the properties of cemented paste backfill (CPB) material is essential for the design of cost-effective, safe and durable cemented backfill structures. The primary properties and characteristics of CPB are categorised by their physical, mechanical, hydraulic, thermal, chemical and microstructural aspects. Each property can be affected by different parameters that are mainly classified as (1) internal factors, which include all the intrinsic and multi-physics properties, such as parameters related to the main components of CPB (e.g. binder, tailings and water) and their changes with curing time. They also include other processes, such as self-desiccation, cement reaction and heat generated by binder hydration, and (2) external factors or processes, which include all phenomena that take place in relation to environmental aspects. For example, the effects of thermal loads (e.g. temperature of the rock mass, self-heating of the rocks) on the strength development of CPB, or the effects of the pressure of the self-weight of backfill and the filling rate on its consolidation behaviour, are external factors. These processes and their interactions affect the behaviour and performance of CPB. Thus, an understanding of the properties of CPB and the factors that affect them is critical for the safe and cost-effective design of CPB structures. In this chapter, a comprehensive review of the current knowledge on the physical, thermal (T), hydraulic (H), mechanical (M), microstructural and chemical properties (C) of CPB as well as the factors that affect them and their interactions is provided. However, the properties for assessing the transportability of CPB (e.g. yield stress, viscosity) are not addressed in this chapter.

A. Ghirian • M. Fall (✉)
Department of Civil Engineering, University of Ottawa, Ottawa, ON, Canada
e-mail: mfall@uottawa.ca

E. Yilmaz, M. Fall (eds.), *Paste Tailings Management*,
DOI 10.1007/978-3-319-39682-8_4

2 Physical Properties of CPB and Influential Factors

The behaviour and performance of CPB are influenced by its physical properties, such as porosity (or void ratio), density (or unit weight), water content and degree of saturation. The physical properties of CPB have been investigated by several previous studies (e.g. le Roux et al. 2002; Fall et al. 2005, 2008; Belem et al. 2006; Yilmaz et al. 2009). However, most of these studies are laboratory investigations and the in situ physical properties of CPB have been only occasionally reported in the literature. This is mainly due to the difficulties associated with in situ sampling and testing, such as lack of access to backfilled stopes, mining activity interruptions, costs of the procedure and safety issues.

The field results obtained by le Roux et al. (2005) (which include a summary of the results obtained by Pierce (1997)) showed that there are variations in the physical properties of mine backfill due to numerous reasons, such as preparation and placement techniques, stress regime, and tailings and cement properties. They reported that at 90 days of curing, the void ratio ranges between 1.10 and 1.40, unit weight from 18.40 to 20.10 kN/m^2 and degree of saturation between 79 and 100% for the studied mine and CPB. Also, they compared the field and laboratory results obtained from tests conducted on the same CPB mix. They found that the in situ void ratios and degree of saturation are approximately 20 and 10% higher and lower, respectively, than the laboratory values.

Laboratory experiments carried out by Belem et al. (2006) on an undrained column that was 3.0 m in height showed that 91-day CPB samples have a void ratio that ranges between 0.85 and 0.97 and degree of saturation between 83 and 94%. In the drained column, however, the void ratio ranges from 0.77 to 0.91 and degree of saturation varies between 75 and 93%, thus generally lower values compared to those of the undrained column. They showed that the variations of the studied physical parameters depend on the location of the column and are not uniform with height.

Experimental studies carried out by Ghirian and Fall (2013) on the physical properties of CPB with the use of column experiments (Fig. 4.1) showed that three important mechanisms can influence the void ratio variation of the studied CPB with time and height: (1) decrease of the void ratio due to the progression of binder hydration. More hydration products fill the empty pores in the CPB matrix and therefore pore refinement takes place as curing time advances (Fig. 4.1a); (2) decrease of the void ratio as a result of drainage (Fig. 4.3a). CPB cured with a lower water-to-cement ratio (*w/c*) produces lower void ratios as less water fills the pores (Fig. 4.2). Moreover, water drainage results in an increase of the effective stress and consolidation of CPB; and (3) shrinkage of the CPB matrix because of evaporation. As can be seen from Fig. 4.1a, a high evaporation rate at the surface results in the development of shrinkage, which is associated with volume changes. A low void ratio (0.86–0.89) at the shrinkage limit can be observed in the 90- to 150-day samples, as shown in Fig. 4.1d.

Moreover, Fig. 4.1b shows that as an overall trend, the wet density (γ_{wet}) decreases and dry density (γ_{dry}) increases as the curing time proceeds. The reduction in the wet density with time is attributed to the water consumption by the binder hydration reaction as well as desaturation due to evaporation. However, due to the pore refine-

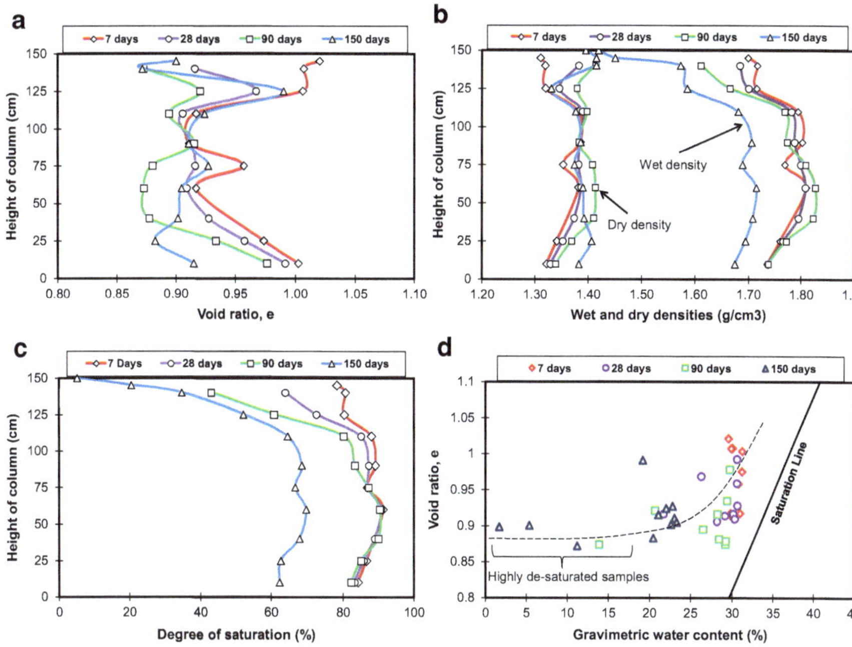

Fig. 4.1 Changes in physical properties in columns with different curing times: (**a**) void ratio, (**b**) wet and dry densities, (**c**) degree of saturation and (**d**) gravimetric water content (shrinkage curve) (Ghirian and Fall 2013)

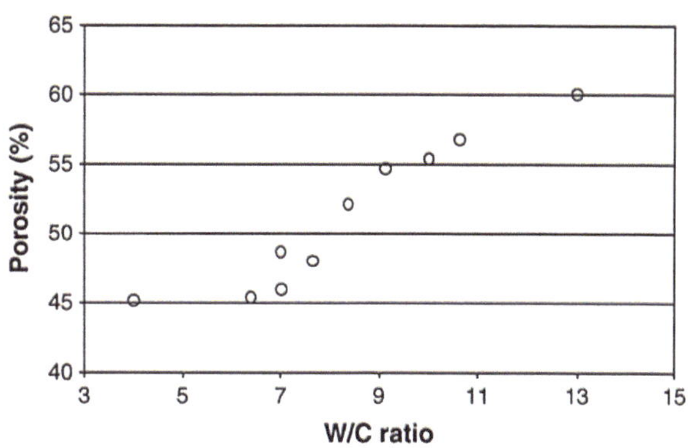

Fig. 4.2 Effects of water-to-cement ratio on porosity of CPB (Fall et al. 2010)

ment and filling of the voids with cement hydration products, an increase in the dry density can be observed as the curing time proceeds. A high rate of surface evaporation results in the lowest wet density value of 1.42 g/cm^3 at 150 days. The wet density shows an almost uniform behaviour in the column. It should also be emphasised that in underground mine environments, evaporation is limited.

In CPB material, the variation of the degree of saturation with time and height can be a function of the binder hydration and evaporation, respectively. As illustrated in Fig. 4.1c, the degree of saturation is reduced as curing time is increased as the result of self-desiccation due to cement hydration. However, the variations within the column are significantly influenced by the effects of evaporation. There is a slight increase of the degree of saturation in the middle layer and then a significant reduction towards the surface. For example, the degree of saturation is 85% in the middle of the column and then reduced to approximately less than 40% when closer to the surface of the column at 90 days.

Figure 4.1d presents the changes in the void ratio with respect to the gravimetric water content (shrinkage curve). A direct relationship between the water content and void ratio can be observed for a degree of saturation that is almost greater than 80%. The excessive desaturation resulted in further reductions of the void ratio until the shrinkage limit is reached, mainly due to cement hydration induced self-desiccation and evaporation, as can be observed in Fig. 4.1d.

Yilmaz et al. (2009) investigated the variations in the physical properties of CPB with respect to binder content (3, 4.5, 7 and 10%), for unconsolidated and consolidated 28-day samples. They reported that for the unconsolidated samples prepared with Portland cement (PC), the porosity ranges between 0.42 and 0.47, degree of saturation between 90 and 97% and gravimetric water content between 38 and 46%. For consolidated samples (cured under pressure), the porosity ranges between 0.41 and 0.45, degree of saturation between 80 and 94% and gravimetric water content between 33 and 42%. This study showed that the values of the physical properties (porosity, void ratio, degree of saturation, bulk density) of CPB decrease as the cement content increases due to the formation of hydration products and the effects of desaturation from cement hydration (in the unconsolidated samples). Also, a comparison between the unconsolidated and consolidated tests showed that drainage due to consolidation reduces the values of the physical properties. Yilmaz et al. (2014) studied the physical properties of CPB samples prepared with a blend of slag-Portland cement (3, 4.5 and 7%) that were cured for 7, 14 and 28 days under different drainage conditions [undrained, gravity drained and consolidated drained (CD)]. They reported that the gravimetric water content varies between 13 and 25%, degree of saturation between 67 and 98% and void ratio between 0.59 and 0.8. They found that the CD samples have a lower void ratio, water content and degree of saturation compared to the undrained samples, regardless of the cement content.

Ghirian and Fall (2015, 2016) investigated the physical properties of CPB samples in laboratory tests with regard to curing stress, curing time, filling rate and drainage conditions. The CPB samples were cured under zero stress (control sample, C–U) and under stress (up to 600 kPa) by using a curing cell system developed at the University of Ottawa, and considering both drained and undrained conditions. Also, three different backfilling rates of 0.155, 0.31 and 0.62 m/h, as well as three curing ages of 1, 3 and 7 days, were adopted in their studies (Fig. 4.3). The results showed that in all of the samples, the porosity (or void ratio) decreases as the curing time is increased. This is due to the ongoing cement hydration process, which produces more hydration products and thus causes the refinement of the pore structure with time (Fall and Samb 2008). Moreover, the samples cured under stress (both drained and undrained conditions, and cured under different filling rates) showed lower porosity compared to the control sample (C–U). This is mainly due to the pore refinement because of the application of pressure. Applied pressure can

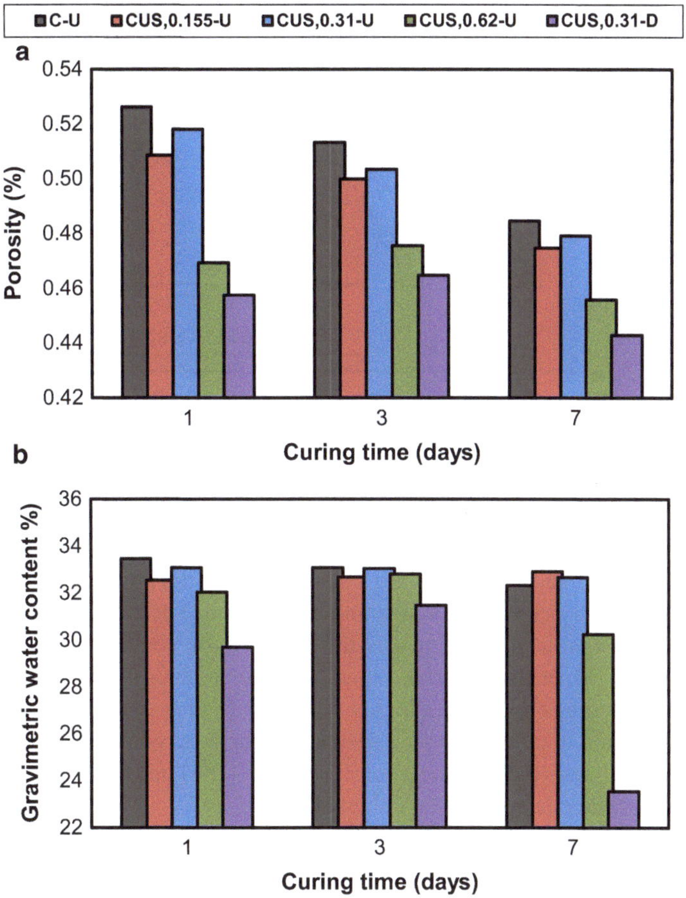

Fig. 4.3 Changes in physical properties: (**a**) porosity, and (**b**) gravimetric water content in load cell experiments (Ghirian and Fall 2016)

increase the packing density of CPB, which decreases the porosity. In the drained condition, however, the CPB samples showed considerable porosity reduction due to the applied pressure and effects of consolidation. Consolidation initially expels unbound water from the CPB pores due to volume changes until setting takes place at a later time of curing, when cementation in CPB and its water retention capacity are more developed.

Figure 4.3b shows that the gravimetric water content of all of the samples decreases with curing time. This is mainly due to the consumption of water by self-desiccation. The water content variation of the control sample and samples cured under stress is similar in the undrained condition, which suggests that the effect of pressure application on the water content is minimal in undrained loading conditions. However, the water content of the samples cured under stress (drained condition) shows a significant reduction as the curing time is increased due to consolidation-induced drainage.

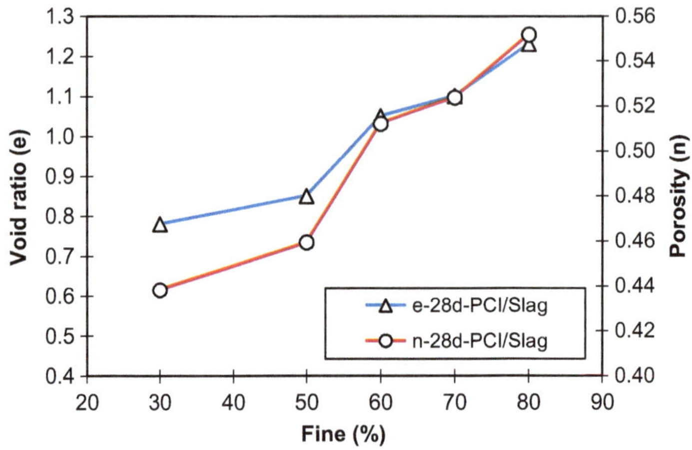

Fig. 4.4 Effects of fine content of tailings on void ratio and porosity of CPB (Modified from Fall et al. 2004)

The physical properties of CPB are also affected by the physical characteristics of the tailings, such as the particle size and fine content. The proportion of fine tailings (<20 μm) significantly influences the porosity (or void ratio) and pore size distribution of CPB, as illustrated in Fig. 4.4 (Fall et al. 2004). A higher proportion of fine particle tailings results in a higher void ratio (Fall et al. 2004). Also, there is a greater reduction in porosity with a decrease in the fines content in CPB samples made with fine (fines <60%) or medium (fines = 35–60%) fine tailings, compared to samples made with coarse tailings (Fall et al. 2004).

In summary, the above findings show that various factors, such as CPB mix components, curing time, curing stress, drainage and consolidation, affect the physical properties of CPB at the laboratory scale. Environmental factors such as evaporation, backfilling rate and stress can also influence the physical properties, but to a certain extent. The obtained results from both field and laboratory tests show that the physical properties of CPB widely differ from one backfill or mine to another. Also, they vary with respect to the height of the backfill. These variations are due to the tailings type, field conditions and placement, as well as the thermo-hydro-mechanical chemical (THMC) processes to which the CPB is subjected in field curing conditions.

3 Mechanical Properties of CPB and Influential Factors

Mechanical stability is one of the most important design criteria of CPB (Fall et al. 2007). The mechanical properties of CPB are usually related to the uniaxial compressive strength (UCS), shear strength properties or parameters, stress–strain behaviour and modulus of elasticity. A good understanding of the mechanical properties is key to assessing the mechanical stability and performance of any CPB structure.

The most common, economical and direct method to determine the mechanical strength and stress–strain behaviour of backfill materials is UCS testing (Fall et al. 2007). Other methods, such as direct shear and triaxial testing, can also be used to determine the shear strength properties and failure criteria of CPB (e.g. le Roux et al. 2005; Rankine and Sivakugan 2007; Fall et al. 2007; Ghirian and Fall 2014). In addition to these conventional testing methods, different indirect techniques have been recently used to assess the mechanical strength of CPB, such as through ultrasonic wave measurements, thermal profiling (maturity index) and measuring the shear wave velocity (e.g. Klein and Simon 2006; Mozaffaridana 2011; Ercikdi et al. 2014).

3.1 Shear Strength Properties or Parameters of CPB

There is limited available information on the shear strength properties or parameters (cohesion and internal friction angle) of CPB in the literature. Triaxial tests have been carried out in previous studies (e.g. le Roux et al. 2005; Fall et al. 2007; Rankine and Sivakugan 2007; Veenstra 2013) to determine the shear strength properties of backfill which showed different values of the internal friction angle (φ) with respect to curing time, binder type and content, and tailings type. For example, Belem et al. (2000) performed triaxial CD compression tests on CPB samples and reported that the friction angle of the 112-day samples varies from 5 to 28°, depending on the binder type and content (3, 4.5 and 6%). They showed that the friction angle decreases as cohesion of the backfill increases. Also, they found that at a certain curing time, the friction angle decreases when the percentage of the cement content is increased. Rankine and Sivakugan (2007) conducted triaxial CD tests on CPB samples and reported that the friction angle is slightly reduced from 38° at 14 days to 37.9° at 28 days, and from 35.7° at 14 days to 32.7° at 28 days for samples with a binder content of 2 and 6%, respectively. Pierce (1997) reported that the cohesion significantly increases when the binder content and curing time are increased. On the other hand, the friction angle decreases when the curing time is increased. le Roux et al. (2005) conducted triaxial tests and reported that the friction angle ranges between 32 and 37° for an undisturbed unsaturated CPB sample at 90 days (field-cured sample), while laboratory-prepared samples from the same CPB mix showed a friction angle of 37°, thus suggesting that there is no significant change in the friction angle due to in situ curing conditions. Direct shear tests conducted by Veenstra (2013) showed that the friction angle is reduced when curing time is increased. Different tailings grain size (from clayey silt to fine sand) and different binder contents were used to prepare the CPB samples. The typical values obtained ranged on average from 38° to 43° at day 1 and from 23° to 36° after 7 days of curing.

A review of the previous work shows that there is a slight decrease in the friction angle with curing time, mostly taking place in the long term. This reason is not well understood, but most likely attributed to different mechanisms, such as the chemical processes related to cement hydration reactions, oxidation of mine backfill (weathering), self-desiccation, drainage, in situ stress and presence of air voids in the backfill (Rankine 2004; Rankine and Sivakugan 2007). The changes in the friction angle

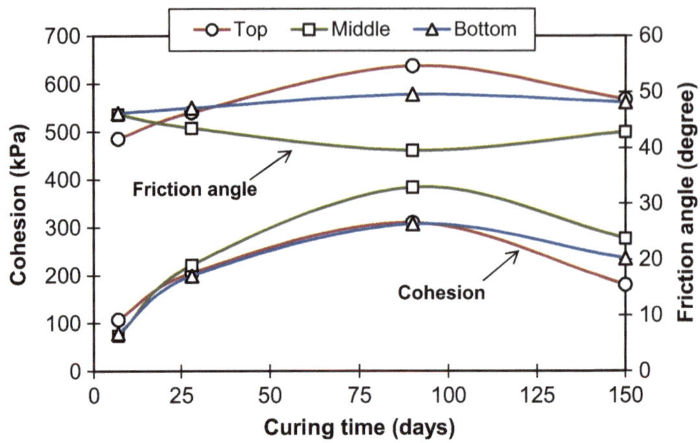

Fig. 4.5 Changes in shear strength parameters in column experiments (Ghirian and Fall 2014)

with increased curing time in the column experiments conducted by Ghirian and Fall (2014) showed that the friction angle of the CPB samples taken from the middle of the column decreases from 46° at 7 days to 39° at 90 days (Fig. 4.5). However, the samples taken from the top and bottom of the column showed different behaviours due to the effects of drying shrinkage and drainage. In these samples, the friction angle is increased with curing time from approximately 42° at 7 days to approximately 52–55° at 90 days followed by a reduction to 48° at 150 days. Ghirian and Fall (2015) also found changes in the friction angle in the short term in their load cell experiments, as illustrated in Fig. 4.6. The friction angle of the undrained CPB samples cured under stress (CUS-U) is slightly decreased as the curing time is increased, while the friction angle of the control sample (cured under zero stress) remains almost constant. It can also be observed that the curing stress does not have a significant influence on the changes in the friction angle.

Cohesion in cemented backfill materials is normally developed through cementation by the bonding that takes place between the tailings particles and cement hydration products. Other sources of cohesion development in CPB can result from desaturation induced by self-desiccation and/or evaporation. Cohesion is a time-dependent property and can develop up to 1.5 MPa, depending on the binder type and content, solid content and curing time (Rankine and Sivakugan 2007; Fall et al. 2007; Veenstra 2013; Ghirian and Fall 2013, 2014). The majority of the previous studies that have examined cohesion development in backfill reported an increase with curing time. For example, Veenstra (2013) reported that cohesion values obtained from direct shear tests increase from 40–130 kPa at 7 days to 240–680 kPa at 21 days for the studied materials. Rankine and Sivakugan (2007) performed a series of undrained unconsolidated (UU) tests and reported that the changes in undrained cohesion with longer curing time for a cement content of 6% in CPB range from 158 to 384 kPa for 1 to 9 months of curing. The reported results in Ghirian and Fall (2013) (Fig. 4.5) for column experiments showed that cohesion increases as curing time is increased from 83 kPa at 7 days up to 400 kPa at 90 days. It should be noted that the cohesion at 90

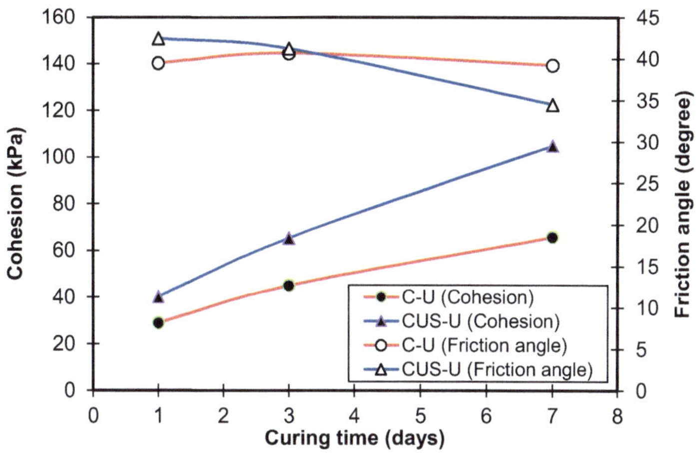

Fig. 4.6 Changes in shear strength parameters in load cell experiments (Ghirian and Fall 2014)

to 150 days decreases despite the continual development of cement hydration due to the effects of deterioration as a result of surface evaporation at the top of the column and associated shrinkage and microcracks. The results of a study by Ghirian and Fall (2015) revealed that cohesion and bonding in the cement matrix can be improved when curing stress is applied, as illustrated in Fig. 4.6.

3.2 Stress–Strain Behaviour of CPB

The stress–strain behaviour of CPB under compression loading was investigated by Fall et al. (2007). They studied the effects of different factors on the stress–strain behaviour and modulus of elasticity of CPB, such as aging, confinement pressure as well as characteristics of the main components of CPB (cement, tailings and water). Figure 4.7 shows that the stress–strain response of CPB is time dependent. Moreover, aging causes stiffening of the material (increase in the modulus of elasticity) which results in a higher peak stress and smaller strain to the peak stress (Fall et al. 2007).

Moreover, the stress–strain behaviour of CPB can be significantly affected by confinement as illustrated in Fig. 4.8. The results of triaxial compression tests showed that as the confinement is increased, the peak stress and strain at failure also increase. Furthermore, at zero or low confinement, distributed microcracks and several macro-cracks developed in the CPB samples, and the stress–strain curves, have a sharper peak at lower strains. However, higher confinement causes elasticity and then prolonged plastic behaviour with higher peak stress and larger strain at failure. This behaviour can be attributed to the friction between the fractured surfaces in the CPB after the initial cracking of the sample. The confining holds the fractured surface together and results in additional resistance of the sample (Fall et al. 2007).

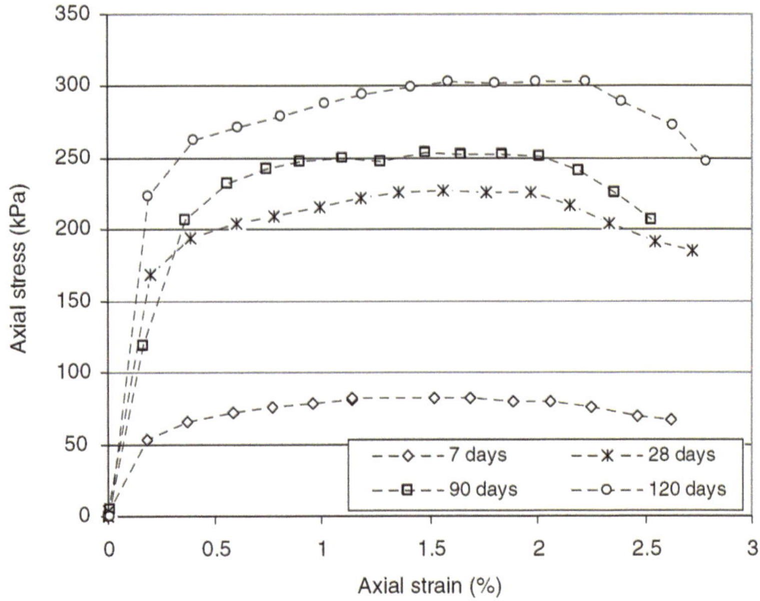

Fig. 4.7 Effects of aging on stress–strain behaviour of CPB samples with 3.0% PCI/slag (Fall et al. 2007)

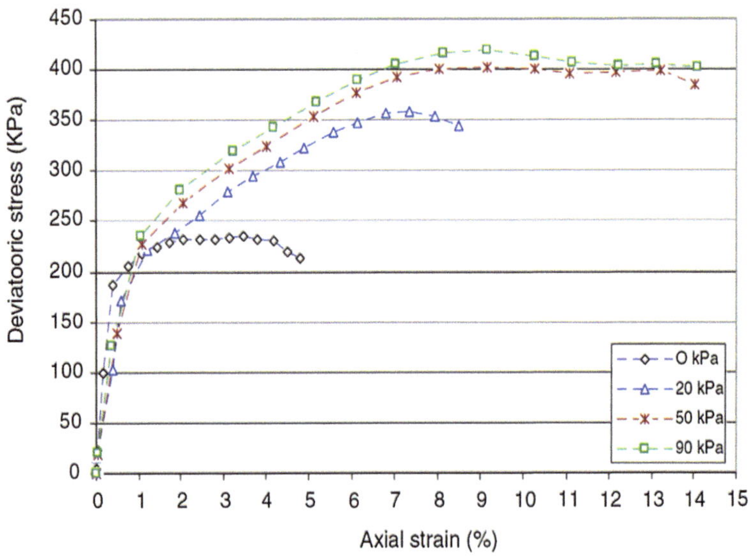

Fig. 4.8 Effect of confinement on stress–strain behaviour of 28-day CPB samples with 7% PCI (Fall et al. 2007)

Figure 4.9 shows that the cement content also considerably influences the stress–strain behaviour of CPB. It can be seen that the peak stress and modulus of elasticity of the CPB samples increase as the binder content is increased for a specific curing time. A higher binder content results in a well-defined peak stress and softening after failure, while a lower binder content causes relatively ductile failure of the sample (Fall et al. 2007).

The *w/c* ratio significantly affects the stress–strain behaviour of CPB as shown in Fig. 4.10. At a given strain, a lower *w/c* can sustain higher stress before and after the peak stress, as well as contribute to higher elasticity in the CPB samples. This is attributed to the fact that CPB prepared with a lower *w/c* has an overall lower porosity and denser microstructure, which will result in an increase in strength and modulus of elasticity (Fall et al. 2007).

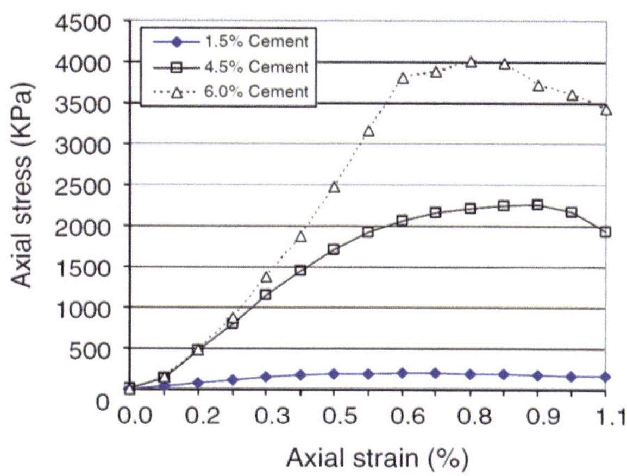

Fig. 4.9 Effect of binder content on stress–strain behaviour of 28-day CPB samples with PCI (Fall et al. 2007)

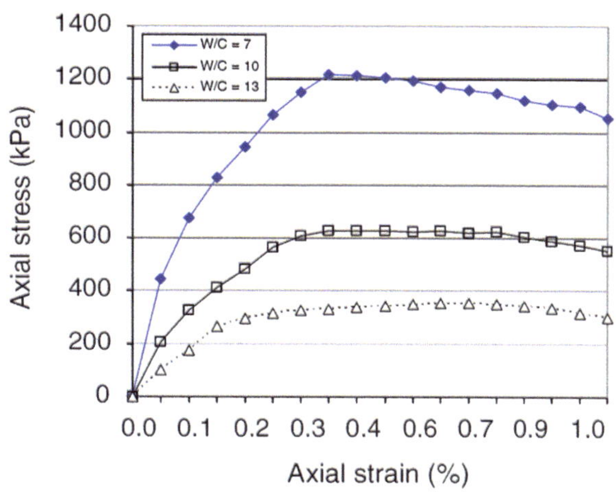

Fig. 4.10 Effect of *w/c* on the stress–strain behaviour of 28-day CPB samples with 4.5% PCI/slag (Fall et al. 2007)

The effect of the fine content of the tailings on the stress–strain behaviour of CPB is presented in Fig. 4.11. It can be observed that the peak stress and modulus of elasticity increase as the fine content of the tailings material is decreased. The failure modes of the samples prepared with different fine content are almost similar. However, a reduction in the fine content to less than 50% of fines does not considerably change the peak stress and modulus of elasticity. The fine content effect can be attributed to the fact that CPB with finer particles requires more water to sustain the required consistency (or slump), thus resulting in a higher *w/c* and thereby lower strength of the CPB (Fall et al. 2007).

3.3 Strength (UCS) of CPB

The required UCS for a CPB structure varies and mostly depends on several factors, such as the application of backfill and the subjected loading conditions. For example, CPB used for void filling typically requires very low compressive strength (e.g. 150–300 kPa), while free-standing or roof support backfill demands a high compressive strength between 1 and 4 MPa (Fall et al. 2005).

Numerous studies have been conducted on the compressive strength of CPB (e.g. Klein and Simon 2006; Pokharel and Fall 2011; Cihangir et al. 2012; Ercikdi et al. 2014). The main outcomes of these studies are that the short-term and long-term mechanical strength of CPB can be significantly affected by variations of several internal and external factors. The internal factors include the physical, chemical and mineralogical characteristics of CPB components such as the tailings, cement and mixing water. The external factors refer to the curing temperature, curing conditions (including curing time and curing stress), drainage condition, consolidation as well as placement technique such as filling rate.

The tailings characteristics, including the mineralogy content, chemical properties (e.g. amount of sulphide), fine content, density and grain size, can affect the mechanical strength of CPB (Fall et al. 2005, 2008). Also, the binder type and content, a blend ratio of two or more binders as well as the chemical properties of the binder can influence the development of the mechanical strength (Kesimal et al. 2005; Fall et al. 2008). The water content (with respect to the *w/c*) and chemistry of the mixing water (e.g. mine processing, lake or saline water) also affect the mechanical properties of CPB.

The effects of the *w/c* ratio and binder content on the compressive strength of CPB are shown in Fig. 4.12. It can be observed that at any given binder content, a reduction in the *w/c* ratios results in an increase in the UCS values (Fall et al. 2008). This effect is mainly because a reduction in the *w/c* ratio decreases the overall porosity of the backfill and consequently increases the UCS value (Fall et al. 2008). Also, an increase in the binder content increases the UCS values. A higher cement content produces more cement hydration products, which in turn increases the cohesion and reduces the porosity and packing density, and thus there is an increase in the UCS. The fine content (particles <20 µm) in tailings can also significantly influence the strength,

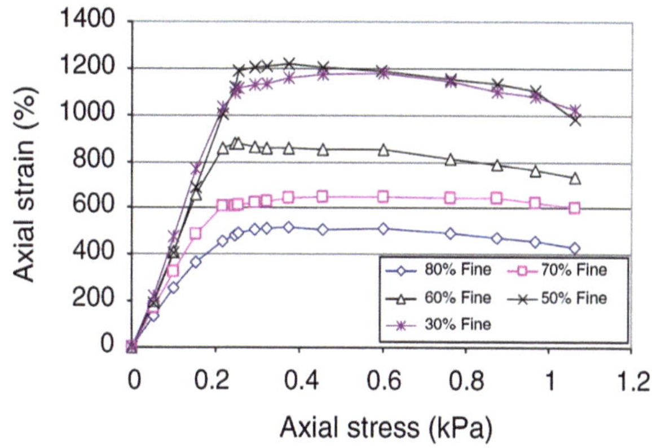

Fig. 4.11 Effect of fine content of tailings on the stress–strain behaviour of 28-day CPB samples with 4.5% PCI/slag (Fall et al. 2007)

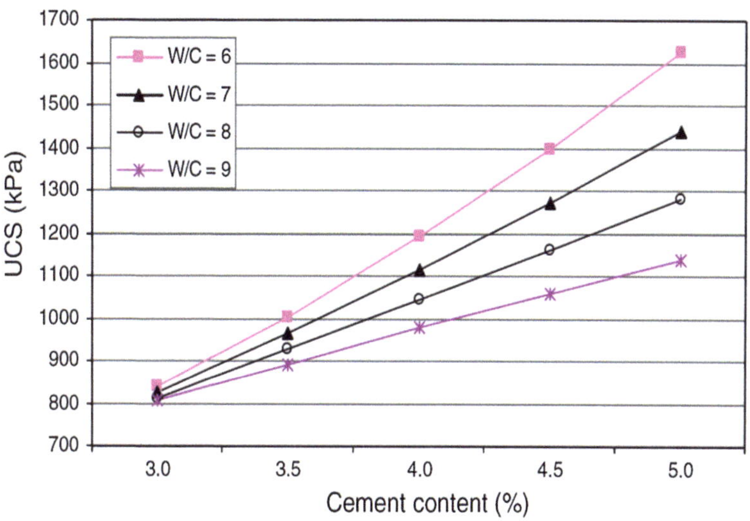

Fig. 4.12 Effect of *w/c* ratio on UCS of 28-day CPB samples with different binder contents (Fall et al. 2008)

as illustrated in Fig. 4.13. In general, a higher compressive strength is obtained when tailings contain 40–45% fine particles (Fall et al. 2008). A reduction in the fine content of the tailings (coarse tailings) reduces the void ratio and leads to the pore refinement of the CPB matrix. This in turn can result in reductions in the *w/c* ratio to sustain the CPB consistency, as well as reduced packing density of the backfill, which can consequently lead to strength gain (Fall et al. 2004, 2005).

Figure 4.14 shows the effect of the tailings density and sulphur content in the tailings on the UCS development of the CPB samples with increased curing time. It can be seen that a higher tailings density results in greater mechanical strength due

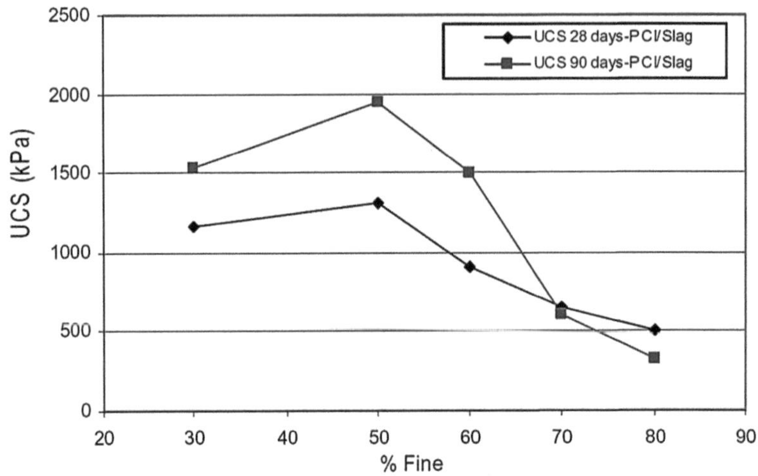

Fig. 4.13 Effect of proportion of fines in tailings on strength development of CPB (Fall et al. 2004)

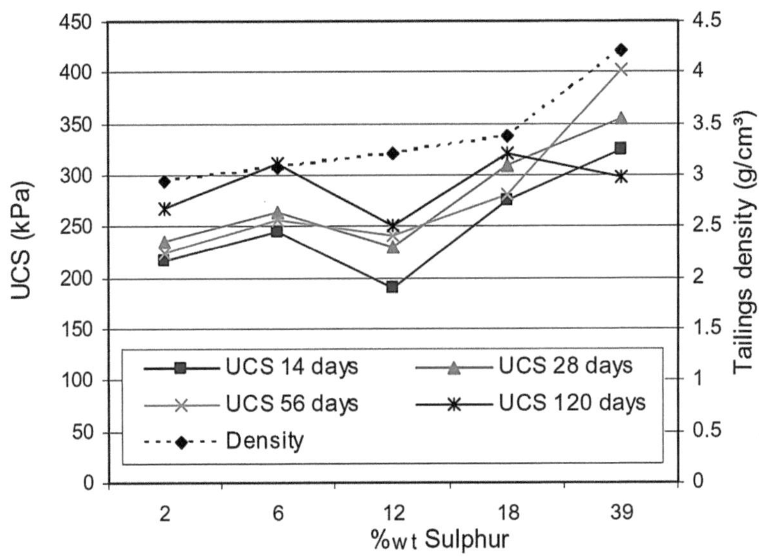

Fig. 4.14 Effect of tailings sulphur content on tailings density and strength development of CPB with PC I/PC V (50/50) (Fall et al. 2004)

to higher binder consumption (Fall et al. 2004). The figure also shows that an increase in the sulphur content results in increased UCS for a given curing time. However, the sample with a large amount of sulphur (39%) at an advanced age (120 days) has the lowest strength compared to the other samples (Fall et al. 2004). This is attributed to the presence of sulphur compounds, such as sulphide minerals and soluble sulphates, which negatively affect the CPB strength due to sulphate attacks (Kesimal et al. 2004; Pokharel and Fall 2011). Similar findings can also be obtained in a study conducted by Li and Fall (2016) on the early age strength development of CPB that contains sulphate; see Fig. 4.15. This figure shows that a higher volume of sulphate results in the development of lower strength of the backfill at the early ages, which is due to the inhibition of cement hydration reactions and fewer hydration products developed. Figure 4.16 shows that the weight loss or peak at 400–450 °C of the CPB sample without sulphate is greater than that with a sulphate content of 25,000 ppm. This indicates that fewer hydration products, such as CH, are produced in CPB with high amounts of sulphate. Other factors, such as the formation of expansive minerals (e.g. ettringite), coarsening of the CPB pore structure and sulphate absorption by calcium silicate hydrate (C–S–H), contribute to the strength loss of CPB (Li and Fall 2016).

The binder characteristics, such as type, content, blend ratio and chemical properties (e.g. the soluble sulphate concentration), have a significant role in the strength development of CPB. Benzaazoua et al. (2002) investigated the UCS development in CPB samples with three different sulphur contents (5, 16 and 32 wt.%) and different binder types and ratios [i.e. ordinary PC (types I and V), slag and fly ash (FA)]. They reported that for tailings with low to medium sulphur contents, only binders with a mix of PC type I and FA, mix of PC type I and slag, and slag provide better UCS in CPB samples compared to binders with a mix of PC types I and V. However, binders with a mix of PC types I and V are more appropriate for tailings with high contents of sulphur and can deliver better compressive strength. Slag-based binders or a mix of PC and slag was not recommended by Benzaazoua et al.

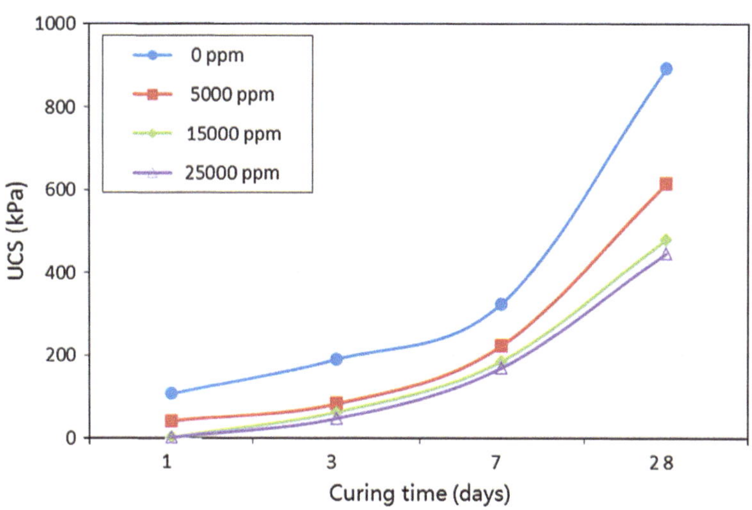

Fig. 4.15 Effect of initial sulphate content on development of UCS of CPB at early ages with 4.5% PC I (Li and Fall 2016)

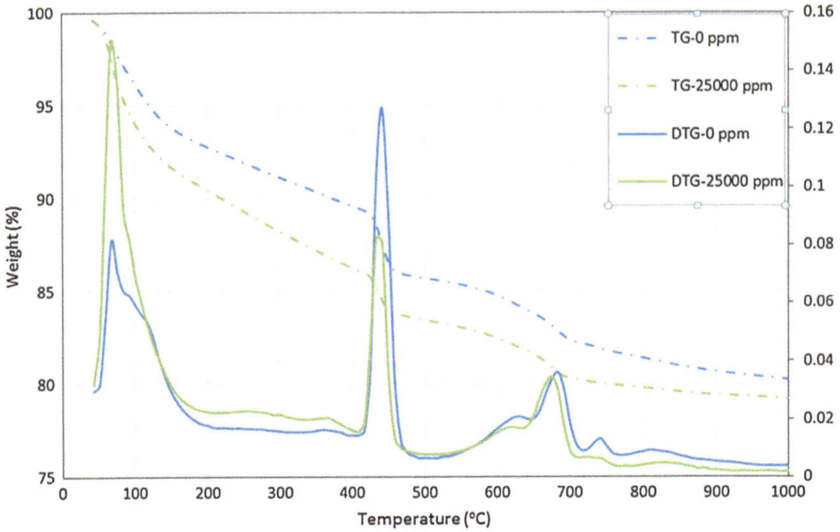

Fig. 4.16 Effect of sulphate content on thermal behaviour of the cemented pastes of 3-day CPB samples (Li and Fall 2016)

(2002) for such tailings. Kesimal et al. (2005) reported that binders that contain lower contents of pozzolanic additives (14%) deliver higher short-term mechanical strength to CPB compared to binders with a large amount of pozzolanic additives (29%). However, a large amount of pozzolanic additives in the binder can provide higher long-term strength to CPB made of sulphide-rich mine tailings.

Benzaazoua et al. (2004) discussed the effect of the blend ratio of the binder on UCS development in CPB samples with different combinations of PC type I (PC I) and type V (PC V), while the mass proportion was kept constant at 4.5%, and mixing water with a sulphate concentration of 2000 ppm was used. The results of the UCS tests showed that ratios of 50/50 and 60/40 provide the highest strength to the CPB samples at 28 days, due to the difference in the rate of strength gain between PC I and V, and the volume of C_3A and CH as a result of the cement hydration reactions (Fall et al. 2005). However, the sample with a 70/30 ratio delivered high strength at 7 days. Ultimately, high concentrations of C_3A and/or C_3S in PC I led to decreases in the sulphate resistance of CPB and therefore reductions in strength at 28 and 56 days. The binder ratio with a more sulphate-resistant cement (PC V) had low early strength, but the highest UCS at an advanced age (Fall et al. 2005, Fig. 4.17).

The mixing water chemistry and properties, including the chemical concentration and water-to-binder ratio, can affect the strength acquisition process. The water chemistry (e.g. mine processing water can contain large amounts of sulphate) can significantly change the chemical composition of cement and the hydration process. It can also affect the formation of primary and secondary cement hydration products, which are responsible for strengthening the backfill and its durability in the long term (Benzaazoua et al. 2002; Fall and Pokharel 2010; Pokharel and Fall 2011, 2013).

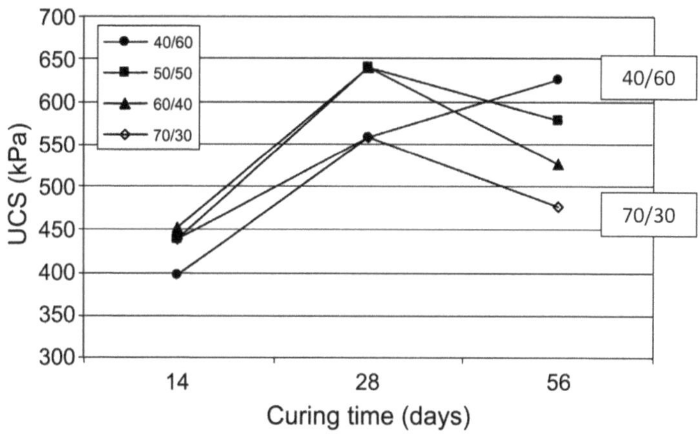

Fig. 4.17 Effect of binder mixing ratio (PC I/PC V) on strength development of CPB with 4.5% binder content and mixing water with 2000 ppm sulphate (Fall and Benzaazoua 2005; modified)

As discussed earlier, in addition to the internal factors (tailings, cement and mixing water properties), several external factors can also influence the UCS development in the backfill. One of the most important external factors is temperature (e.g. curing and initial backfill temperatures) which can significantly influence the short-term and long-term mechanical strength, pore structure and performance of CPB materials (e.g. Fall and Samb 2008; Fall et al. 2007, 2010).

Fall et al. (2010) studied the effect of the curing temperature on the short-term and long-term UCS development, as well as the tensile strength. Figure 4.18 shows that UCS increases as the curing temperature is increased. The reason is that a higher temperature causes a rapid hydration reaction in the cement, and thereby the formation of more hydration products (e.g. C–S–H, CH) in the backfill. This can then lead to pore refinement and a denser CPB matrix. Figure 4.19 shows that regardless of the binder type, the CPB strength increases with curing time up to 150 days, except for the PC I/slag samples cured at temperatures above 35 °C. This loss of strength of the PC I/slag samples is due to the "crossover effect", in which curing at a higher temperature produces a coarser pore structure in the cementitious materials (Fall et al. 2010). Then, the hydration products are not uniformly distributed in the pores, thus resulting in higher porosity and lower strength (Alexander and Taplin 1962).

Fall et al. (2010) also investigated the combined effects of fine content and temperature on UCS development. It is observed in Fig. 4.20 that for a low curing temperature (≤20°C), the CPB strength is reduced with increased fine content. The increased fine content results in an increase in the porosity, which eventually produces CPB with lower mechanical strength. However, after 28 days of curing at high temperatures (≥35 °C), CPB with 35% fine tailings has a higher strength than CPB with 45% fine tailings because the former has the optimal number of pores to accommodate the hydration products formed (Fall et al. 2010).

Fall and Pokharel (2010) studied the coupled effect of temperature and sulphate on CPB strength. This interaction of temperature with sulphate in cement reactions

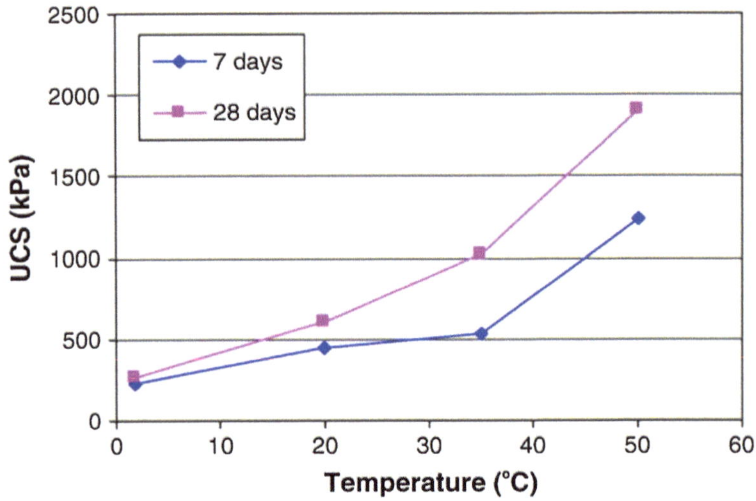

Fig. 4.18 Effect of temperature on UCS development of CPB with 4.5% PC I and *w/c* = 7.6 (Fall et al. 2010)

can have positive or negative effects on the UCS depending on the amount of sulphate and the curing temperature, as illustrated in Fig. 4.21. Fall and Pokharel (2010) reported that temperatures over 20 °C and sulphate concentrations up to 15,000 ppm contribute to the UCS development in CPB samples cured for 28 and 150 days. However, the samples with an initial sulphate content of 25,000 ppm generally show much lower strength, especially at a curing temperature of 50 °C, regardless of the curing time. Higher curing temperatures (≥35°C) and initial sulphate contents (≥15,000 ppm) lead to absorption of a larger amount of sulphate ions by the C–S–H, which may lead to the formation of lower quality C–S–H, and thus results in reduced CPB strength.

CPB structures can be subjected to high temperatures due to the self-heating of sulphidic rock masses (oxidation of the sulphide minerals) or accidental fires (Fall and Samb 2008, 2009). This loading condition under an extreme temperature can significantly impact the UCS development in CPB. Figure 4.22 shows that the UCS increases as the temperature is also increased, up to 200 °C. However, there is lower UCS from 200 to 600 °C due to the thermal decomposition of the hydrates as well as the development of microcracks at the interface between the tailings particles and cement matrix as a result of excess vapour pressure in the CPB pores (Orejarena and Fall 2008). However, Orejarena and Fall (2008) indicated that the adverse thermal effects can be reduced by using more binder in the backfill.

One of the most important external factors that significantly affects the UCS of CPB structures are the curing conditions, including the curing stress (or self-weight pressure) and drainage conditions, as well as the filling rate of the backfill. The effect of curing under stress on UCS development is presented in Fig. 4.23. This figure shows that samples cured under stress in an undrained condition show higher

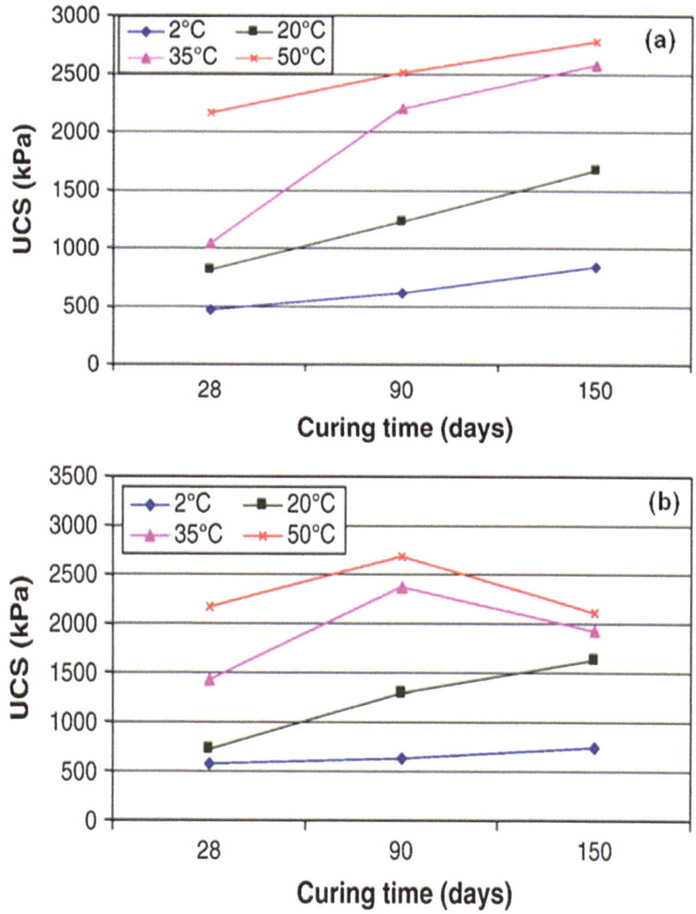

Fig. 4.19 Effect of curing temperature on the development of long-term strength of CPB with: (**a**) PC I, (**b**) PC I/slag (50/50), *w/c* = 7.6 and cement = 4.5% (Fall et al. 2010)

early-age mechanical strength compared to those cured under zero stress, due to the effect of the application of pressure on the microstructure or pore structural changes. Curing under stress causes particle rearrangement which then increases the packing density of the CPB material. Then, there is a reduction in the total porosity and void ratio, which can lead to compressive strength gain (Ghirian and Fall 2015). In contrast to the undrained condition, consolidation can take place if CPB samples are cured under stress in a drained condition. The effect of consolidation and drainage (cured under zero stress) on UCS development is shown in Fig. 4.24. It can be observed that the drained samples (cured with and without stress) exhibit higher mechanical strength compared to the sample cured under zero stress and the undrained sample. The UCS of the consolidated sample (drained and cured under stress) has the highest value for all of the curing times. Also, drainage itself, without the

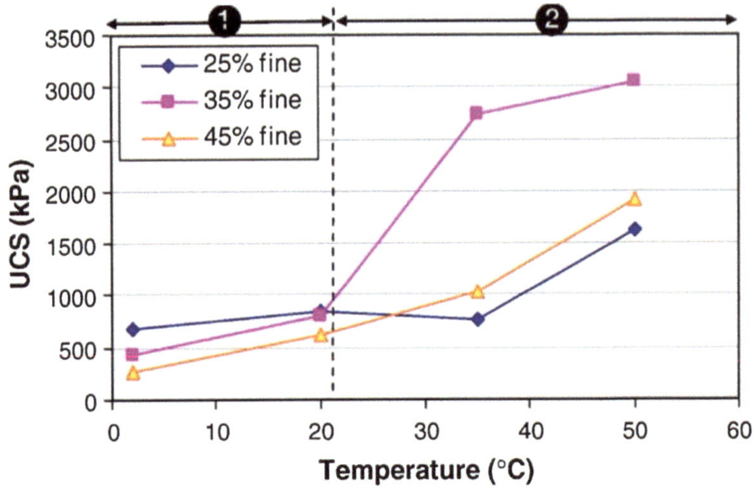

Fig. 4.20 Combined effect of fine content of tailings and curing temperature on strength of CPB at 28 days of curing with w/c = 7.6 and 4.5% PC I (Fall et al. 2010)

application of curing stress, significantly improves the mechanical strength of the backfill. The gain in strength of the consolidated samples is from the pore pressure dissipation due to the compression of the CPB pores, and as a result, there is development of effective stress in the backfill. This provides stronger cementation and bonding in the CPB matrix. Moreover, water drainage can result in curing of CPB under a lower *w/c* ratio. As explained earlier, a lower *w/c* ratio helps with the formation of denser CPB pores, eventually delivering higher strength to the backfill (Ghirian and Fall 2016). Figure 4.25 presents the effect of the filling rate on the UCS development in the undrained condition. It can be observed that CPB with a higher filling rate exhibits somewhat higher UCS values, except for the CUS 0.62-U sample on the first day of curing. The 0.62 m/h corresponds to the filling rate, which is fast and thus generates high pore pressure in CPB at a very early age, which results in almost zero effective stress development in the CPB. This, in turn, causes low mechanical strength development in CPB (Ghirian and Fall 2016).

4 Hydraulic Properties of CPB and Influential Factors

The primary hydraulic properties and processes in CPB materials include hydraulic conductivity (saturated and unsaturated) of the backfill and generation of pore water pressure (PWP; positive or negative) in the cemented backfill. The hydraulic conductivity of CPB can be expressed in two states: saturated and unsaturated conditions.

Some of the previous research works have examined the saturated hydraulic conductivity (e.g. Godbout and Bussière 2007; Fall et al. 2009; Pokharel and Fall 2013; Ghirian and Fall 2013), as well as the unsaturated hydraulic conductivity (e.g.

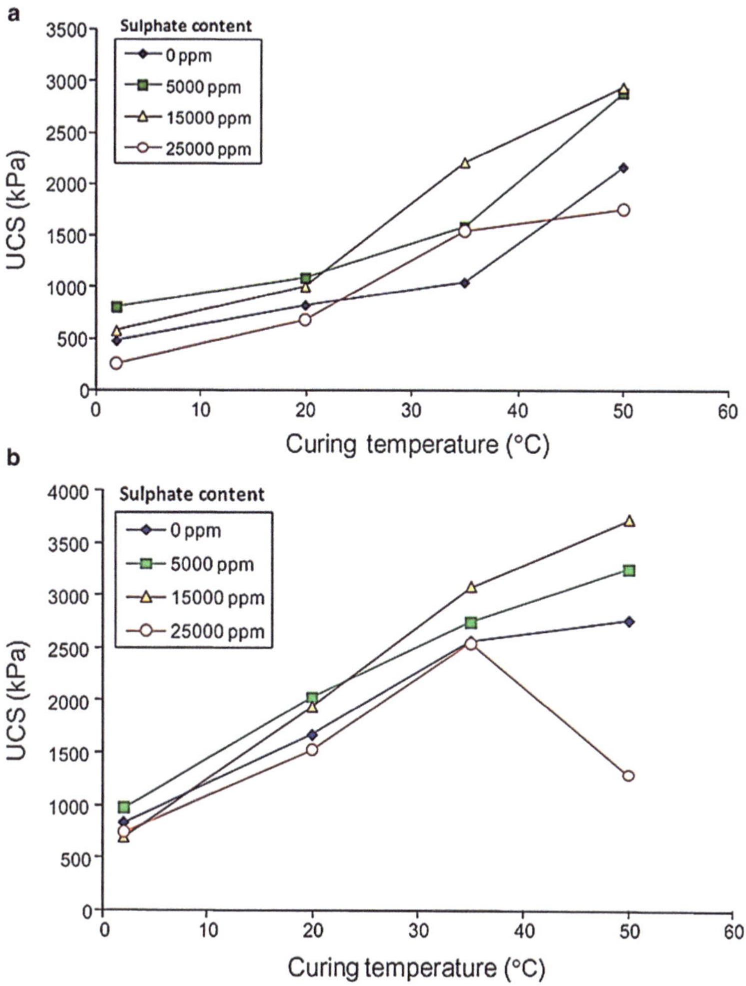

Fig. 4.21 Coupled effect of temperature and initial sulphate content on UCS of (**a**) 28-day and (**b**) 150-day CPB samples with $w/c = 7.6$ and 4.5% PC I (Fall and Pokharel 2010)

Witteman and Simms 2011; Abdul-Hussain and Fall 2011) of CPB materials. Saturated hydraulic conductivity (k_{sat}) is necessary to examine the consolidation behaviour and drainability as well as assess the transportability of backfill in a fluid state, such as groundwater flow in the backfill and/or between the CPB and its surrounding environment. Information on the unsaturated hydraulic conductivity (k_{unsat}) and properties [e.g. water retention capacity (WRC) and air entry value (AEV)] is necessary to assess the environmental performance of backfill structures, such as the generation of acid rock drainage and leakage potential of the mine backfill (specifically, backfill with sulphate-bearing tailings). A review of the available and recent body of research work related to the hydraulic properties of CPB materials is presented in this section.

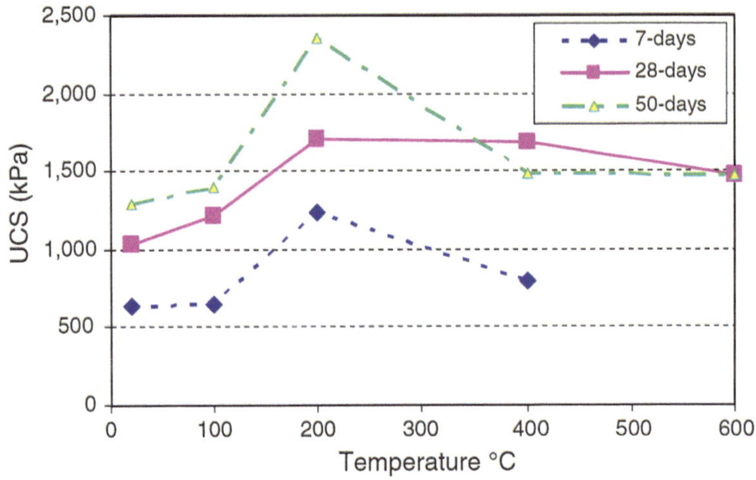

Fig. 4.22 Effect of high temperature on strength development of CPB with PC I (4.5% and *w/c* = 7.0) (Orejarena and Fall 2008)

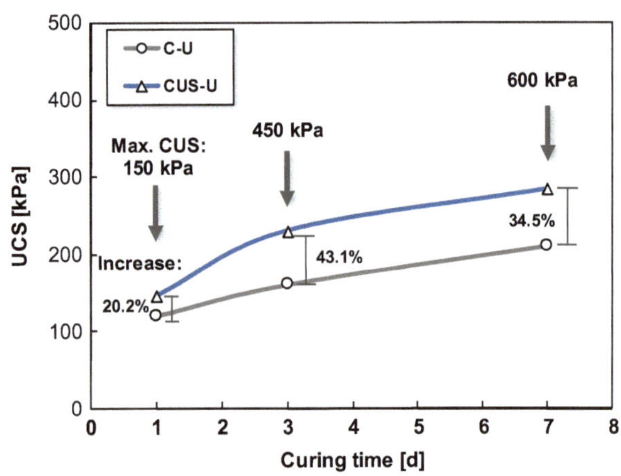

Fig. 4.23 Effect of curing under stress on UCS development (Ghirian and Fall 2015)

The saturated hydraulic conductivity of CPB is a property that changes with curing time. The changes can be observed in the hydraulic properties obtained from the column experiments performed by Ghirian and Fall (2013). Figure 4.26 shows that the saturated hydraulic conductivity values decrease as the curing time is increased, due to the refinement of the pores as a result of the cement hydration process. Also, Ghirian and Fall (2013) observed that the variations in the saturated hydraulic conductivity are not uniform inside the columns. The saturated hydraulic conductivity of the CPB samples located at the top of the column at the more advanced ages is significantly reduced. This is due to the effect of the microcracks from surface evaporation, which leads to the creation of preferential liquid paths in the CPB, and thus higher permeability (Ghirian and Fall 2013).

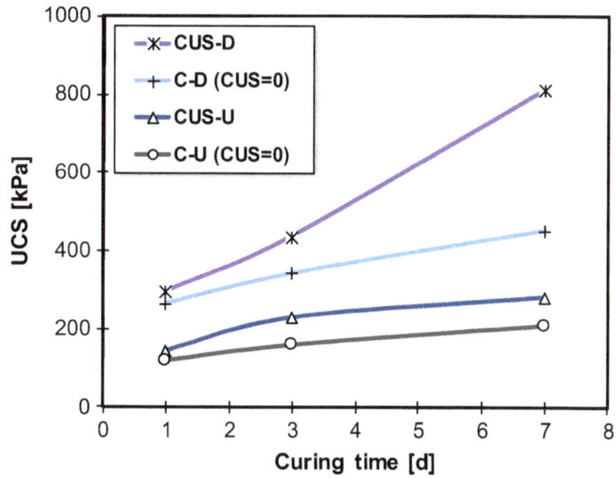

Fig. 4.24 Effect of curing under stress and drainage (consolidation) on UCS development at the early ages (Ghirian and Fall 2016)

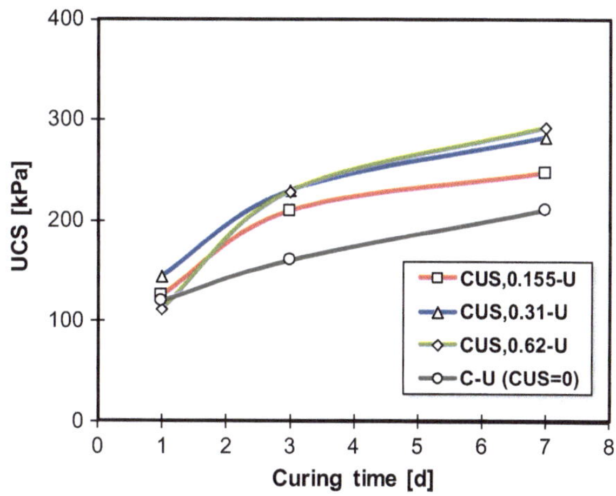

Fig. 4.25 Effect of filling rate on UCS development at the early ages (Ghirian and Fall 2016)

There are many factors that affect the saturated hydraulic conductivity of CPB. Internal factors such as the fine content of the tailings, *w/c* ratio, binder content and type, and sulphate content, as well as external factors such as the curing temperature and curing stress, can affect the hydraulic properties of CPB. Figure 4.27 shows the effect of the fine content of tailings on the changes in the saturated hydraulic conductivity, where medium tailings with 45% fine particles reduce the hydraulic conductivity of CPB, compared to coarse tailings (≤32% fine content), for a given consistency. This is due to a reduction of the packing density of the tailings materials and coarsening of the pore structure as the fine content is reduced (Fall et al. 2009).

Figure 4.28 shows that the permeability of CPB is reduced with a lower *w/c*. This is more evident at the early ages of curing compared to the advanced ages because a lower *w/c* leads to reduced porosity and faster binder hydration. These then cause pore refinement and result in the reduction of the transportability of backfill in the

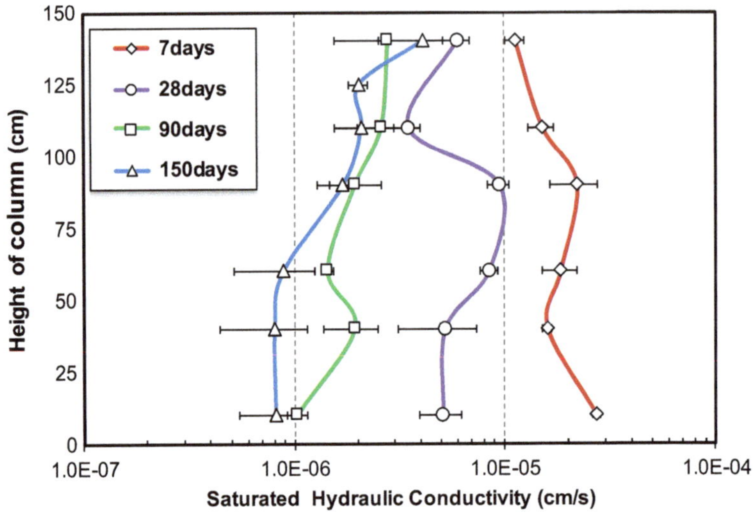

Fig. 4.26 Changes in saturated hydraulic conductivity in column experiments (Ghirian and Fall 2013)

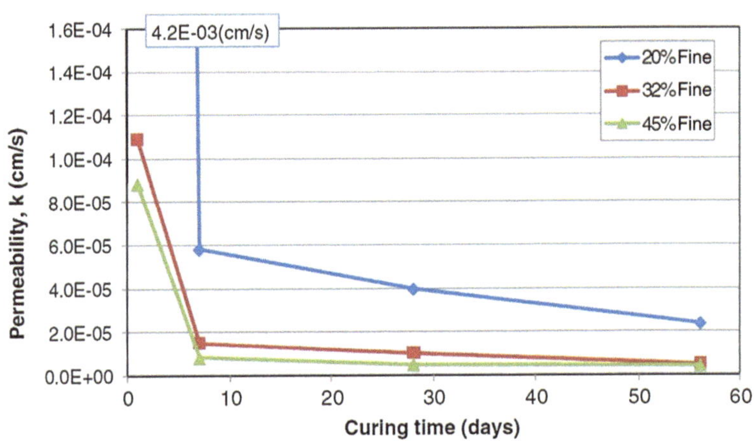

Fig. 4.27 Effect of fine content on saturated hydraulic conductivity of CPB with 4.5% PC I and slump = 18 cm (Fall et al. 2009)

fluid state (Fall et al. 2009). Pokharel and Fall (2013) studied the combined effects of temperature and sulphate content on the saturated hydraulic conductivity of CPB. Generally, a higher curing temperature results in lower hydraulic conductivity, regardless of the sulphate content (except for samples with 25,000 ppm of sulphate and cured at 50 °C); see Fig. 4.29. High curing temperatures increase cement hydration and thereby more hydration products are generated that fill the pores in the CPB matrix. Also, a moderate sulphate content and temperature produce secondary hydration products. These products then fill the CPB pores and reduce the

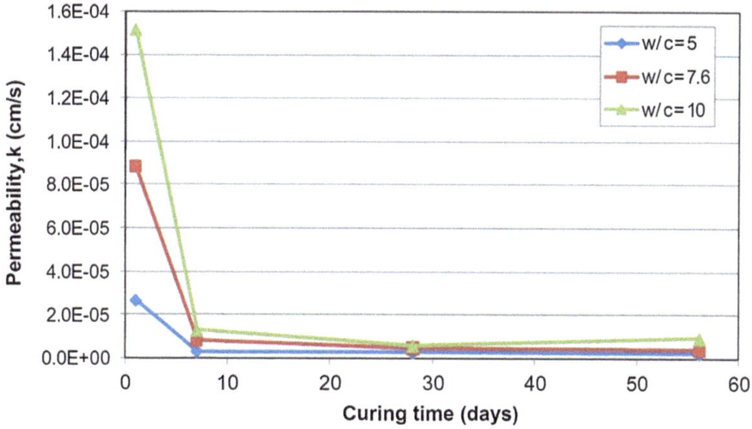

Fig. 4.28 Effect of *w/c* ratio on saturated hydraulic conductivity of CPB with 4.5% PC I and slump = 18 cm (Fall et al. 2009)

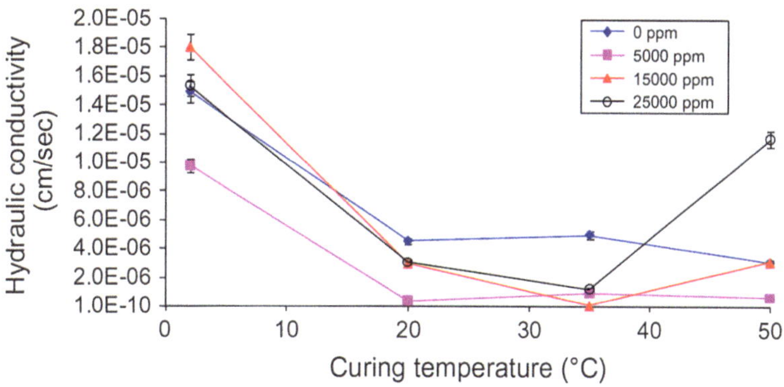

Fig. 4.29 Coupled effects of sulphate and curing temperature on hydraulic conductivity of 90-day CPB samples with 4.5% PC I and slump = 18 cm (Pokharel and Fall 2013)

permeability. However, there is an increase in the hydraulic conductivity of the samples with a high sulphate content at 50 °C, due to the adsorption of the sulphate by the C–S–H gel, as well as destabilisation of the ettringite at a high curing temperature, which results in less precipitation of ettringite in the CPB pores and thereby greater permeability (Pokharel and Fall 2013).

An important external factor that can affect the changes in the saturated hydraulic conductivity is the curing stress. Ghirian and Fall (2015) reported that at any curing time, the application of stress has a significant effect on reducing the hydraulic conductivity at the early ages; see Fig. 4.30. This is due to the particle rearrangement and porosity reduction in the CPB matrix caused by the curing stress, which in turn results in reduced permeability of the CPB sample. However, mechanical damage induced by high levels of applied stress (more than 80% of the UCS) can

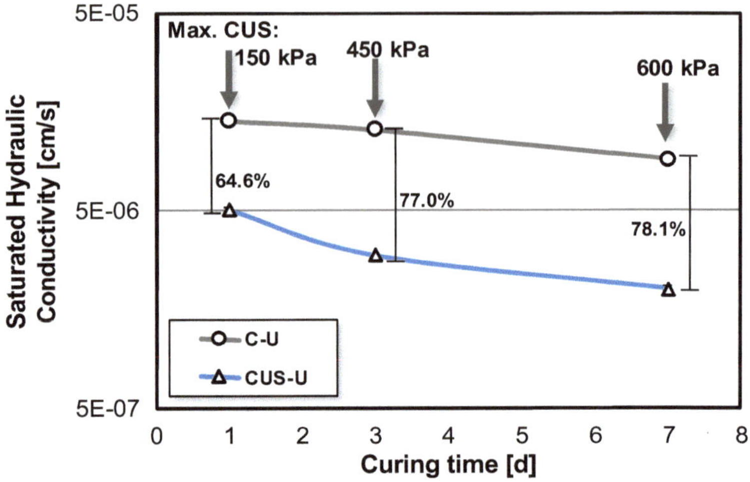

Fig. 4.30 Effect of curing under stress on saturated hydraulic conductivity of CPB with 4.5% PC I and $w/c = 7.6$ (Ghirian and Fall 2015)

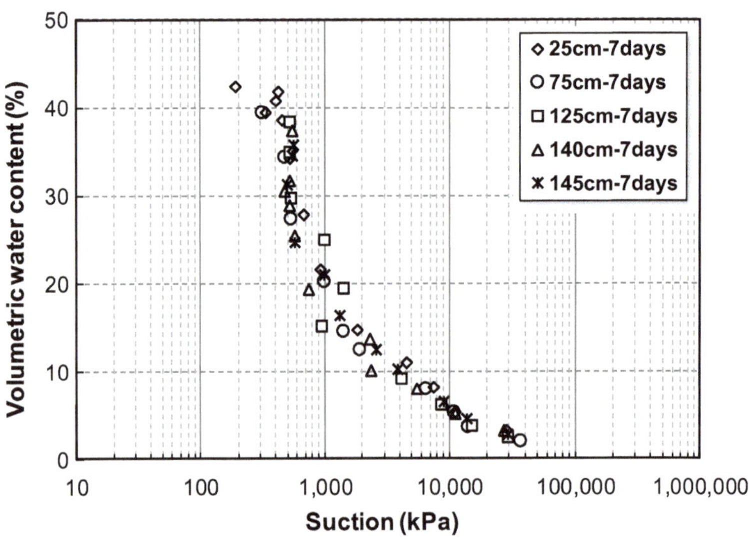

Fig. 4.31 Water retention properties of 7-day CPB samples in column experiments (Ghirian and Fall 2013)

increase the hydraulic conductivity, mainly due to the formation of microcracks in the CPB matrix under high external stress (Fall et al. 2009).

In unsaturated conditions, factors such as the degree of saturation, water retention capacity (WRC; water content versus suction) and AEV control the unsaturated hydraulic properties of CPB. Figure 4.31 illustrates the WRC of 7-day CPB samples with respect to height in column experiments. WRC is the capacity of an unsaturated porous medium to retain water, which can be determined from the variation of the water content with matric suction. Also, the changes in the AEV with curing time are shown in Fig. 4.32. The AEV is the matric suction at which air enters the large pores in a porous medium and causes desaturation. It can be seen that the AEVs increase with curing time and degree of hydration because the pore structure of CPB is refined during the hydration process. Moreover, a higher binder content leads to higher AEVs for a given curing time, except at 90 days (Abdul-Hussain and Fall 2011).

Figure 4.33 illustrates the computed unsaturated hydraulic conductivity of the CPB sample cured for 3 days with different binder contents. The technique in van

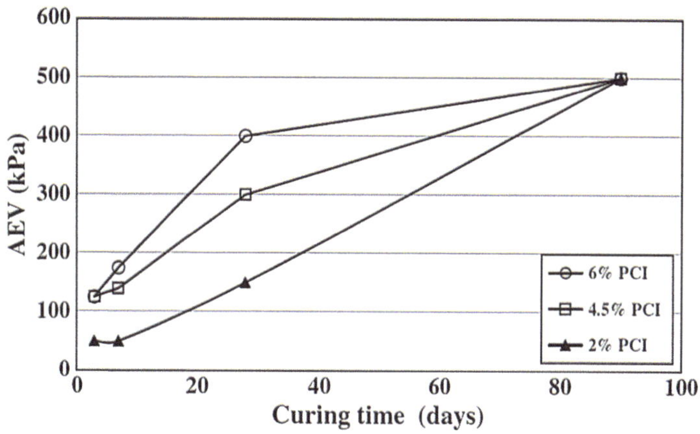

Fig. 4.32 Changes in air entry values of CPB samples with curing time (Abdul-Hussain and Fall 2011)

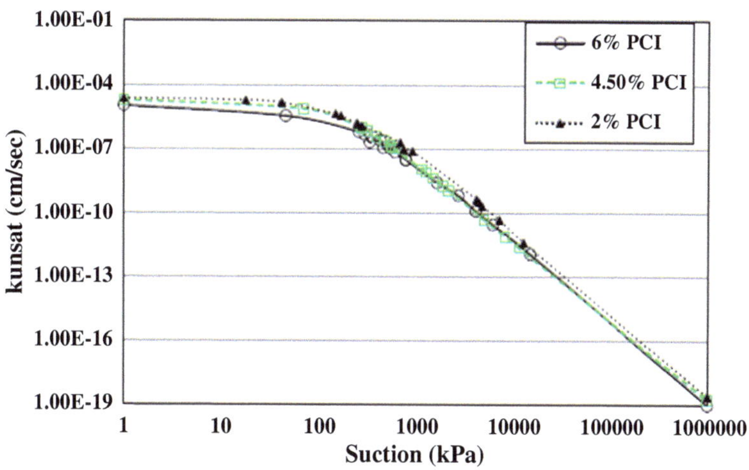

Fig. 4.33 Unsaturated hydraulic conductivity of 3-day CPB sample with different binder contents (Abdul-Hussain and Fall 2011)

Genuchten (1980) is used by Abdul-Hussain and Fall (2011) to examine the changes in hydraulic conductivity versus suction based on the obtained WRC. It can be observed that a higher binder content results in slightly lower k_{unsat} for a given suction because as the binder content is increased, more cement hydration products are produced, thus resulting in pore refinement as well as changes in the WRC of CPB (Abdul-Hussain and Fall 2011).

An understanding of the development and changes in the positive and negative (suction) PWP in CPB is essential for a reliable assessment of the mechanical strength and stability of CPB from the early to the advanced ages, liquefaction potential of CPB, stability of barricades as well as quantification of the deformation behaviour and stress distribution in the CPB. The PWP that develops in CPB influences the magnitude of the effective stresses. The principle of effective stress is one of the most important concepts of soil mechanics. Therefore, effective stress has considerable impacts on the behaviour of porous media, such as soil, CPB and rocks, because it affects their mechanical behaviour.

Experimental measurements of the changes in the PWP in CPB have been rarely addressed in previous studies. Helinski (2007) conducted a centrifuge experiment to investigate the relationships among consolidation, pore pressure changes (due to cement hydration) and total stress. Despite the fact that the apparatus monitored the stress and pore pressure changes in the backfill, it could not fully couple the studied parameters. Simms and Grabinsky (2009) modified a triaxial cell with a miniature tensiometer to measure the changes in the negative pore pressure (suction) during consolidated undrained (CU) triaxial tests. CPB samples cured for 2 days were subjected to triaxial shear testing, and the stress–strain and suction were monitored during shearing.

Despite the very limited number of experimental studies on pore pressure measurements, the field instrumentation of backfill has received increasingly more attention among researchers (e.g. Yumlu 2008; Thompson et al. 2009; Grabinsky 2010; Veenstra et al. 2011; Thompson et al. 2012). Field monitoring has been conducted by installing different sensors, such as piezometers and pressure cells, in different locations of mine stopes (different stope heights and proximity to barricades). Different parameters, such as PWP, and vertical and horizontal total stresses are measured up to 150 days of curing. The related literature has also discussed the effects of the filling sequence (effects of a plug) and filling rate. These studies consequently provide an understanding of the fundamental behaviours of CPB with regard to changes in the pore pressure, effective stress, cement hydration reaction, self-desiccation and hardening process in the field stopes. Yet a review of the previous works in the field shows that there is still a lack of understanding of the effects of THMC factors on pore pressure and the stress state. Also, external factors, such as drainage conditions, arching effects, curing temperature and chemical composition of CPB components, have not been controlled and therefore their effects on the studied parameters are not well understood.

To address these issues, Ghirian and Fall (2013) carried out column experiments that investigated the changes in suction (negative pore pressure) with curing time; see Fig. 4.34. The rate of suction development quickly increases with curing time up to approximately 80 h after loading of the column. Beyond this point, there is a gradual

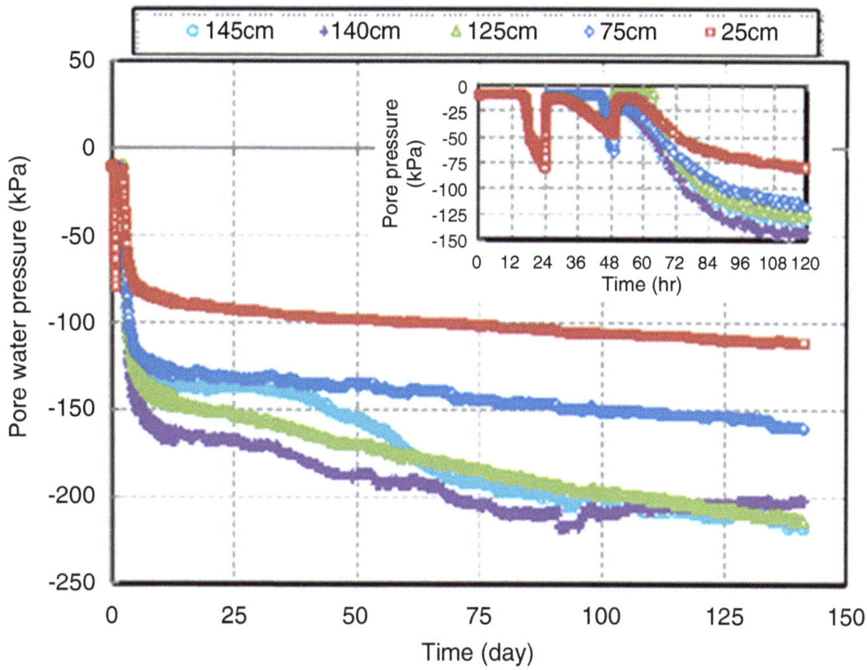

Fig. 4.34 Changes in suction with curing time in column experiments (Ghirian and Fall 2013)

change in the suction with time. The suction is primarily due to self-desiccation as a result of the cement hydration reaction. The faster rate of hydration at the early ages results in the rapid development of suction. Also, it can be noticed that the magnitude of the suction differs with respect to the height of the column. For example, the suction that develops at the bottom of the column is lower compared to the middle (and top) of the column which is due to the self-weight of CPB (i.e. self-weight-induced effective stress), as well as the drainage of water from the middle to the bottom of the column. Also, a significantly higher magnitude of suction was measured by the suction meters located near the top of the column at approximately 40 days to the end of the experiment because the development of suction resulted from the effects of self-desiccation due to the binder reaction and surface evaporation.

One of the main factors that can influence the development of suction in CPB is the sulphate content in the mine tailings and/or mine process water, as reported by Li and Fall (2016). Figure 4.35 shows that the sulphate content significantly influences the self-desiccation of CPB at the early ages. Increases in the initial sulphate content result in reduced suction in the samples (Li and Fall 2016). This is evident when the changes in suction are compared between the samples with no sulphate and 25,000 ppm of sulphate. Also, it can be observed that a higher sulphate content delays the onset of suction. For example, suction developed in the sample with no sulphate almost immediately after placement, but the suction of the sample with a high sulphate content started to change 1 week after the testing started. This is due to the

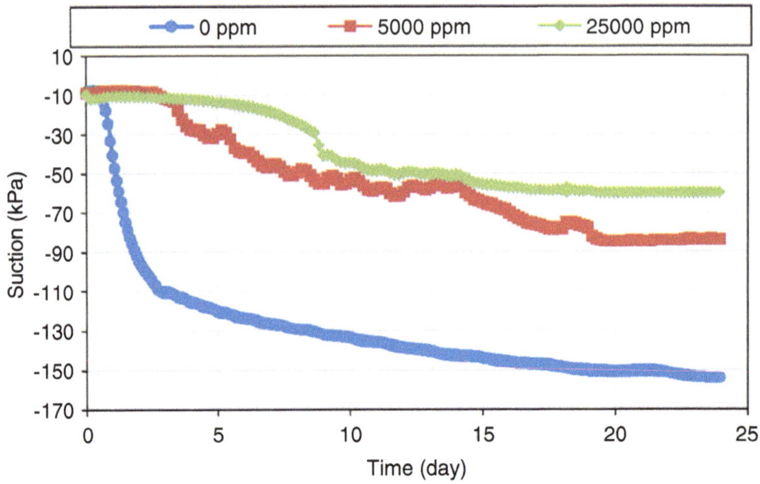

Fig. 4.35 Effect of sulphate on self-desiccation of CPB at early ages (Li and Fall 2016)

adverse effect of sulphate content (tailings chemistry) on cement hydration. The sulphate acts as an inhibitor and reduces the rate and degree of hydration, thereby resulting in less consumption of water and self-desiccation (Li and Fall 2016).

Other factors such as binder content, binder type and curing temperature can also affect the development of suction due to self-desiccation, as illustrated in Fig. 4.36. Figure 4.36a shows that an increase in the binder content will lead to increased development of suction in the CPB sample because this increases the binder hydration, so that a large volume of pore water is consumed, thus resulting in greater self-desiccation (Wu et al. 2014). Also, Fig. 4.36b shows that the binder type affects the degree of self-desiccation in the CPB samples. For example, the sample with PC I has a greater self-desiccation compared to the sample with PC I and slag or FA, which is most likely due to the fact that the rate of hydration changes with the type of binder, thus resulting in differences in the degree of self-desiccation (development of suction). Curing temperature can also affect the development of suction as higher temperatures increase hydration reactions, which in turn increases the rate of pore water consumption. This rapid consumption of the pore water increases self-desiccation (Wu et al. 2014). Therefore, it can be concluded that a combination of different factors (such as the coupled effects of curing temperature, binder content, sulphate content) can result in different degrees of self-desiccation (greater or less self-desiccation) in CPB materials, which in turn results in variations in the hydro-mechanical properties of CPB structures.

The effect of different external factors such as the curing stress, filling rate and drainage condition on the development of PWP is illustrated in Fig. 4.37. In mine backfills, the filling rate directly affects the PWP, stress regime and applied forces on the barricade. It can be observed that a faster filling rate (e.g. 0.61 m/h) produces a higher PWP in the samples. Another issue is mechanical damage induced by rapidly applied stress. Cemented paste subjected to excessive curing stress has a weak

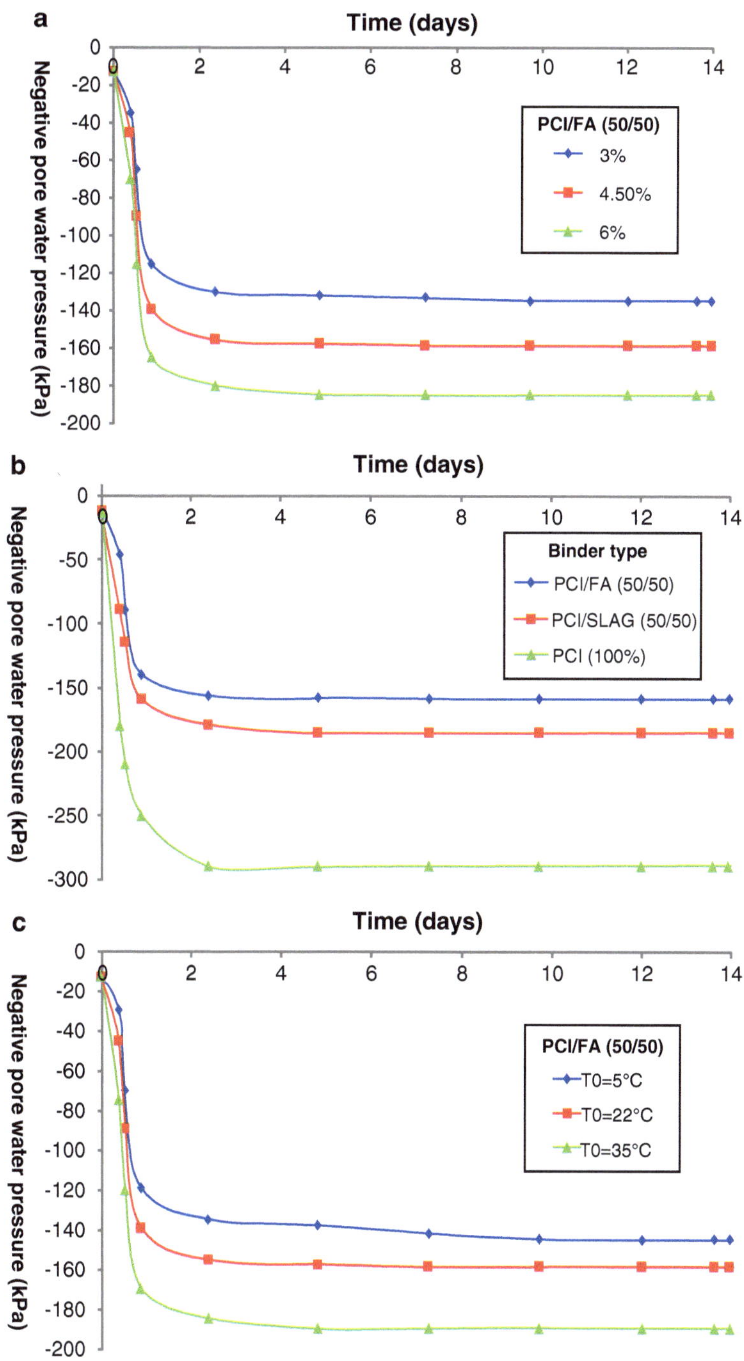

Fig. 4.36 Effects of (**a**) binder content, (**b**) binder type and (**c**) curing temperature on suction development due to self-desiccation in CPB (Wu et al. 2014)

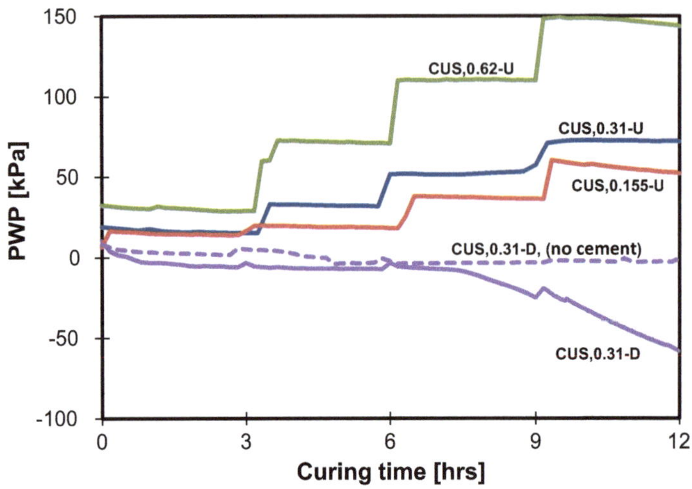

Fig. 4.37 Changes in pore water pressure of CPB samples cured under stress. Filling rates and drainage conditions obtained from load cell experiments (Ghirian and Fall 2016); (U: undrained conditions; D: drained condition); 0.31: filling rate of 0.31 m/h

microstructure at the early ages and thereby the points of contact between the cement hydration products and tailings/hydration particles during hydration are prone to damage due to the excessive stress. Also, it can be observed that the drainage condition has significant effects. For instance, the PWP is reduced for CUS 0.31-D (sample cured in drained condition), which is cured under stress (consolidated sample). The drainage condition also changes the positive PWP in CUS 0.31-D to a negative PWP (suction). These changes in suction can significantly increase the effective stress, which in turn leads to higher development of mechanical strength of the CPB. The results show the importance of drainage conditions on the stability of backfill and barricades (Ghirian and Fall 2016).

A review of the previous studies on the hydraulic properties of CPB shows that the hydraulic characteristics, such as saturated hydraulic conductivity, suction development (due to self-desiccation, drainage and evaporation), pore pressure changes and unsaturated properties of CPB, are important factors in the design and performance assessment of backfill structures and barricades.

5 Thermal Properties and Temperature Development in CPB and Influential Factors

The thermal properties and thermal processes of CPB materials are grouped into two main categories, including (1) the intrinsic thermal properties of CPB, such as the thermal conductivity, and (2) external thermal factors, such as the curing temperature and heat of binder hydration. Both are necessary to gain a better understanding of the thermal behaviour of cemented backfill structures. Thermal conductivity is required for the thermal analysis of CPB structures and heat transfer between the CPB and surrounding environment (rock mass, mine, etc.) (Celestin and Fall 2009).

There are some factors that do not significantly affect the thermal conductivity of CPB, as indicated in Celestin and Fall (2009), such as the binder type, binder content, curing age, *w/c* ratio and sulphate concentration of the tailings. However, there are many factors that affect the thermal conductivity of CPB, especially the tailings mineralogy (e.g. quartz content) (Celestin and Fall 2009; Ghirian and Fall 2013), and others, such as the fine content, curing temperature, porosity and degree of saturation. Figure 4.38 shows that the thermal conductivity of CPB increases with increased quartz content because the thermal conductivity of quartz is much higher than that of the many other available minerals found in tailings (e.g. $K = 7.7$ W/m °C for quartz, 2.25 W/m °C for feldspar) and other components of CPB (water and binder). Therefore, for a given CPB recipe, the tailings materials which contain less quartz or minerals with a high thermal conductivity will produce a less conductive material (Celestin and Fall 2009).

The thermal conductivity of the CPB also increases as the fine content of the tailings is decreased, as shown in Fig. 4.39, because an increased fine content reduces the packing density of the tailings, which in turn contributes to increasing the overall porosity of the hardened cement matrix (Celestin and Fall 2009).

However, the thermal conductivity of CPB decreases with an increase in the curing temperature, as shown in Fig. 4.40, due to the effect of the degree of saturation. CPB samples that are cured at a higher temperature (35 °C, 50 °C) have a higher rate of water consumption due to faster hydration. Furthermore, the samples most likely become more quickly desaturated with higher curing temperatures due to some

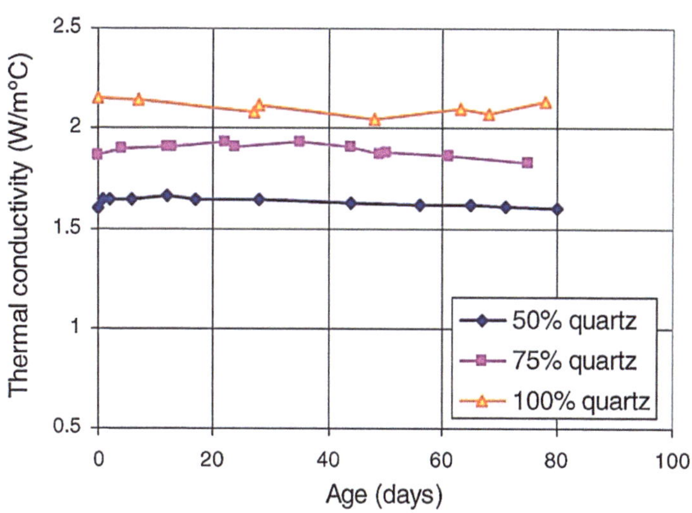

Fig. 4.38 Effect of mineralogical composition (proportion of quartz) of tailings on thermal conductivity of CPB (Celestin and Fall 2009)

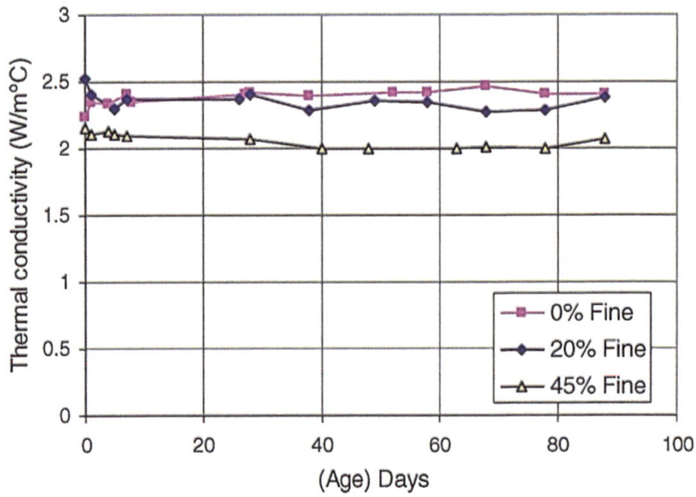

Fig. 4.39 Effect of tailings grain size with constant slump on thermal conductivity of CPB (Celestin and Fall 2009)

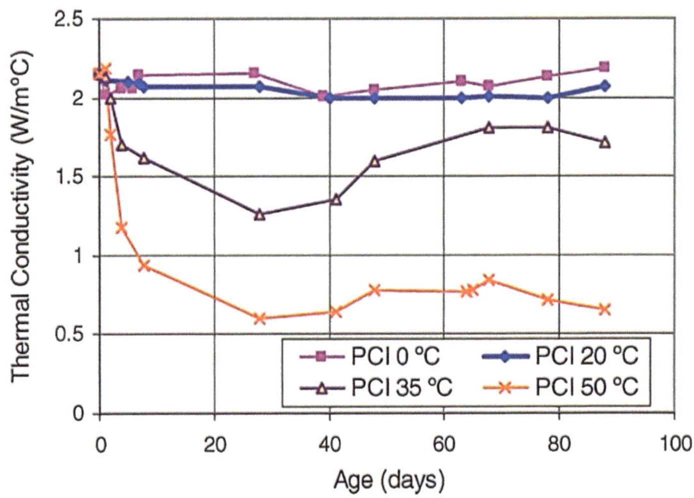

Fig. 4.40 Effect of curing temperature on thermal conductivity of CPB (Celestin and Fall 2009)

evaporation of the water from the pores. The combined effects of water loss caused by self-desiccation and evaporation reduce the degree of saturation. The degree of saturation of the samples cured at 2 °C and 50 °C is determined to be about 94 and 88%, respectively (Celestin and Fall 2009). Through desaturation, air substitutes the water in the pores, and since water has a thermal conductivity about 25 times that of air, it is obvious that when the air content is increased, the thermal conductivity of the CPB is reduced. This is shown in Fig. 4.41 (Ghirian and Fall 2013), which demonstrates that the degree of saturation significantly affects the thermal conductivity of CPB structures.

There are various sources of temperature variation (or thermal loads) in mine backfill operations. They comprise the heat generated by binder hydration, heat in the host rock and deep mines, the self-heating of sulphidic rock (and tailings), initial temperature of the CPB mix components and heat generated by blasting and mine fires. The rock temperature (i.e. hot rock temperature in deep mines and cold rock temperature in permafrost mines) mostly depends on the type of rock, depth of the mine and geographical location (Fall et al. 2010; Fall and Pokharel 2010). The self-heating of rocks/tailings is dependent on the type and quantity of sulphide and pyrrhotite minerals as well as accessibility to oxygen and water for oxidation (Bernier and Li 2003). Among the different sources of temperature variations in mine backfills, the heat produced by binder hydration is the most significant source of variation. Since CPB structures are usually massive, the temperature due to binder hydration can increase to 50 °C or even higher (Fall et al. 2010; Thompson et al. 2012). The amount of heat generated in the backfill depends on factors such as the *w/c* ratio, mixing water chemistry, and binder type and quantity. It can also be affected by external factors such as the stope size, binder content, filling rate and placement temperature (Nasir and Fall 2009).

This is demonstrated by column experiments carried out in Ghirian and Fall (2013), in which the changes in temperature with curing time for different heights of the column are presented in Fig. 4.42. The column is 1.5 m in height and filled with CPB. The temperature changes at the top, middle and bottom of the column were measured with a thermometer. For these three different depths of the column, the heat of hydration rapidly reaches the highest value after approximately 20 h of curing, and then remained constant for at least another 72 h. Afterwards, the temperature gradually decreased up to 7 days. It can also be seen in Fig. 4.42 that the temperature measurement taken close to surface of the column shows the lowest temperature compared to the rest of the column because a large amount of heat is lost through the evaporation process, which in turn reduces the temperature. Eventually, after moisture and thermal equilibrium is reached between the CPB and surrounding environment, the temperature at the surface becomes ambient. Details on the results and experimental method can be found in Ghirian and Fall (2013).

The literature review here shows that the thermal properties and behaviour of CPB materials are affected by several factors and processes, such as the physical properties of the tailings and degree of saturation. Moreover, the temperature variations due to binder hydration can be affected by the binder type and content. Temperature variations also significantly affect the hydromechanical and chemical properties or behaviour of CPB.

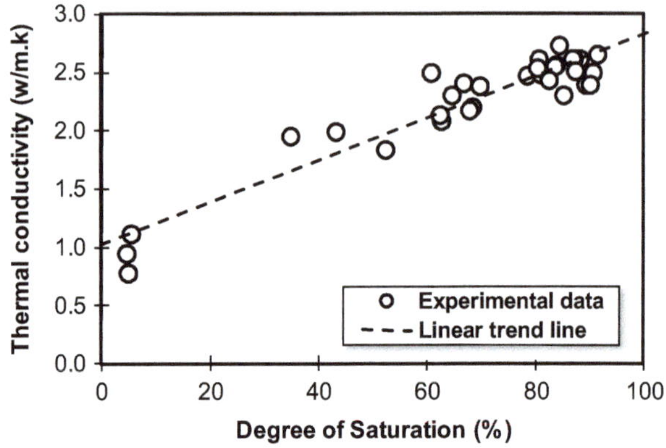

Fig. 4.41 Effect of degree of saturation on thermal conductivity of CPB (Ghirian and Fall 2013)

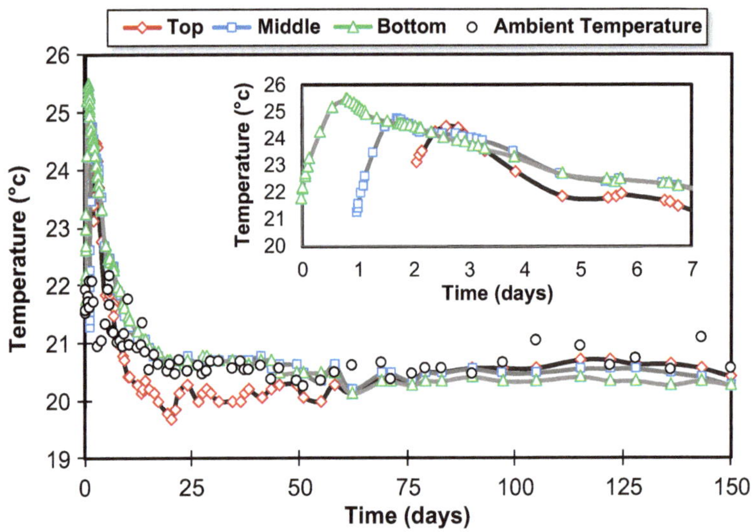

Fig. 4.42 Changes in temperature with curing time in column experiments (Ghirian and Fall 2013)

6 Chemical Properties of CPB and Influential Factors

The chemical characteristics and chemical processes of CPB are primarily controlled by the chemistry and mineralogy of its constituents, including the tailings, binder and mixing water. External factors such as the curing temperature (which has a strong effect) and curing stress (has a weak effect) can also influence the chemical properties or behaviour of CPB.

The chemical characteristics or chemical processes of CPB, in turn, can significantly affect the thermo-hydro-mechanical properties and processes in CPB structures. Therefore, understanding these coupled THMC processes is necessary when the stability and performance of a backfill structure are assessed. Also, this understanding provides valuable knowledge on the binder hydration reactions and processes in CPB, which can be used to examine the changes in the CPB materials and their behaviour when subjected to THMC loads and processes.

To do so, there are different ways and techniques to measure and monitor the chemical processes in CPB and the properties of CPB. For example, measuring the pore water chemistry (e.g. ion concentration analysis) can show the changes in the chemistry of a CPB sample. This approach is found in Ghirian and Fall (2014) in which a high-pressure pore water extractor apparatus was developed, and then used to analyse the pore water to study the different anion and cation concentrations in the solution, pH, etc. An example of the results is presented in Fig. 4.43, which are obtained from column experiments in Ghirian and Fall (2014). The figure shows the changes in the pore water solution of CPB with different chemistries which was mechanically extracted from different heights of the column, as well as the different curing times. More details can be found in Ghirian and Fall (2013, 2014).

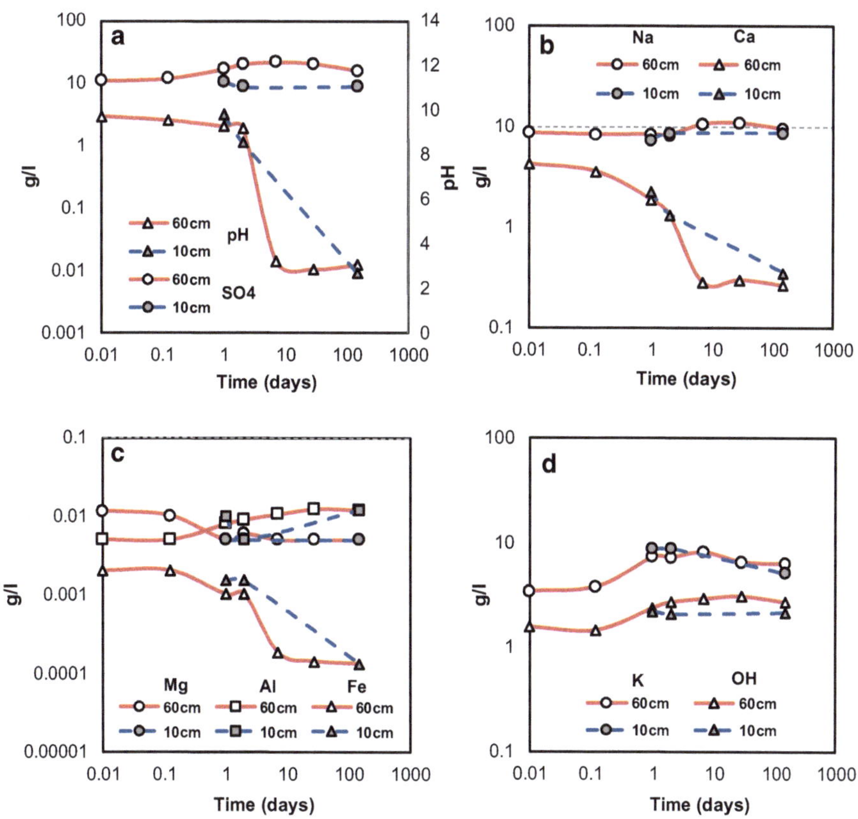

Fig. 4.43 Changes in ion concentration of pore fluid with curing time (Ghirian and Fall 2014)

Other techniques, such as analyses of cement dissolution and process water obtained from mine tailings, can also be subjected to chemical analysis when the effects of chemical processes on CPB properties and performance are being investigated (Benzaazoua et al. 2004). Some researchers used metal leaching tests to study the potential of metal generation of CPB made with arsenopyrite-rich tailings. These techniques can be used when the long-term geo environmental performance of a CPB structure needs to be investigated.

In addition to laboratory testing, other indirect techniques can also be used to study the chemical properties of CPB and chemical processes in CPB material, particularly when continuous monitoring of the chemical reactions with curing time is required. These also provide information on the setting time and rate of hydration of the CPB, as well as qualitative information on the total depletion rate of ions from the CPB pore solution. Successful field and laboratory applications of these techniques have been reported in Thottarath (2010) and Ghirian and Fall (2015), respectively. The change in the ion concentrations in the pore fluid of CPB as a result of cement hydration reactions can be measured by the electrical conductivity (EC) of the backfill. Figure 4.44 shows the changes in the EC of CPB with curing time and different sulphate contents (Li and Fall 2016). It can be observed that soon after mixing, the EC starts to gradually increase up to the peak value, regardless of the initial amount of sulphate. This increase in the EC is due to the ion concentration increase in the pore fluid as well as temperature increase as a result of exothermic cement reactions. The peak of the EC corresponds to the initial setting (transforming from the paste phase to the formation of the solid skeleton) of the backfill (Ghirian and Fall 2015). However, a longer time is required to reach the peak as the sulphate content is increased, because sulphate delays the hydration of the binder (Li and Fall 2016). Afterwards, the EC values start to decrease with time which is due to the reduction in the volume of unbound water as a result of self-desiccation and fewer connected capillary pores, which in turn increase the path of the ion flow (Ghirian and Fall 2015).

As explained above, changes in the chemical properties of CPB and chemical processes in CPB can significantly affect the other properties of CPB. For example, direct or indirect chemical processes can take place that affect the mechanical strength of backfill. In the former, CPB prepared with sulphide-rich tailings (e.g. pyrite) can be oxidised in the presence of oxygen (also called weathering). The degree of oxidation is mainly a function of the percentage of pyrite and degree of saturation. The weathering causes release of the metal ions and acid mine drainage into the environment (Ouellet et al. 2003). In the latter, the initial chemical compositions (e.g. sulphides, sulphate) in the CPB ingredients (i.e. tailings, binder and mixing water) can negatively affect the strength development of CPB due to sulphate attacks (Fall and Pokharel 2010; Li and Fall 2016). For example, sulphate in the initial CPB matrix can inhibit cement hydration reactions at the early ages and therefore reduce the strength (Fall and Benzaazoua 2005; Pokharel and Fall 2011). In advanced ages, the formation of secondary expansive minerals in the backfill pores, such as ettringite and gypsum, can cause internal cracks and eventually lead to strength deterioration (Fall and Benzaazoua 2005).

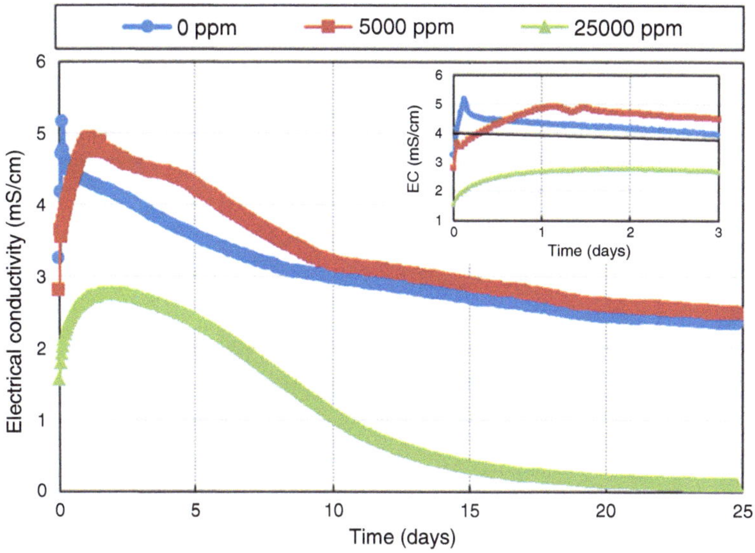

Fig. 4.44 Changes in electrical conductivity with curing time and different sulphate contents (Li and Fall 2016)

In the tailings chemical or mineralogical composition, the presence of sulphide minerals (e.g. pyrite) and sulphate ions as well as the chemical characteristics of the mixing water, which may have soluble sulphates, can affect the strength acquisition of backfill (Kesimal et al. 2005). Benzaazoua et al. (2004) used different types of binders (ordinary Portland cement (OPC), blast furnace slag and FA) with different chemical compositions, five types of tailings with different percentages of pyrite and sulphur, and mixing water with different percentages of soluble sulphate to prepare CPB samples. They found that the chemical properties of the three main components of the CPB are interrelated and play an important role in mechanical strength acquisition. Also, the chemical composition and concentration of dissolved ions in the pore water are the main factors that influence the hardening process in cemented materials (Benzaazoua et al. 2004).

7 Microstructural Properties of CPB and Influential Factors

The microstructural properties of CPB, such as the porosity, pore size distribution, interconnectivity of the pore system, and types and mechanisms of formation of hydration products, significantly influence the other properties of CPB materials, such as strength, durability and permeability (Fall and Samb 2008). An understanding of the microstructural properties of CPB is necessary to determine the main mechanisms and processes that take place during the hardening process of CPB.

There are different testing methods and techniques to examine the CPB microstructural properties, such as mercury intrusion porosimetry (MIP), thermal analysis

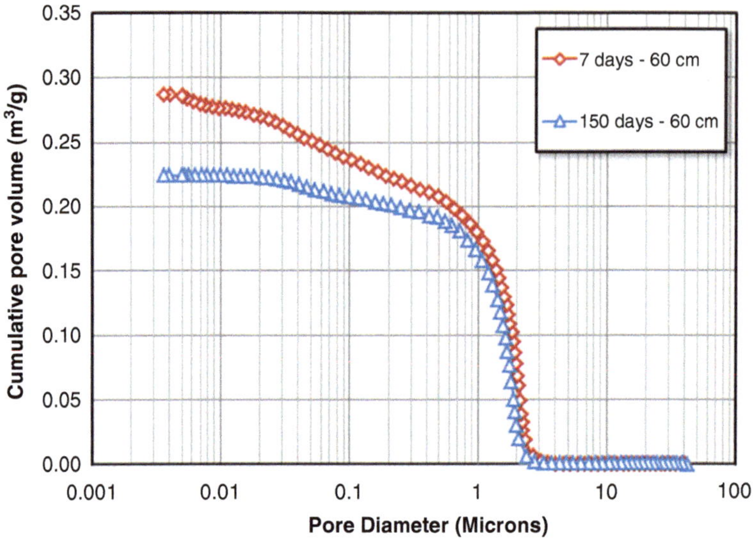

Fig. 4.45 MIP test results of column experiments with 7- and 150-day CPB samples (Ghirian and Fall 2013)

[differential thermal analysis/thermal gravimetric analysis (DTA/TGA)], X-ray diffraction (XRD) and scanning electron microscopy (SEM). MIP is used to study the pore size distribution and total porosity of the CPB materials, while DTA/TGA and XRD are used to study the mineralogy of CPB as well as the type and quantity of binder hydration products. SEM is a useful technique to visually identify the density and type of hydration product, and examine the pore structure, pore connectivity and microcracks in the CPB matrix. Figure 4.45 shows the results of MIP tests performed on 7- and 150-day CPB samples. The cumulative pore volume as a function of the pore diameter is compared. A sample cured for 150 days has a much lower cumulative pore volume compared to the 7-day sample, which means that the former has a finer pore size distribution (Ghirian and Fall 2013). Figure 4.46 shows the changes in the pore structure of the samples cured for 7 and 150 days with time. The refinement of the CPB pores at an advanced age (150 days) is mainly due to the increases in cement hydration products such as C–S–H, CH and ettringite with time as shown in the SEM images. Figure 4.47 presents the mineralogical composition of cement paste samples (after undergoing XRD) performed on 28-day cement paste with a sulphate content of 25,000 ppm and without sulphate. It can be observed that more ettringite is produced in the cured sample with a sulphate content of 25,000 ppm compared to that with no sulphate. Also, the plot shows that the peak intensities of C–S–H and CH are higher in the cemented paste with no sulphate as opposed to the samples with sulphate, which means that greater proportions of C–S–H and CH have formed in the sample with no sulphate. Therefore, this can lead to higher strength in the sample (Li and Fall 2016).

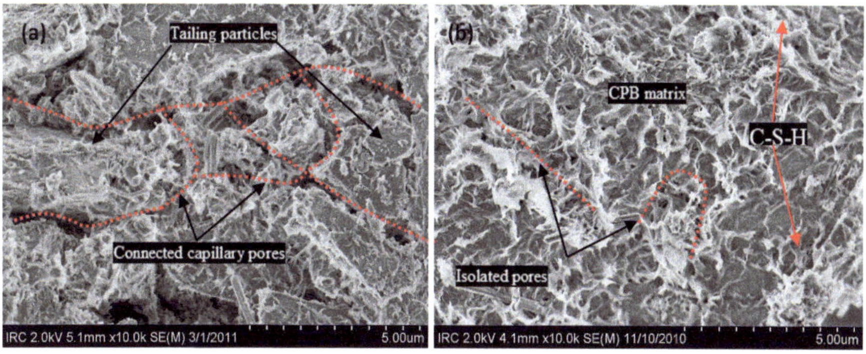

Fig. 4.46 SEM images of CPB microstructure after (**a**) 7 days and (**b**) 90 days of curing (Ghirian and Fall 2013)

Figure 4.48 shows the results of a thermal analysis performed on 7-day cement paste samples cured at different temperatures (20 °C, 35 °C). Generally, typical peaks can be seen in the temperature range of 50–150 °C, and at 450 °C and 750 °C. These peaks indicate the presence of C–S–H gel, ettringite, CH and calcite (CaCO3). Furthermore, it can be observed that a high curing temperature significantly increases the hydration rate of the cement paste. This is demonstrated by the increase in the amount of C–S–H, ettringite, CH and CaCO3 formed with increased curing temperature. This means that weight losses at the temperature range of 50–150 °C and at 450 °C and 750 °C are greater with higher curing temperatures (Fall and Samb 2008).

The results of these studies show that the mix components (such as *w/c* ratio, cement type and proportion, tailings grain size), curing time, curing temperature, sulphate content and curing stress significantly affect the pore structure of CPB materials.

8 Primary Coupled THMC Processes and Factors in CPB

In this section, the importance of coupled THMC processes or factors in understanding the properties and behaviour of CPB is discussed and highlighted, which contribute to the cost-effective designs of CPB structures. The coupled processes that mainly occur in the geotechnical systems of CPB poured into mine stopes can be classified as thermal (T), hydraulic (H), mechanical (M) and chemical (T) processes or factors. These factors are related and coupled (interact). The magnitude of the effect of coupling in one direction may be different from that in the opposite direction. Therefore, coupled factors have either weak or strong interactions. Strong interactions control the geotechnical performance of backfill structures. Figure 4.49 schematically illustrates the primary THMC processes in CPB structures as well as the strong and weak interactions between these processes.

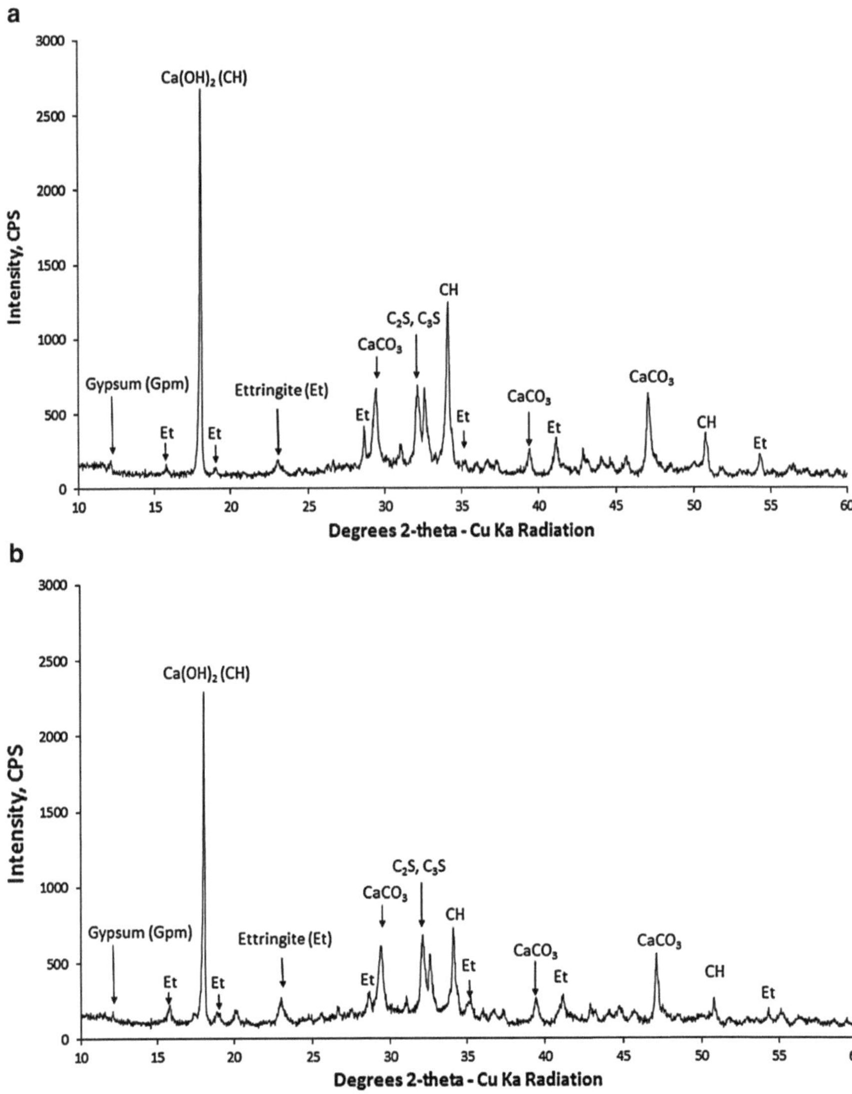

Fig. 4.47 XRD plots of cured cement paste samples with sulphate content of (**a**) 0 ppm and (**b**) 25,000 ppm at 28 days (Li and Fall 2016)

The primary THMC interactions are summarised below. It should be noted that although these are the most important interactions that can take place in a backfill structure, there are also others that are relevant.

- $T \rightarrow H$: The primary source of temperature increase in mine backfill is from binder hydration reactions. A higher temperature induces the rapid development of suction and reduction of the PWP in mine backfills. Also, temperature increases can lead to a reduction in the transportability of backfill in the fluid state because the rates of binder hydration and pore refinement of CPB are both increased.

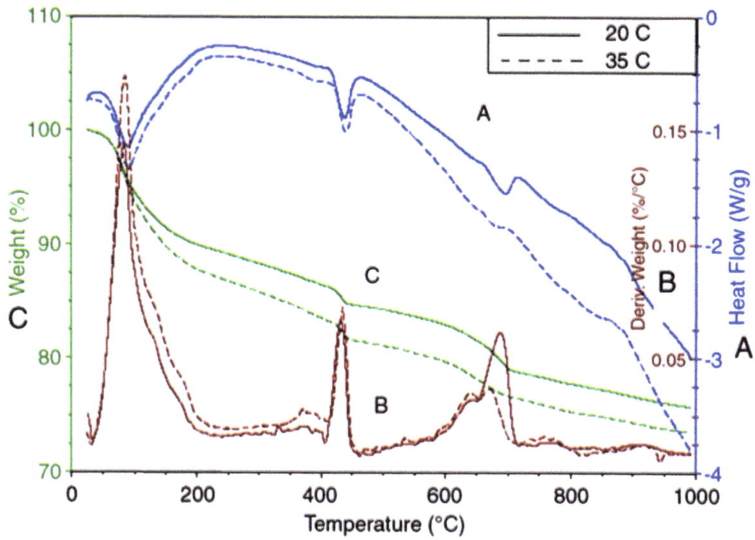

Fig. 4.48 Thermal analysis on CPB samples with $w/c = 2$ (Fall and Samb 2008)

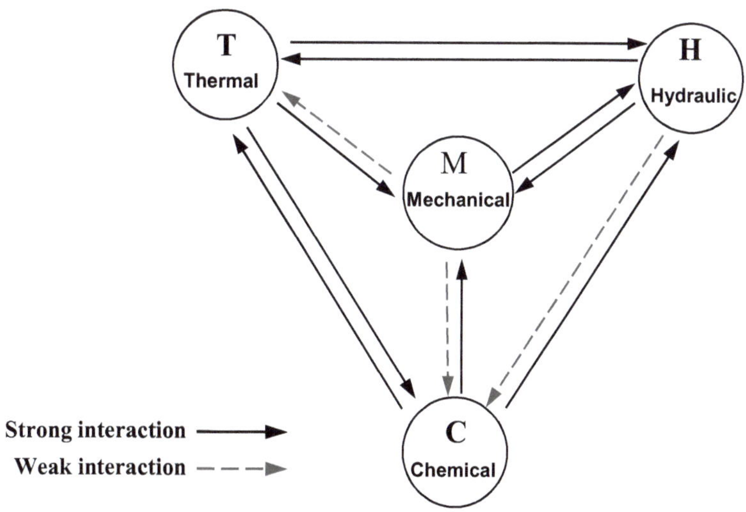

Fig. 4.49 Primary multi-physics processes in CPB material and their interactions

- $T \rightarrow M$: Temperature increases in mine backfill will increase the rate of binder reaction, which in turn leads to the production of more cement hydration products. More hydration products result in a finer and denser CPB pore structure, which consequently delivers higher mechanical and shear strength properties to CPB.
- $T \rightarrow C$: Temperature increases in a CPB structure can induce a faster rate of binder hydration. This is a strongly coupled interaction that influences the chemical concentration of pore water as well.

- $H \rightarrow T$: Hydraulic factors, such as the degree of saturation, directly influence thermal factors, such as the thermal conductivity of backfill. Reductions in the degree of saturation (or desaturation—increase of air voids in the CPB matrix) due to self-desiccation and/or evaporation reduce the thermal conductivity of backfill.
- $H \rightarrow M$: Suction development as a result of self-desiccation can affect the mechanical properties. Self-desiccation induces suction as a result of binder reactions and thereby results in the decrease of the PWP and increase in the effective stress (strength increase).

Evaporation as a hydraulic factor also results in the development of surface shrinkage and associated microcracks. The presence of microcracks then reduces the mechanical strength of CPB by creating preferential failure planes of weakness in the CPB structure. This process is considered to be weak interaction as long as the CPB remains in a saturated or near-saturation state, or the air-backfill interface cannot be identified in the mine stope.

- $H \rightarrow C$: Hydraulic factors, such as drainage, evaporation and self-desiccation, can indirectly influence ion concentrations in pore fluid. Removing water from the CPB pore structure results in increases in ion concentration by reversing the dilution effect. This is a weak interaction.
- $M \rightarrow T$: Not a significant interaction.
- $M \rightarrow H$: Consolidation is a strongly coupled interaction that can take place in mine backfill. Backfill is consolidated under the pressure of its own self-weight in drained conditions. During this process, the water expelled from the CPB pores lead to a reduction in the PWP. A reduced PWP increases the effective stress in the CPB, which is similar to curing backfill under stress, and significantly contributes to the strength improvement of CPB. Consolidation also reduces the porosity of backfill, which results in reduction of permeability.

Mechanical damage can induce microcracks. Crack propagation at a micro-scale might be found in CPB cured under (excessive) sustained stress. The connected microcracks act as preferential liquid paths in CPB, which results in decreases in tortuosity, and increases in the permeability.

- $M \rightarrow C$: Curing under stress induces a slightly faster rate of binder hydration. However, due to the relatively low amount of stress in the backfill, the effect of stress on the rate of hydration is not significant. This can be considered as a weak interaction in the backfill.
- $C \rightarrow T$: Binder hydration reactions are an exothermic reaction. Therefore, cement hydration produces heat and increases the temperature in mine backfill. This is one of the most important and strong interactions in CPB.
- $C \rightarrow H$: Binder hydration reactions, as a chemical factor, directly influence the porosity of mine backfill. Binder reactions produce hydration products, such as CH and CSH, and therefore cause refinement of the pore structure of backfill. A lower porosity results in reduced water permeability in backfill. This is therefore a strong reaction in CPB.
- $C \rightarrow M$: As explained above, binder hydration reactions cause pore refinement. A lower porosity also increases the mechanical strength of backfill. Furthermore,

as binder hydration progresses, cement hydration products (e.g. C–S–H, CH) are produced, which facilitate cohesive strength in CPB. This is therefore a strong interaction.

THMC-coupled processes in CPB material mostly take place as changes in the intrinsic properties. However, several external factors can also induce and/or initiate a process. A comprehensive list of the main coupled THMC factors and the main intrinsic and external factors that can affect them is shown in Table 4.1. Some related sources for each individual interaction are also provided in the table for reference purposes. Also, Fig. 4.50 illustrates the primary and important THMC processes and factors that control and affect the CPB properties and performance.

9 Conclusion

A comprehensive review of the properties of CPB materials and the factors that can affect them has been provided in this chapter. A summary of the obtained results from testing carried out in previous studies and recent laboratory and field experiments has been outlined. The findings are relevant for understanding how these properties change in backfill, as well as how they can affect the overall behaviour and performance of cemented backfill and the degree that they do so. The properties of CPB are mainly affected by thermal, hydraulic, mechanical and chemical factors and processes, as well as physical and microstructural characteristics. Physical (such as porosity, void ratio, degree of saturation, density), mechanical (such as compressive and shear parameters, stress–strain behaviour); hydraulic (such as saturated and unsaturated permeabilities, suction development due to self-desiccation, pore pressure), and thermal (such as thermal conductivity, heat of binder hydration)

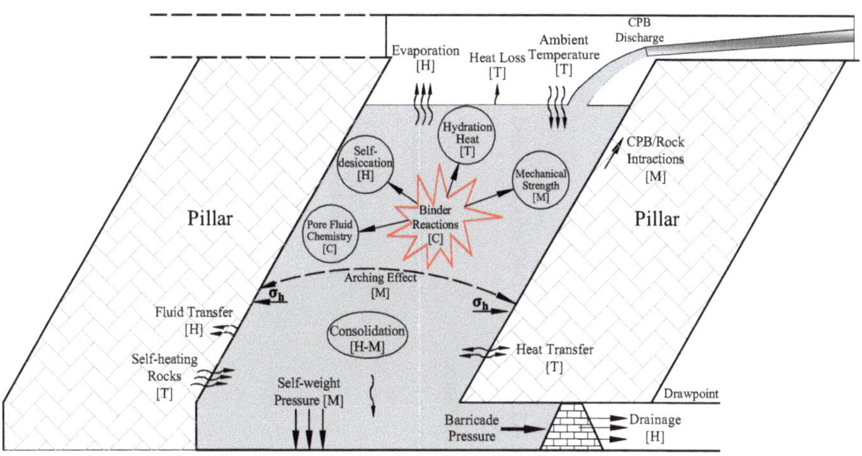

Fig. 4.50 Schematic diagram of different THMC-coupled processes in a CPB structure (static condition) (Ghirian and Fall 2015)

Table 4.1 Coupled THMC processes and factors that affect behaviour of CPB structures

Process/factor		Source	Influential property/parameter[a]		Source
			Intrinsic property	Parameter	
Thermal (T)	Intrinsic thermal properties	Thermal conductivity and heat capacity	Tailings type and fine content; CPB porosity and degree of saturation	–	Celestin and Fall (2009), Ghirian and Fall (2013)
	Thermal loading	Heat of binder hydration	Chemical reactions, binder content and type	Stope size, filling rate	Nasir and Fall (2009), Fall and Pokharel (2010), Fall et al. (2010); Fall and Samb (2009), Pokharel and Fall (2013), Wu et al. (2012)
		Self-heating mechanism	Host rock mineralogy, oxidation and chemical reaction		
		Environment temperature (cold/hot mining)	–	Geographic location, mine depth	Orejarena and Fall (2008)

Hydraulic (H)	Intrinsic hydraulic properties/factors	Saturated and unsaturated hydraulic conductivities	Porosity; intrinsic permeability; degree of saturation; water retention properties	Binder content and type; tailings fine content; temperature	Abdul-Hussain and Fall (2011), Fall et al. (2009), Ghirian and Fall (2013)
		Suction	Self-desiccation, cement reaction rate; temperature	Drainage	Ghirian and Fall (2013), Li and Fall (2016)
		Pore pressure	Consolidation	Drainage, consolidation	Ghirian and Fall (2015)
		Evaporation	Ambient humidity and temperature	Geographical location; depth of mine; exposed backfill	Ghirian and Fall (2013)
	External hydraulic loading	Groundwater flow (after mine closure)	–	–	Levens et al. (1996)
Mechanical (M)	Intrinsic mechanical properties/factors	UCS; shear strength; deformation	Porosity; tailings fine content; binder type/content; density of tailings; sulphate content; *w/c* ratio; self-desiccation	Curing stress; external heat	Fall et al. (2004), Benzaazoua et al. (2004), Fall et al. (2005, 2007, 2008, 2010), Fall and Pokharel (2010), Li and Fall (2016)
		Consolidation	Hydraulic conductivity; self-desiccation	Drainage; filling rate	Ghirian and Fall (2014, 2016)
		Shrinkage and cracks	Evaporation	Sustained stress	Ghirian and Fall (2014)
	Mechanical loading	Self-weight pressure	Backfill density	Drainage; filling rate	Yilmaz et al. (2009), Ghirian and Fall (2015)
		Stress regime and arching	Consolidation	Stope geometry; drainage condition	Yilmaz et al. (2015)
		Traffic loads	–	Stope design	Belem and Benzaazoua (2008)

(continued)

Table 4.1 (continued)

Process/factor		Source	Influential property/parameter[a]		Source
			Intrinsic property	Parameter	
Chemical (C)	Intrinsic chemical properties/factors	Pore fluid chemistry	Chemical reactions; binder content; binder type, sulphate content	–	Ghirian and Fall (2014)
		Electrical conductivity	Cement reactions; tailings chemistry, temperature	–	Thottarath (2010); Li and Fall (2016)
		Tailings and chemical composition of mixing water	Tailings/water mineralogy and chemistry	–	Benzaazoua et al. (2004)
		Rate of chemical reaction	Stress; binder content and type; temperature	–	
	Chemical loading	Sulphate effect	Sulphate concentration; temperature	–	Li and Fall (2016)
		Tailings reactivity; leaching	Tailings mineralogy and chemistry	–	Fall and Benzaazoua (2005); Pokharel and Fall (2011), Coussy et al. (2011)

[a]Only key influencing factors shown, other factors may also have effects

factors, as well as the microstructural properties of CPB, belong to the most important internal factors that control CPB behaviour. Furthermore, it has been shown that several external factors (or at the field scale) also influence the properties of CPB, such as self-weight pressure, external heat, drainage of the stope and self-heating rock masses. It has been shown that all of these THMC processes and factors are related and control the CPB behaviour. Information on the CPB properties and the factors that influence them will provide a better understanding of the THMC-coupled processes in CPB and their behaviour, and contribute to the designing of safe, economical and durable backfill structures.

References

Abdul-Hussain N, Fall M (2011) Unsaturated hydraulic properties of cemented tailings backfill that contains sodium silicate. Eng Geol 123(4):288–301

Alexander KM, Taplin JH (1962) Concrete strength, cement hydration and the maturity rule. Austr J Appl Sci 13:277–284

Belem T, Benzaazoua M (2008) Design and application of underground mine paste backfill technology. Geotech Geol Eng 26(2):147–174

Belem T, Benzaazoua M, Bussière B (2000) Mechanical behaviour of cemented paste backfill. In: Proceedings of the Canadian geotechnical society conference geotechnical engineering at the Dawn of the Third Millennium 15–18 October, Montréal, vol. l, p 373–380

Belem T, El Aatar O, Bussière B, Benzaazoua M, Fall M, Yilmaz E (2006) Characterization of self-weight consolidated paste backfill. In: Proceedings of 9th international seminar on paste and thickened tailings—paste'06, Limerick, Ireland. 3–7 April 2006, p 333–345

Benzaazoua M, Belem T, Bussière B (2002) Chemical factors that influence the performance of mine sulphidic paste backfill. Cem Concr Res 32(7):1133–1144

Benzaazoua M, Fall M, Belem T (2004) A contribution to understanding the hardening process of cemented paste fill. Miner Eng 17(2):141–152

Bernier LR, Li M (2003) High temperature oxidation (heating) of sulfidic paste backfill: a mineralogical and chemical perspective. In: Proceedings of Sudbury 2003: mining and environment III conference, held May 25–28 in Sudbury

Celestin J, Fall M (2009) Thermal conductivity of cemented paste backfill material and factors affecting it. Int J Min Reclam Environ 23(4):274–290

Cihangir F, Ercikdi B, Kesimal A, Turan A, Deveci H (2012) Utilisation of alkali-activated blast furnace slag in paste backfill of high-sulphide mill tailings: effect of binder type and dosage. Miner Eng 30:33–43

Coussy S, Benzaazoua M, Blanc D, Moszkowicz P, Bussière B (2011) Arsenic stability in arsenopyrite-rich cemented paste backfills: a leaching test-based assessment. J Hazard Mater 185:1467–1476

Ercikdi B, Yilmaz T, Karaman K, Kulekci G (2014) Assessment of strength properties of cemented paste backfill by ultrasonic pulse velocity test. Ultrasonics 54(5):1386–1394

Fall M, Benzaazoua M (2005) Modeling the effect of sulphate on strength development of paste backfill and binder mixture optimization. Cem Concr Res 35(2):301–314

Fall M, Pokharel M (2010) Coupled effect of sulphate and temperature on the strength development of cemented backfill tailings: Portland cement paste backfill. Cem Concr Compos 32(10):819–828

Fall M, Samb SS (2008) Pore structure of cemented tailings materials under natural or accidental thermal loads. Mater Charact 59(5):598–605

Fall M, Samb SS (2009) Effect of high temperature on strength and microstructural properties of cemented paste backfill. Fire Saf J 44(4):642–651

Fall M, Benzaazoua M, Ouellet S (2004) Effect of tailings of paste backfill properties. In: International symposium Mine Fill 2004, Beijing, 19–21 September 2004

Fall M, Benzaazoua M, Ouellet S (2005) Experimental characterization of the influence of tailings fineness and density on the quality of cemented paste backfill. Miner Eng 18(1):41–44

Fall M, Belem T, Samb S, Benzaazoua M (2007) Experimental characterization of the stress-strain behaviour of cemented paste backfill in compression. J Mater Sci 42(11):3914–3992

Fall M, Benzaazoua M, Saa EG (2008) Mix proportioning of underground cemented tailings backfill. Tunn Undergr Space Technol 28(1):80–90

Fall M, Adrien D, Celestin J, Pokharel M, Toure M (2009) Saturated hydraulic conductivity of cemented paste backfill. Miner Eng 22:1307–1317

Fall M, Celestin J, Pokharel M, Touré M (2010) A contribution to understanding the effects of curing temperature on the mechanical properties of mine cemented tailings backfill. Eng Geol 114(3–4):397–413

Ghirian A, Fall M (2013) Coupled thermo-hydro-mechanical–chemical behaviour of cemented paste backfill in column experiments. Part I: physical, hydraulic and thermal processes and characteristics. Eng Geol 164:195–207

Ghirian A, Fall M (2014) Coupled thermo-hydro-mechanical–chemical behaviour of cemented paste backfill in column experiments. Part II: mechanical, chemical and microstructural processes and characteristics. Eng Geol 170:11–23

Ghirian A, Fall M (2015) Coupled behaviour of cemented paste backfill at early ages. Geotech Geol Eng 33(5):1141–1166

Ghirian A, Fall M (2016) Strength evolution and deformation behaviour of cemented paste backfill at early ages: effect of curing stress, filling strategy and drainage. Int J Mining Sci Technol 26(5):809–817

Godbout J, Bussière B (2007) Evolution of cemented paste backfill saturated hydraulic conductivity at early curing time. In: Proceedings of 60th Canadian geotechnical conference and the 8th Joint CGS/IAH-CNC groundwater conference, 21–24 October 2007, Ottawa, p 2230–2236

Grabinsky MW (2010) In situ monitoring for ground truthing paste backfill designs. In: Proceedings of the 13th international seminar on paste and thickened tailings, 3–6 May 2010, Australian Centre for Geomechanics, Toronto, p 85–98

Helinski M (2007) Mechanics of mine backfill. Ph. D. thesis, School of Civil and Resource Engineering, The University of Western Australia

Kesimal A, Yilmaz E, Erikdi B (2004) Evaluation of paste backfill mixtures consisting of sulphide-rich mill tailings and varying cement contents. Cem Concr Res 34(10):1817–1822

Kesimal A, Yilmaz E, Ercikdi B, Deveci H (2005) Effect of properties of tailings and binder on the short-and long-term strength and stability of cemented paste backfill. Mater Lett 59(28):3703–3709

Klein KA, Simon D (2006) Effect of specimen composition on the strength development in cemented paste backfill. Can Geotech J 43:310–324

le Roux KA, Bawden WF, Grabinsky MWF (2002) Comparison of the material properties of in situ and laboratory prepared cemented paste backfill. In: Annual conference, BC, Canada, p 201–209

le Roux KA, Bawden WF, Grabinsky MW (2005) Field properties of cemented paste backfill at the Golden Giant mine. Min Technol Inst Min Metall Trans Sect A 114(2):65–80

Levens RL, Marcy AD, Boldt CMK (1996) Environmental impacts of cemented mine waste backfill. RI 9599. United States Bureau of Mines, 23p

Li W, Fall M (2016) Sulphate effect on the early age strength and self-desiccation of cemented paste backfill. Construct Build Mater 106:295–304

Mozaffaridana M (2011) Using thermal profiles of cemented paste backfill to predict strength. Master Thesis. University of Toronto, p 138

Nasir O, Fall M (2009) Modeling the heat development in hydrating CPB structures. Comput Geotech 36:1207–1218

Orejarena L, Fall M (2008) Mechanical response of a mine composite material to extreme heat. Bull Eng Geol Environ 67(3):387–396

Ouellet S., Bussière B, Benzaazoua M, Aubertin M, Fall M, Belem T (2003) Sulphide reactivity within cemented paste backfill: oxygen consumption test results. In: Proceedings of 56th Canadian geotechnical conference; 28 Sept–1 Oct 2003 in Winnipeg, p 176–183

Pierce, M.E. 1997. Laboratory and numerical analysis of the strength and deformation behaviour of paste backfill, Master's thesis. Department of Mining Engineering, Queen's University, Kingston

Pokharel M, Fall M (2011) Coupled thermo-chemical effects on the strength development on Slag-Paste backfill materials. ASCE J Mater Civ Eng 23(5):511–525

Pokharel M, Fall M (2013) Combined influence of sulphate and temperature on the saturated hydraulic conductivity of hardened cemented paste backfill. Cem Concr Compos 38:21–28

Rankine RM (2004) The geotechnical characterisation and stability analysis of BHP Billiton's Cannington Mine paste fill. PhD Thesis, James Cook University, Townsville

Rankine R, Sivakugan N (2007) Geotechnical properties of cemented paste backfill from Cannington Mine, Australia. Geotech Geol Eng 25:383–393

Simms P, Grabinsky M (2009) Direct measurement of matric suction in triaxial tests on early-age cemented paste backfill. Can Geotech J 46(1):93–101

Thompson BD, Grabinsky MW, Bawden WF (2009) In-situ measurements of cemented paste backfill in long-hole stope. In: Proceedings of the 3rd CANUS rock mechanics symposium, Toronto

Thompson BD, Bawden WF, Grabinsky MW (2012) In-situ measurements of cemented paste backfill at the Cayeli Mine. Can Geotech J 49(7):755–772

Thottarath S (2010) Electromagnetic characterization of cemented paste backfill in the field and laboratory. Master thesis, p 105

van Genuchten MT (1980) A closed-form equation for predicting the hydraulic conductivity of unsaturated soils. Soil Sci Soc Am 44:892–898

Veenstra RL (2013) A design procedure for determining the in situ stresses of early age cemented paste backfill. PhD thesis, University of Toronto, p 275

Veenstra RL, Bawden WF, Grabinsky MW, Thompson BD (2011) Matching stope scale numerical modelling results of early age cemented paste backfill to in-situ instrumentation results. In: 14th Pan-American conference on soil mechanics and geotechnical engineering (PCSMGE), the 64th Canadian geotechnical conference (CGC), Toronto

Witteman M, Simms P (2011) Unsaturated flow in hydrating porous media: application to cemented paste backfill. In: Proceedings of the 2011 Pan-Am CGS geotechnical conference, Toronto

Wu D, Fall M, Cai S-J (2012) Coupled modeling of temperature distribution and evolution in cemented tailings backfill structures that contains mineral admixtures. J Geomag Geoelec 30(4):935–961

Wu D, Fall M, Cai S-J (2014) Numerical modelling of thermally and hydraulically coupled processes in hydrating cemented tailings backfill columns. Int J Min Reclam Environ 28(3):173–199

Yilmaz E, Benzaazoua M, Belem T, Bussiere B (2009) Effect of curing under pressure on compressive strength development of cemented paste backfill. Miner Eng 22(9–10):772–785

Yilmaz E, Benzaazoua M, Belem T (2014) Effects of curing and stress conditions on hydromechanical, geotechnical and geochemical properties of cemented paste backfill. Eng Geol 168:23–37

Yilmaz E, Belem T, Bussiere B, Mbonimpa M, Benzaazoua M (2015) Curing time effect on consolidation behaviour of cemented paste backfill containing different cement types and contents. Construct Build Mater 75:99–111

Yumlu M (2008) Barricade pressure monitoring in paste backfill. Gospodarka Surowcami Mineralnymi 24(4/3):233–244

Chapter 5
Design and Characterization of Underground Paste Backfill

Erol Yilmaz and Murat Guresci

1 Introduction

The mining and mineral industries make considerable commercial revenues, which are vibrant for the countries' economic growths and vital for the well-being of populations in the world. But, the principal activities of these industries, covering exploration, blasting, mining, milling, and metallurgical extraction, generate large quantities of mining wastes such as waste rocks, mill tailings, and water treatment sludge (Yilmaz 2015a). The mine wastes manufactured and stored at mining sites are problematic since they may embrace detrimental substances, such as heavy metals, metalloids, leachates, acids, and chemical reagents. Consequently, the treatment, management and monitoring of these problematic wastes (e.g. waste rocks and mill tailings) have become a foremost issue for most modern mining operations around the world (Bussière 2007; Lottermoser 2010). Mill tailings are the finely ground host rocks and gangue minerals left over after the recoverable metals and minerals have been extracted from mined ore in a mine processing plant. Waste rocks are solid materials excavated from a mine to reach the mineralized rock and/or to conduct underground openings required for mine operation such as ventilation shafts and access tunnel (Chan et al. 2008).

Over the decades, the volumes of wastes being created by mining and mineral industries have grown dramatically as the demand for minerals and metals has increased and lower grades of ore are being mined by advances in extraction and processing technology (Hudson-Edwards et al. 2011). In the 1960s tens of thousands of tons of tailings were generated each day and by 2000 this figure had increased to hundreds of thousands (Jakubick and McKenna 2003). In addition, there are individual mines manufacturing in excess of 200,000 tons of tailings per

E. Yilmaz (✉) • M. Guresci
Cayeli Bakir Isletmeleri A.S., P.O. Box 42, Madenli Beldesi, Cayeli, Rize TR53200, Turkey
e-mail: yilmazer@fqml.com

E. Yilmaz, M. Fall (eds.), *Paste Tailings Management*,
DOI 10.1007/978-3-319-39682-8_5

Table 5.1 Principal underground backfill methods and their characteristics

Characteristics	Rock fill	Slurry fill	Paste fill
Placement state	100 wt% solids	60–75 wt% solids	75–85 wt% solids
Transport system	Raise, mobile equipment, separate cement system	Borehole or pipeline via gravity	Borehole or pipeline via gravity
Cement addition	Cemented or uncemented	Cemented or uncemented	Cemented alone
Water:cement (*w*/*c*) ratio	Low *w*/*c* ratio, higher cement strength	High *w*/*c* ratio, very low cement strength	Low-to-high *w*/*c* ratio, low-to-high cement strength
Placement rate	100–400tons/h	100–200tons/h	50–200tons/h
Segregation	Stockpile segregation, reduced strength	Slurry settlement segregation, low strengths	No segregation
Stiffness	High stiffness	Low stiffness	Low-to-high stiffness
Strength	High	Low	Moderate
Grain size	>20 cm	>20 μm (≥90 wt%)	<20 μm (≥15 wt%)
Capital costs	High	Low	High
Operating costs	High	Moderate	Low
Barricade	Not necessary	Expensive	Less expensive
Tight filling	Hard-to-tight fill	Cannot tight fill	Easy-to-tight fill

day. To prevent, rectify, lessen, or eliminate the potential environmental impacts of mining wastes, numerous techniques and methods have been developed in the mining industry. One practical mine waste management method for underground mines involves refilling, which is depositing wastes as mine backfill into underground voids created by ore extraction (Yilmaz 2015b). At underground mine sites, mine backfills can be utilized adeptly as a construction material to create a floor wall or roof cover for mining activities, a major ground support tool to provide a safe work environment, and an effective means of mine waste disposal.

2 Underground Backfill Methods

The type of backfill used by an underground mine operation is dependent on several factors: the configuration of the mining process, the stope sequences, and excavation sizes determined by the mining method, the depth and orientation of the orebody, and the materials available to use as backfill, focusing on tailings management requirements over the life of the orebody. (Landriault et al. 2000). Underground mine fills are categorized mainly as rock fill, hydraulic or slurry fill, and paste fill. The choice between these three types is site specific and depends on the particular requirements of each mining operation. Table 5.1 lists a summary of different characteristics of the three most important types of most modern mine backfill used in

Fig. 5.1 Mine backfill methods: (a) rock, (b) slurry, and (c) paste

hard rock mining (Hassani and Archibald 1998; Henderson et al. 2005; Landriault 2006). The different backfill systems have different capital and operating costs attached to them.

Cemented rock fill (CRF, Fig. 5.1a) consists typically of classified or unclassified waste rocks mixed with a cement slurry to improve the bond strength between the rock fragments. CRF contains a mixture of coarse (<150 mm) and fine aggregate (< 10 mm fraction). The ideal gradation is one which minimizes the voids. Cement slurry concentration is approximately 55 wt%. Hydraulic or slurry fill (Fig. 5.1b) is prepared by dewatering the mill tailings stream to a pulp density of about 65–70 wt% solids and then passing it through hydrocyclones to remove the "slimes" and retain the sand fraction for backfill. Paste backfill (Fig. 5.1c) is a high-density backfill (>70 wt% solids) which is made up of full tailings, hydraulic cement, and process water (Archibald et al. 2003; Potvin et al. 2005).

Among the other types of mine backfill (i.e., rock fill, hydraulic, or slurry fill), cemented paste backfill (CPB) is extensively used in many underground mines worldwide, reducing volumes of surface-disposed mine wastes. This environmental benefit has spurred the acceptance of CPB as an economical alternative to both rock fills and hydraulic fills (Wilkins et al. 2004; Stone 2007; Belem and Benzaazoua 2008). Unlike hydraulic fill, CPB can be composed of tailings without fine particle removal, thereby maximizing the amount of tailings that can be used regardless of size distribution. CPB offers several advantages over other backfill types in operational, financial, and environmental terms such as speed of delivery to a mined out stope, and higher specific mechanical strength development. Supplementary information on design and application of CPB technology for underground metalliferous mines around the world can be found elsewhere (Kesimal et al. 2003; Benzaazoua et al. 2004; Grabinsky and Bawden 2007; Fall et al. 2008; Yilmaz 2015b).

3 Integrated Paste Technology

Paste technology may be considered as a modern technique in order to handle and dispose of tailings from the mining industry in an environmentally sound way. Surface paste disposal (SPD) is a novel alternative used by the mining industry for safe storage of tailings at the surface (Verburg 2002; Johnson and Slotte 2004). SPD consists in the deposition of mine tailings at a higher percent solid (>70%) in relatively thin layers that are allowed to drain and dry before the next layer is put in place. The term "paste" typically signifies that mine tailings will not segregate into fine and coarse particles during the transport and disposal, and will exhibit minimal bleeding after the final placement of tailings. SPD provides a number of operational and environmental advantages, such as a better water management, no necessity for complex retaining impoundments, a reduced footprint of the tailings disposal area, and the possibility to utilize progressive reclamation (Sofra and Boger 2002; De Souza et al. 2003; Ulrich and Coffin 2013; Yilmaz 2015a).

In addition to surface paste disposal, the paste can also be mixed with cements and used to stabilize underground metalliferous mines. This means that an engineered mixture of paste and cement is pumped into mined-out stopes to form a rock solid material. Cemented paste backfill (CPB) supports the walls of adjacent adits as mining progresses. Implementing paste backfill system, mines can be run more efficiently and safely (le Roux et al. 2005; Thompson et al. 2012). Optimizing the paste's consistence can be crucial since the cost of binders is essential for the competitiveness of backfill. The mix variables must always be optimized to provide a backfill material that exactly meets the demands of the mine (Kesimal et al. 2005). Figure 5.2 illustrates a schematic view of integrated paste tailings technology intended for hard rock mining operations: surface paste disposal and cemented paste backfill.

In addition, strict environmental regulations pertaining to the tailings disposal, lower tolerance to the release of heavy metals into the environment, and high cost

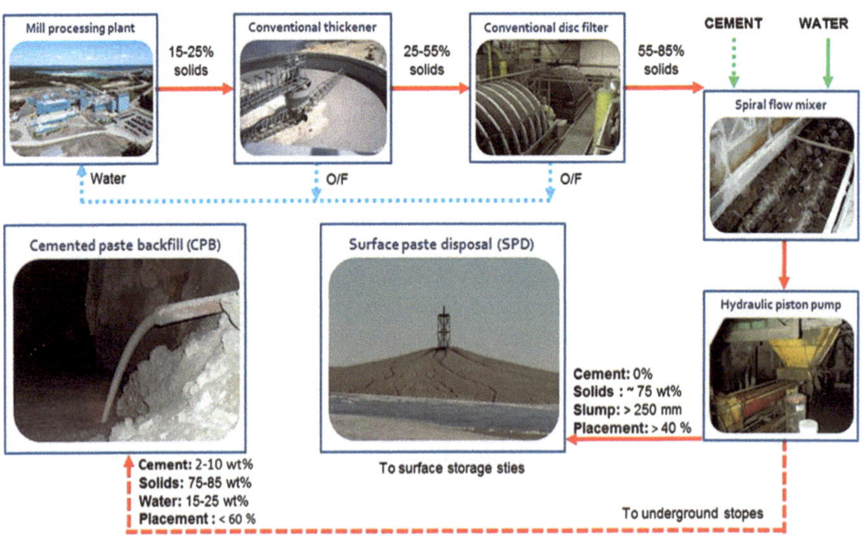

Fig. 5.2 Schematic view of integrated paste technology for sustainable mining (Yilmaz et al. 2014b)

linked with the treatment of acid mine drainage (AMD) effluent are other reasons for adopting paste technology (Awoh et al. 2013; Bouzahzah et al. 2014; Plante et al. 2014). Underground paste backfill uses the finer fraction of reactive tailings, accordingly avoiding its surface impoundment. The costly environmental problems and long-term liability of the AMD treatment from surface ponds is avoided at later stages since the reduction in the amount of reactive tailings decreases their environmental impact as well as future capital expenditure.

4 Cemented Paste Backfill

The paste can be considered as a granular material mixed with sufficient cement and water to fill the interstices between the particles so that the material behaves as a fluid. Moreover, it has been experimentally shown that the tailings used in a paste mix should contain at least 15 wt% of particles finer than 20 μm, which act as lubricant (colloidal water retention) and so facilitate plug flow in pipelines (Landriault 2006). The material properties of paste mixes, such as water retention and flow resistance, are not only controlled by grain size distribution, but also their chemical and mineralogical composition. Thus, each backfill mix must be tested separately to govern the paste behavior. The paste plant and the pipeline distribution system need to be designed according to the properties and behavior of paste mixtures (Kesimal et al. 2004; Grice 2005; Cooke 2006; Ghirian and Fall 2013, 2014). In addition, there is no critical flow velocity and the paste mix can remain idle in an underground pipeline system for varying periods of time without plugging the line. Sufficient shear yield stress is needed to remobilize paste that is idle. The longer the paste is left idle, the greater becomes the yield stress to reinitiate flow. If the yield stress is relatively high, and there is insufficient energy available to remobilize the paste, the line will become plugged (Revell 2004; Ouellet et al. 2006; Ouattara et al. 2010; Helinski et al. 2010). Many operations use the concrete slump tests as a means of accurately measuring the yield stress of paste backfill mixes (Pashias et al. 1996). There is a direct link between yield stress and slump. As paste's slump increases, the yield stress is reduced, and the paste will flow further with less energy input. Figure 5.3 shows a typical relation between yield stress and slump for a paste backfill sample.

One of the most important characteristics of any paste backfill is the grain size distribution. The aim of grain size optimization is to produce a paste backfill that develops dense packing during placement. This is usually achieved by a well-graded aggregate which allows attaining optimum porosity and thus reducing binder consumption and mine operating costs (Zou and Nadarajah 2006). It has also been shown that pipeline pressures and wear are sensitive to the percentage of minus 20 μm size material within backfill (Brackebusch 1994; Nonnen 2001; McGuinnes and Cooke 2011). The three GSD categories of tailings are typically considered in the CPB mix design for modern underground mines all over the world: coarse, medium, and fine tailings (Landriault 2006). Coarse-grained tailings contain

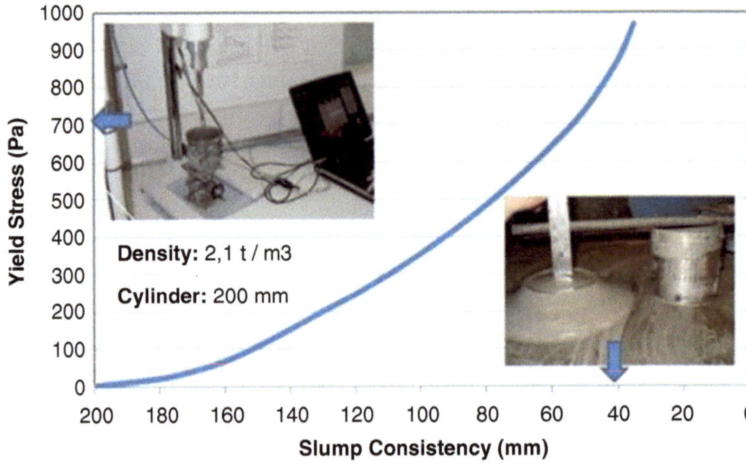

Fig. 5.3 Measurement of yield stress via cylinder slump and rheometer (modified after Grice 2005)

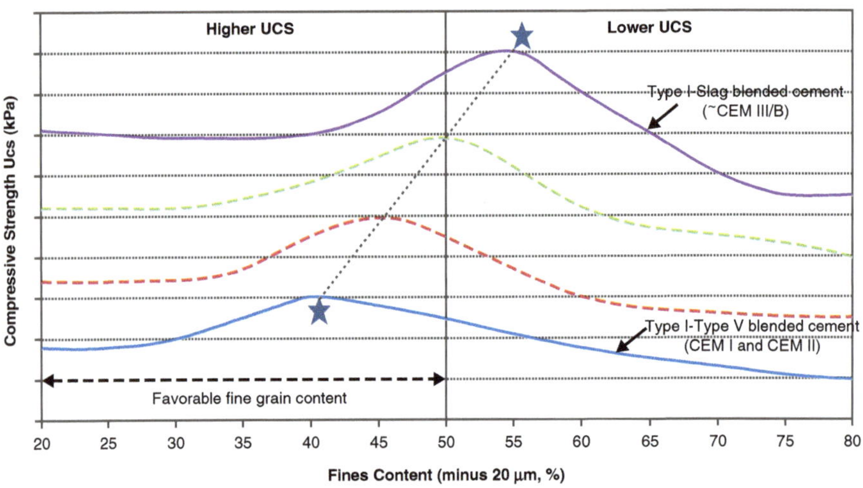

Fig. 5.4 The change in compressive strength with grain size (modified after Fall et al. 2005)

15–35 wt% particles finer than 20 μm. For a given binder content, high solids content creates a good fill strength. Medium-grained tailings contain 35–60 wt% particles finer than 20 μm, and produce a good CPB mix, but with lower strength than those obtained from coarse-grained tailings. Fine-grained tailings contain 60–90 wt% particles finer than 20 μm (Fig. 5.4). Depending on their mineralogy, fine tailings show high water retention, which create the backfill with good flow properties but relatively lower strength. By adding coarse material to the tailings, the strength of paste backfill is increased due to the lower porosity (Fall et al. 2005; Yilmaz et al. 2009, 2014a).

Tailings used within CPB must be analyzed for mineralogical composition (zinc, lead, pyrite, and pyrrhotite), which can affect the binder chemical reactions (Fried et al. 2007; Godbout et al. 2009). Some minerals can induce strength development retardation, strength reduction, and long-term strength loss due to internal sulfate attack (Bernier et al. 1999; Bertrand et al. 2000; Hassani et al. 2001; Benkendorff 2006). CPB made of this type of tailings should be laboratory tested for its strength and stability performance before it can be used for CPB in underground mines. Also the health and safety of workers must be considered when designing CPB for mines. When pyritic tailings are used to prepare CPB mixtures, their exothermic properties must be examined in detail to avoid self-heating (Blowes et al. 1995). For instance, pyrrhotite and pyrite can react chemically in underground water and oxygen conditions. Such reactions can cause internal heating to temperatures that may ignite the CPB's sulfur content and hence cause a self-sustaining underground fire that produces toxic sulfide gas (Ouellet et al. 2006; Tarig and Nehdi 2007; Weatherwax et al. 2010; Pokharel and Fall 2011).

4.1 Principal Ingredients

Cemented paste backfill (CPB) is an engineered mixture of the total or deslimed filtrated wet tailings, binder, and mixing water (Fig. 5.5). The percentage of filtered tailings solids ranges from 75 to 85 wt%. The binder (generally varies 2–7 wt%) is made up of a general-use Portland cement (GU) or a blend of GU cement and either granulated blast furnace slag or fly ash. The role of binding agents is to develop cohesion and strength within CPB so that the exposed backfill faces will be self-supporting and stable when adjacent stopes are extracted. Mixing water is added to reach the targeted slump, which should vary in the range of 15–25 cm to ease paste transport to underground stopes by either gravity or pumping (Benzaazoua et al. 2004; Moghaddam and Hassani 2007; Fall et al. 2008; Cihangir et al. 2011; Yilmaz 2015b).

Physical, chemical, and mineralogical properties of the CPB ingredients (tailings, binder, and water) have a significant effect on the strength and durability of backfill (Benzaazoua et al. 2004; Kesimal et al. 2005; Klein and Simon 2006; Fall et al. 2008). For instance, grain size distribution (GSD) of tailings determines in part some of the CPB final bulk properties. Overall, a backfill sample with a small spread of grain sizes is referred to as "poorly graded," whereas a sample with a widespread would be "well graded." Numerous authors (Kesimal et al. 2003; Benzaazoua et al. 2004; Fall et al. 2005; Belem and Benzaazoua 2008; Sivakugan et al. 2013; Yilmaz et al. 2009, 2014a, b) have shown that for a given water/cement *w/c* ratio, a CPB with coarse- and medium-grained tailings is more liable to produce higher UCS than that of fine-grained tailings, mainly due to the lower void ratio. GSD is directly related to the flow properties, permeability, and pumpability of CPB. Thus, the higher the fines content (<20 μm), the lower the permeability coefficient (Fall et al. 2005). A suitable binder or combination of binders is mostly required to attain the CPB's desired strength properties and resistance to liquefaction. Figure 5.6

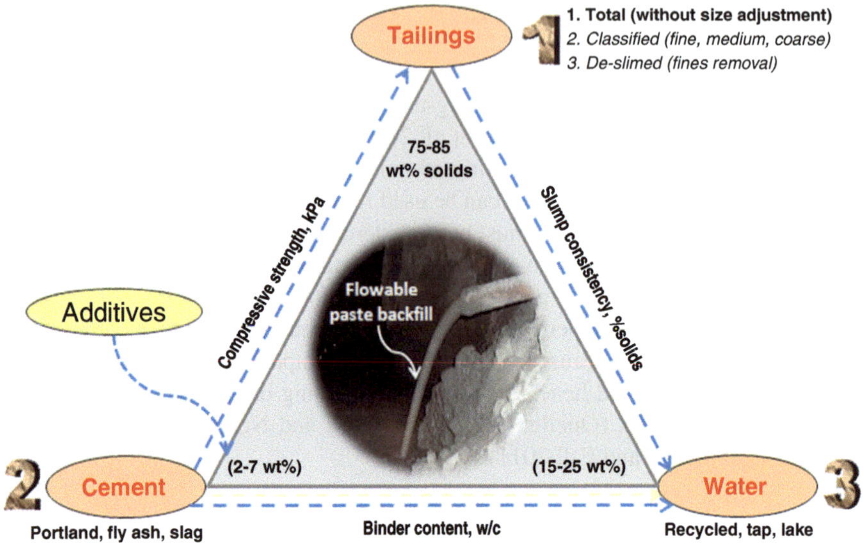

Fig. 5.5 Schematic diagram showing the principal ingredients of cemented paste backfill

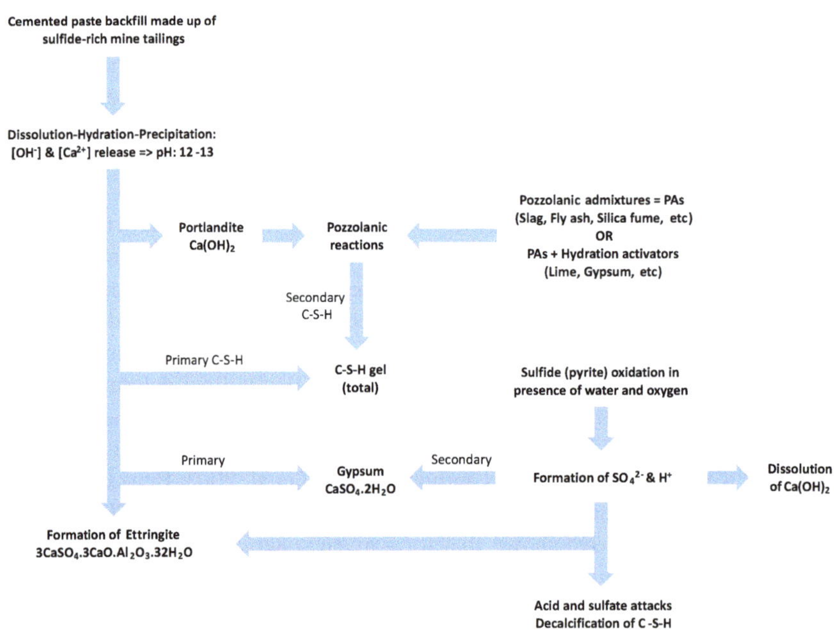

Fig. 5.6 Schematic diagram showing the hydration process of cemented paste backfill

represents a schematic diagram of the development of hydration process, mod of potential interactions of hydration products within sulfide-rich tailings, and the potential role of pozzolanic additives.

The mineralogy of tailings and binders can also affect backfill strength by influencing certain chemical reactions that take place during curing (Bernier et al. 1999; Bertrand et al. 2000; Benzaazoua et al. 2004; Cihangir et al. 2012). When paste backfill is created from sulfide-rich mill tailings, certain events may occur, such as pyrite oxidation in the presence of oxygen and water, forming acid and sulfate. Sometimes, these chemical interactions between sulfate ions and cement hydrate products can cause internal heating to temperatures. All these events may affect harmfully paste backfill strength (Hassani et al. 2001; Kesimal et al. 2005; Fall et al. 2005; Yilmaz et al. 2009, 2014a, b, Yilmaz et al., 2015a). Minerals such as sericite, micas, and clay can reduce the CPB's strength and stability predominantly due to their water-absorbent mineral layers. The sulfide-rich minerals raise the solids-specific gravity, while the strength of abrasiveness of silica minerals causes serious pipeline wear (Revell 2004; McGuinnes and Cooke 2011; Wang et al. 2011).

Mixing water may also affect paste backfill mechanical strength with regard to the *w/c* ratio and cement hydration mechanisms (Kesimal et al. 2005; Ouellet et al. 2006; Fall et al. 2008; Yilmaz et al. 2014a, b). One can state that pH of mixing water and the amount of dissolved chloride and sulfide salts are of greater importance. The acidic water with a pH lower than 6.5 and sulfate salts can react with the cement components (e.g., C_3A) and hydration products (e.g., C-S-H) and lead to the long-term strength and stability losses of cemented paste backfill. The water-to-cement *w/c* varies depending on target slump and strength values for a backfill mixture. The higher amount of mixing water results in the higher slump consistency, a more porous mix, and lower compressive strength. Figure 5.7 shows a schematic

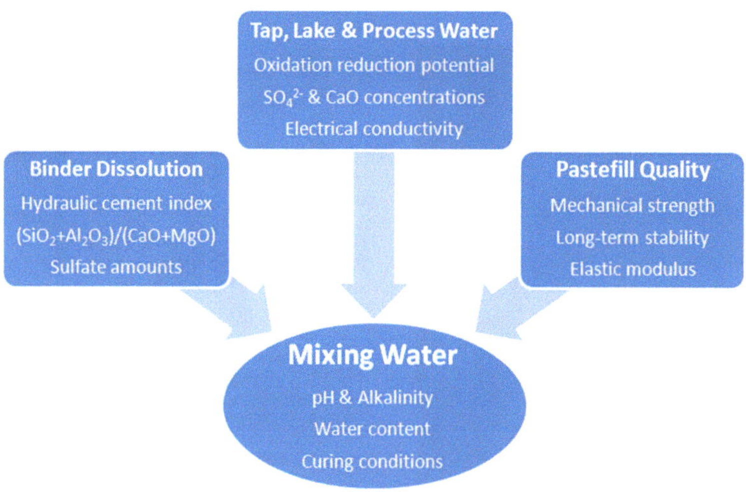

Fig. 5.7 Interactions between water and binder during hydration

illustration of chemical interactions between mixing water and hydraulic binder during hydration processes with or without pressure.

4.2 Preparation of Paste Mixtures

Sulfide-bearing tailings generated from the mineral processing plant are usually discharged as a dilute slurry (20–35 wt% solids). The tailings slurry is dewatered by a conventional gravity tailings thickener or high-density deep-cone thickener up to a solids content of 60–70 wt.%. If the tailings contain too much slimes (fine contents less than 20 μm), part of the tailings can be cycloned, consequently increasing the thickening and filtration rates. The underflow from a conventional thickener should be a steady slurry which does not reveal segregation of grain sizes or fast settling of larger grains. The tailings slurry can then be easily pumped to paste storage tank by using centrifugal pumps. Figure 5.8 illustrates a photo of cyclone typical paste thickener integrated with filter cake and a typical paste backfill preparation flow sheet.

A number of filters, such as disc and drum vacuum filters, belt vacuum filters, and belt filter presses, can be used as the final dewatering step in preparing a paste backfill mix. The primary criterion for filter selection is capital and operating costs. The moisture of filter cake may vary from 10 to 20% solids by weight. If the thickener underflow always keeps the same density (based on the moisture content and grain size distribution of the tailings), then the filtration step can be avoided and the obtained slurry may be directly sent to mixer in order to produce a paste backfill mixture. The ingredients of a paste mixture (filtered tailings, hydraulic binder, and mixing water) must be mixed systematically for obtaining the homogenous pastes which may provide superior paste flows without plugging during the pipeline transportation.

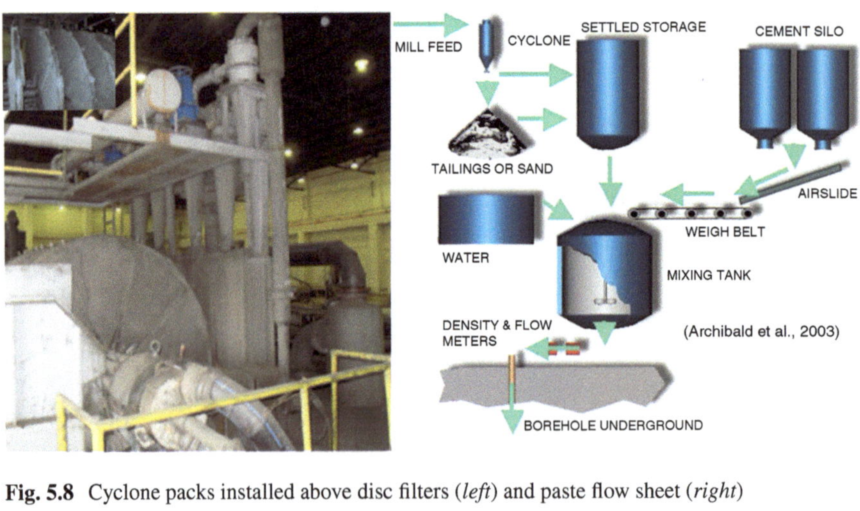

Fig. 5.8 Cyclone packs installed above disc filters (*left*) and paste flow sheet (*right*)

The filtrate cake discharged from vacuum disc filters and drops to a reversible belt conveyor, which may sometimes feed to the conditioning mixer. The filter cake together with tailings slurry from paste storage tank or agitating slurry tank is mixed in a conditioner mixer before being discharged into the paste mixer. At the mixer additional makeup water and cement are added and mixed to the desired slump (typically varies between 18 and 23 cm). After mixing, paste backfill mixes can be delivered to underground stopes in several ways: either by truck, by pumping and/or gravity, or as dense slurry or paste through boreholes and pipelines.

4.3 Advantages of Paste Backfill Technique

CPB requires the presence of slimes to maximize the solids content of the backfill mass, to act as a lubricant for pipeline flow, and to reduce friction losses during pipeline transport. This significantly reduces the need for high volumes of transport water; any water that is used to manufacture the paste backfill tends not to bleed out of emplaced fill. This greatly reduces or eliminates the drainage need of the backfill from underground mines. CPB offers a number of technical, operational, and financial advantages over other backfilling types as follows:

- Paste is a homogenous, non-segregating, flowable, and a low-porosity or high-density backfill material, which avoids plugging in the pipeline.
- Total tailings (with coarse- and fine-graded tailings) can be used for the preparation of paste backfills, not necessary to remove fine contents from tailings.
- A relatively high strength can be achieved with an equal or even less cement contents for a given CPB mix recipe, thereby resulting in a reduction in overall binding costs.
- An overall mining cycle can be reduced drastically due to the relatively high strength gain of early-aged paste backfills.
- Better mine mechanical control enables more selective mining or ore excavation and, due to this, less mill tailings production.
- Increasing the ore production from the underground mine could give the opportunity to store more tailings underground as paste backfill.
- Paste backfill provides higher tailings use, which reduces the need of surface tailings disposal, and therefore its environmental impacts.
- Paste backfill keeps cleaner operation as slime drainage from the stope is eliminated (not necessary to drain water from stopes being filled with paste backfill).
- Backfill dilution of adjacent ore can be reduced potentially due to increased strengths for similar cement contents.
- Mine operating and maintenance costs (less damage to roads, sumps, pumps, support elements, etc.) reduce significantly.
- Paste backfill allows the use of fully mechanized mining system, thereby increasing safety, production, and reducing accidents and injuries.
- Underground open and/or blind stopes can be filled continually with CPB materials, avoiding liquefaction and washout of the barricade.

- The construction of the stope barricades can be simple due to less water retention within paste.
- Part of tailings can be used for sprayed concrete purposes in underground mines while using the majority of the tailings for underground paste backfill aims.
- Pipeline or borehole system provides faster paste backfill placement. This provides the hardening and curing process, leading to high compressive strengths at early ages.

5 Paste Backfill Plant Overview

A typical paste plant consists of thickener, storage tank, filters with or without hydrocyclone, belt conveyor, mixer, cement silos, and pumps. This section examines the production stages of underground paste backfill. The delivery and placement processes will then be discussed in more detail. Figure 5.9 illustrates a typical

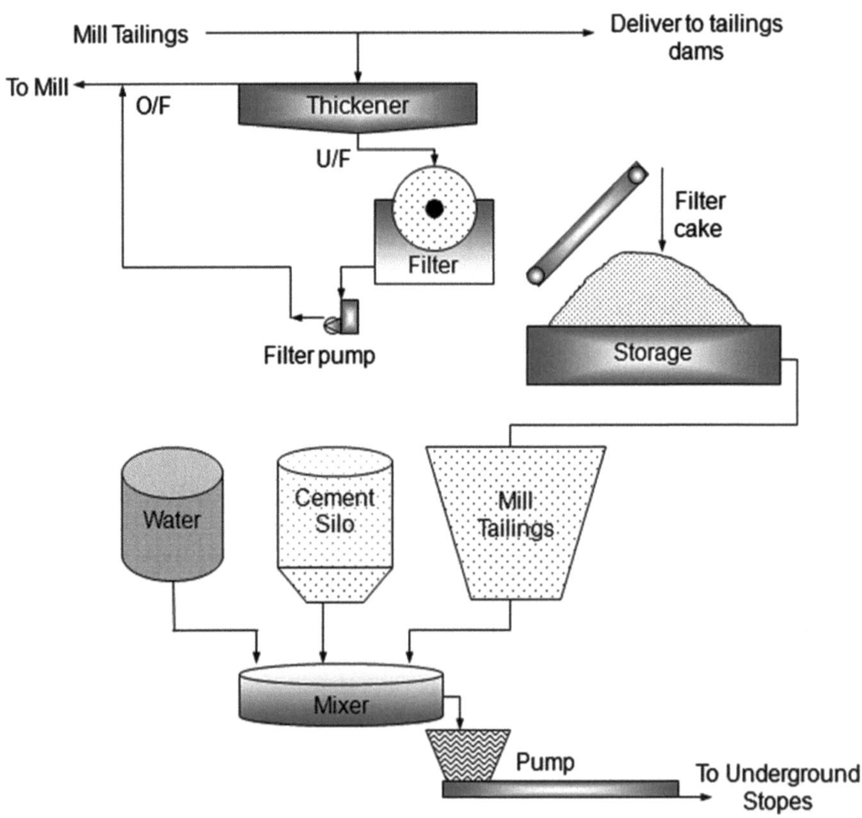

Fig. 5.9 A typical paste backfill plant flow sheet

paste plant flow sheet which covers all fundamental equipment required for the manufacture of paste mixtures.

5.1 *Paste Thickeners*

The slurry tailings generated during ore processing cannot be used for CPB preparation as long as they are not dewatered mechanically or naturally to attain a certain solid concentration in an appropriate thickener. There are two types of thickeners which have been used lengthily as a cost-effective means of producing the paste. Conventional paste thickeners are the raked vessel which uses flocculent and concentrate a suspension of solids, producing a consistent high-density underflow (with a yield stress of more than 200 kPa). Deep-cone paste thickener was developed to optimize flocculation with tank designed to allow high gravity compression of the tailings, providing a considerable increase in underflow solids concentration. Figure 5.10 shows a schematic diagram of both conventional thickener and deep-cone paste thickener. The height of settled solids in a deep-cone thickener is much higher than the one obtained from a conventional thickener, resulting in a high-throughput rate.

Deep-cone paste thickener is a fairly new, highly accepted technology in the mining industry for dewatering tailings solids to a higher concentration (above 65 wt% solids) than achievable with conventional thickeners (Slottee and Johnson

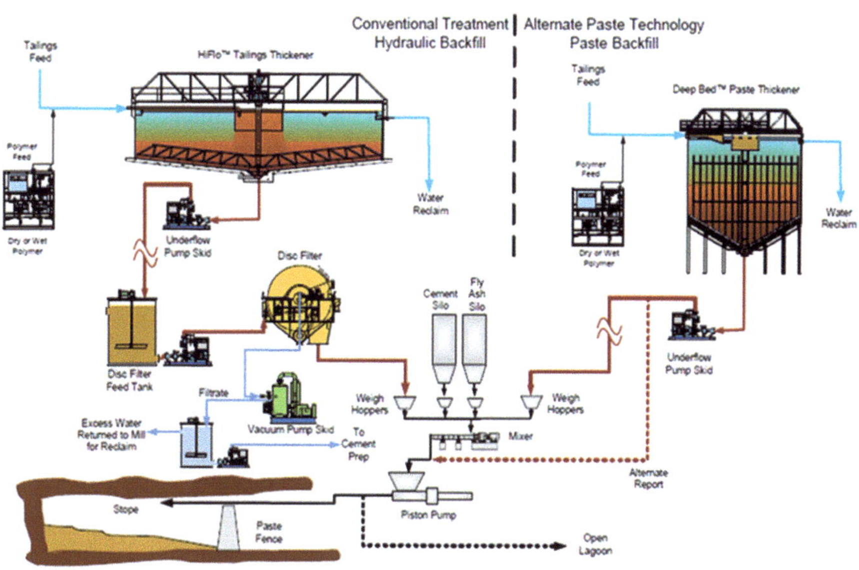

Fig. 5.10 A schematic view of a paste backfill flow sheet associated with alternative paste thickeners: deep-cone-type and conventional thickeners (adapted from WesTech (2015))

2009). The thickening process of paste tailings is typically controlled by the two functions: flocculation and settled solids. If these conditions are managed properly, thickener will produce the paste having a steady underflow density, rake torque, and overflow clarity. Most process operators agree that one of their major challenges is to properly monitor the bed level and mass of their thickeners, which in tandem with other process factors allow the optimization of thickener efficiency.

5.2 Vacuum Disc Filters

Filtering is most often done by using vacuum disc filters, drum filters, or belt filters. The selection of a filter is frequently based on a consideration of capital and operating costs. Disc filter has been used commonly in the mining industry because of its large filtration area and low capital cost. But, it requires changing filter cloths constantly, which increases filtration-related operating costs (Henderson et al. 2005). Figure 5.11 shows two different filter cakes discharging from identical disc filter but using full plant and cycloned tailings slurries.

The removal of ultrafine particles (i.e., the finest 10% of the material or material less than 5 μm in size) from processing tailings can help improve the material's ability to be dewatered as well as reduce the amount of binder that would be required to achieve a target strength (Wilson and Calverd 2011). Nevertheless, the material would still need a certain amount of fines to remain a stable paste backfill and allow its transport within a pipeline without coarser particles settling out. Landriault (2006) has shown from the vacuum filtration results that higher cake loading rates can be attained with the cycloned tailings. Changing the grain size and removing some of ultrafine particles can have a primary effect on the filtration rates. With the coarser grain size there are more voids in the tailings resulting in less blinding of filter cloths, both of which will allow more water to pass over the matrix. In addition to lowering capital cost, reducing the number of vacuum disc filters also lowers the overall operating and maintenance costs.

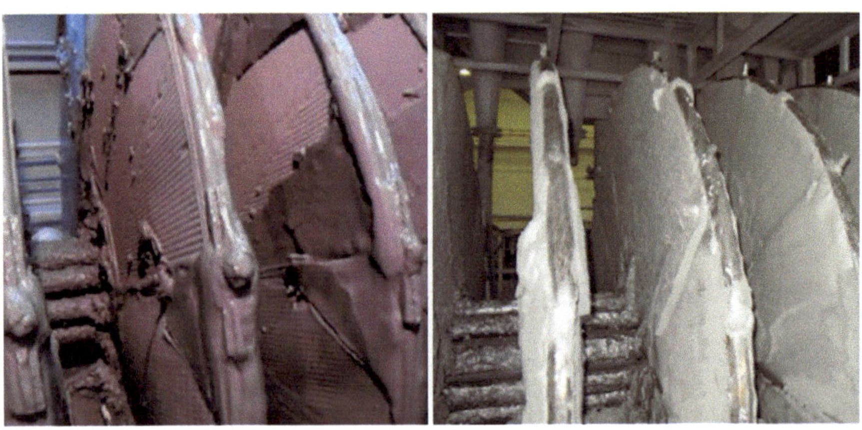

Fig. 5.11 Disc filters for dewatering full plant (*left*) and cycloned (*right*) tailings slurries

5.3 Mixing Process

Mixing is one of the most important stages in manufacturing a high-quality and homogenous paste backfill, and frequently done by either batch or continuous process. Batch systems have separately weighed material being fed into the batch mixer before mixing and discharged as discrete batches while continuous systems have each paste ingredient being fed continually into the mixer. One can state that a batch process is much easier to control than a continuous system. Screw mixers are used for continuous systems while high-intensity mixers (screw or paddle mixers) of the type utilized in the concrete industry can be operated in batch or continuous modes. Figure 5.12 shows photos of conventional and plowshare mixers for paste backfill preparation. Plowshare mixers operate on the principle of a mechanically generated fluid bed with three-dimensional movement of the product enabling homogenous mixing of the paste ingredients.

The filter cake is discharged on a reversible conveyor belt first and then fed to a conditioning surge hopper where the solids concentration is adjusted adding makeup water. The weight of filter cake passing along the conveyor will be continuously monitored, allowing control of the amount of thickened tailings sent to the filter bypass to achieve a desirable slump for CPB mixtures. The mixture slump will be fine-tuned by adding more or less slurry to achieve the target power draw on the mixer motors to obtain a desirable slump. Dirige and Archibald (2014) have shown that for identical binder components and curing time, the plowshare mixing process exhibits significant potential for generating improvement in terms of long-term CPB strength and stiffness relative to conventional mixing processes. On average,

Fig. 5.12 Photos of (**a**) screw-type, (**b**) paddle-type, (**c**) plowshare or plow mixers, and (**d**) special shovels of plowshare mixer (modified from Dirige et al., 2008.)

CPB mixes that were prepared using the plowshare mixer exhibited approximately 15% strength gain for specimens cured at 28 days, 40% strength gain for 56-day cure, and 25% strength gain for 112-day cure relative to conventional screw mixer-prepared samples with equivalent binder contents. The stiffness of almost all the blends prepared using the plowshare mixer is higher than for equivalent blends mixed using the conventional screw mixer. Elastic modulus was indicated to be nearly 28% greater for 28-day cured samples, 66% greater for 56-day cured samples, and 36% greater for 112-day cured samples.

5.4 Binder Management

The binding agents play a vital role in cohesion and strength within paste backfill. General-use Portland cement is commonly used for the preparation of CPB mixtures. Supplementary cement materials, such as ground granulated blast furnace slag, pulverized fly ash, and natural pozzolans, or waste materials, such as cement kiln dust, and finely ground industrial and waste glass can be added to the paste materials as a partial replacement of Portland cement in binary or ternary blends. The binder quantity used within the backfill is most often expressed as a percentage of the total dry mass of solids. Binders are usually fed to the mixer via a screw conveyor under a high-capacity cement storage silo adjacent to the paste plant (Fig. 5.13). The relatively large cement storage silos guarantee continuous material flow when the paste plant is operating continuously at the highest binder content. The binder silo is equipped with a dust collector and fan that will exhaust and transport compressed air through a ducting exhaust system. A monitoring system will continuously monitor the level in cement storage silo. A high-level switch and a low-level switch provide

Fig. 5.13 A 350-ton cement storage silo considered for paste backfill plants

indication of binder levels. The silos are also equipped with a pressure-relief valve which opens if the silo is pressurized excessively. The cement silo cone will be fluidized using aeration and will discharge through a rotary valve to a binder weigh hopper via a screw conveyor. The rotary valve and screw conveyor control the flow rate of binder into the weigh hopper to achieve the desired weight on the load cell. The weigh hopper discharges the binder to the mixer.

The binder costs (approximately US$1 per % per ton of the backfill) comprise 50–80% of the operating costs in a paste backfill plant (De Souza et al. 2003). Accordingly, the optimization of binder dosages and/or the use of suitable and cost-effective binders to achieve the required performance of paste backfill are important in the mining industry. While calculating the cost of optimum binder dosage without reducing especially long-term strength of paste backfill mixes, the type and content of binder must be selected firstly, which will enable achieving the intended mechanical strength by reducing the costs associated with binder use. Even a slight reduction in the binder dosage gives rise to a considerable cost saving. Thus, it is essential to conduct a series of tests using different CPB mixes, to study the effects of binder type and dosage on paste backfill characteristics and performance. In this regard, most investigations to date have been conducted on paste backfill by many researchers (Ouellet et al. 2006; Belem and Benzaazoua 2008; Cihangir et al. 2012; Ercikdi et al. 2014, Yilmaz 2015a, b).

6 Backfill Reticulation System

After mixing, the paste can be either discharged into the feed hopper of a pump and pipeline system or discharged over a borehole to underground stopes. Depending on the topography of the site, there are two ways of designing the underground distribution system: a gravity-driven system where pumping is not required, and service or standby piston pumps where the paste will be delivered through the distribution pipeline to the stopes. Indeed, the paste plants were designed initially to use positive displacement pumping systems, but later switched to gravity-driven flow systems (boreholes) due to the fact that it requires low pumping pressures and hence offers less friction losses in the delivery pipeline. The following section outlines the integral items of the paste backfill reticulation systems.

6.1 Positive Displacement Pumps

The paste material is discharged from the screw or spiral mixer to the hopper prior to entering the underground distribution system. The paste hopper is indeed essential to guarantee proper operation of the positive displacement pumps. Load cells on the paste hopper ensure that the level of the paste in the hopper is maintained above a minimum level in order to prevent the risk of air entering the distribution system.

Fig. 5.14 Paste backfill being pumped underground using Putzmeister pump

There are interchangeable positive displacement pumps with s-swing tube used for paste backfill operations in the mining industry (Fig. 5.14).

The pressure peaks in the pipeline can be reduced importantly by these pumps. While the first delivery cylinder is pushing the paste into the pipeline, the second piston sucks the paste into the delivery cylinder fast, and the seat valve switches over and recompresses to the pipeline pressure. At the end of the stroke of cylinder one, the speed of this piston is reduced and the other starts pushing. As a result of this operation, the big pressure peaks of a standard double-piston pump are eliminated, and the lifetime of the pipeline is increased.

Note that the high-pressure positive displacement pumps are installed typically for clearing line blockages and for cleaning out the pipeline. In most modern underground mines around the world, however, this is not apt and a positive displacement pump operates to transport the mixed cemented paste backfill materials. These pumps are hydraulically powered equipment which offer the flexibility with pressures of up to 100 bars and grain sizes of up to 65 mm, and well suited for full tailings mixtures.

6.2 Borehole Flows

In most mining operations, there are two boreholes drilled from the surface backfill plant to underground voids for the paste delivery. This is mostly because mine operators want to keep delivering the paste to underground even if one of them becomes blocked or is used for line change. Based on the needed delivery flow rate, boreholes may have an inner diameter of up to 25 cm. Paste backfill will flow under gravity at

Fig. 5.15 Photos of the borehole system for paste backfill transport to underground mines

an angle greater than 30°. This makes the selection of the plant location important for mines. In designing a cemented paste backfill system, the plant should be located where a series of boreholes and short lateral runs could carry the paste to all stopes under gravity (Brackebusch 1994). The transportation of paste backfill by means of pipeline is governed solely by gravity. This is due to the high vertical-to-horizontal ratio of the distribution system. Figure 5.15 shows photos of paste backfill boreholes equipped with energy dissipater.

The high-pressure gradients within the horizontal section of the pipeline can be overcome by the gravity head contributed by the vertical drop. There are however some facts that vertical lined borehole may be prone to blockage due to liner damage caused by wear and vacuum effects. If the boreholes are designed at an inclination of ~70°, the line wear-related blockage can be reduced significantly. For gravity injection, the delivery auger is placed immediately above the PVC casing and the paste is poured through a screened funnel into the top of the PVC casing. The metal screen has holes 10 cm by 5 cm and is used to avoid obstruction of the borehole with debris or chunk material. At the borehole distribution point, video feed can be established to verify proper operation for paste backfill flows.

6.3 *Pipeline Networks*

Each mine has its own backfill distribution system which needs to be operated without any blockages, hence preventing production losses and exposing mine personnel to operational hazards. The piping network should not fail due to excessive pressures or worn piping. More importantly, the piping must wear at an expectable rate to allow for the efficient planning and execution of pipe replacement schedules. The size and type of paste backfill pipes are strongly reliant on the network of the distribution system. The use of poly pipe for lateral distribution is a point of operational

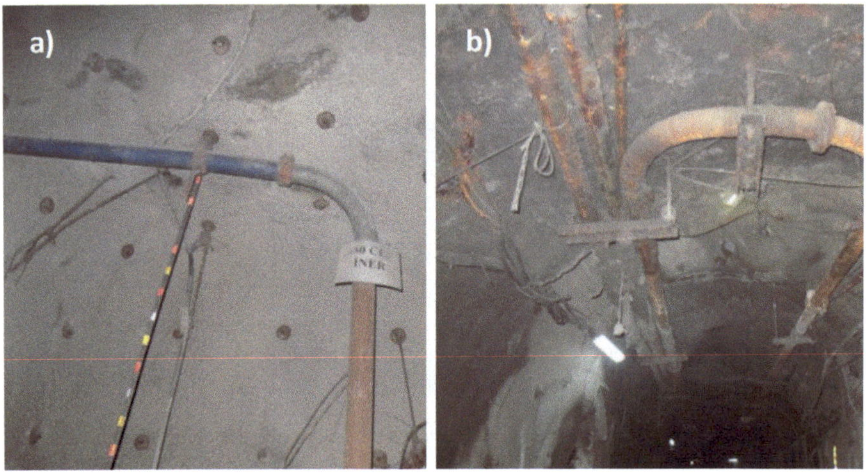

Fig. 5.16 Lateral poly pipe to steel pipe connection (**a**) and backfill pipe loop (**b**)

concern since it restricts the operating pressure limit of the line and poses safety hazards associated with the bursting or rupture of pipes. Figure 5.16 shows photos of paste backfill poly and steel pipes equipped in the underground mines.

When designing a CPB distribution system, several transport and material parameters should be considered: paste velocity, density, grain size distribution, flow regime, binder addition, and pipe diameter. A rule of thumb in the industry is that paste backfills with 175 mm slump (having a yield stress lower than 250 kPa) can be delivered to the underground stopes at a placement rate of 35 m^3 per hour via a 125 mm diameter pipeline. As the placement rate of backfill is increased to 70 m^3 per hour, a 225 mm diameter system is required since the pipe diameter has an effect on a backfill distribution system. The distribution systems for paste backfill depend on constant slow-flowing feed. Friction losses during transport are higher for paste backfill than hydraulic backfill due to the fines within paste (Cooke 2006). The fines help keep the particles in suspension. A minimum of 15% fines (<20 μm) should be present within backfill to keep the particles from settling (Landriault 2006). In practice, this means that flow in a pipeline full of paste can be reinitiated after a period of time because the solids are still in suspension. The stope is most often filled from the upper sill drift through 125 mm underground pipeline system.

6.4 Flushing Mechanism

Flushing is one of the important daily paste plant operations and practiced with compressed air and water to clean up the pipelines or boreholes at the end of each backfill run or prior to a planned paste plant shutdown. Flushing is achieved successfully with a number of positive displacement pumps used in the concrete or

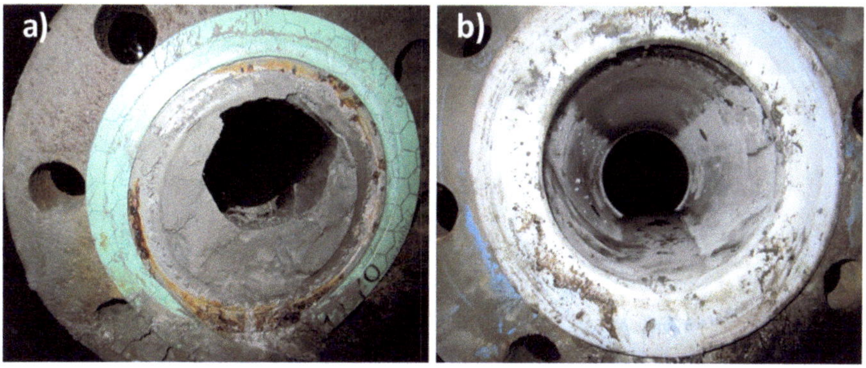

Fig. 5.17 Photos of (**a**) buildup of paste backfill on pipe and (**b**) cleaned pipe after flushing

slurry industry to solve problems associated with blockages or line narrowing (Cooke 2006; Goosen et al. 2011). As a routine paste delivery process, some of the pipes (mainly ones located in the horizontal line) can be blocked with denser residues or chunks and may cause pressure increases in pipeline or boreholes and even at times their failures due to the combined effects of pipes worn and high pressures caused. Figure 5.17 shows some photos of blocked or cleaned paste backfill pipes.

Based on the production rate and stope change, delivery lines are flushed two or three times per day with water (typically of 5–10-min duration) first and then a compressed air in the paste plant (typically 700 kPa). To reach the unobstructed flow conditions in the line, a total volume of 1000–3000 liters of water is used during individual flushing. The horizontal pipelines are flushed until the clean water flow is observed through camera from the end of the pipe. Paste plant operator must see clearly liquid paste flow, and hear air to ensure that no blockage exists in the line. Once paste backfilling is finished, the flushing process by means of water and compressed air is again conducted. In case of power failure, there is an obvious risk of the line blockage if the line is not cleaned properly in a timely manner. To end up this problem, a plant emergency generator should be connected directly to a positive-displacement piston pump so that the line can be cleaned by flushing without causing serious problems for underground mine operations.

7 Backfill Placement and Curing Conditions

After determining all the required transport parameters, the manufactured paste backfills are delivered to stopes either through gravity/pumping or a gravity system. CPB is then placed by end-of-pipe deposition from the overcut access (Fig. 5.18). In most cases, firstly a plug fill of few meters high (up to 7 m) is poured into the stope and then the residual fill is placed. The cement content in the plug fill varies between 5 and 7 wt% while the cement content in the residual fill varies between 2

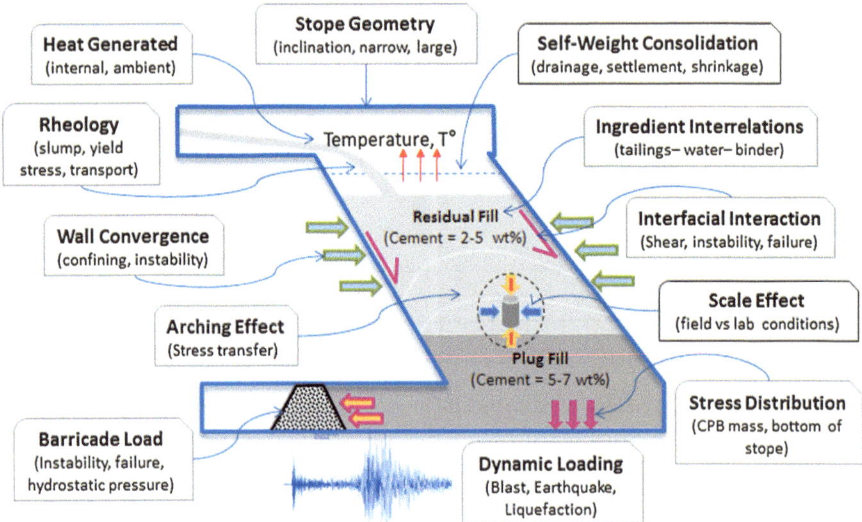

Fig. 5.18 Overview of factors affecting the behavior of paste backfills (Yilmaz et al. 2015b)

and 5 wt%. The plug fill is usually left 2–7 days for curing prior to the residual fill to avoid excess pressure on barricade.

Visual inspection of reticulation boreholes and pipeline flows should be controlled by using cameras to evaluate their conditions and to take corrective actions, if needed. The operating pressures in vertical or horizontal pipelines can be monitored by installing pressure gauges. Accordingly, the pressures in each part of the distribution system can be designed according to piping requirements. In paste plant operations, numerous sensors are utilized to remotely measure whether flow exists, or leakages or blockages have occurred in the pipe. Video feed at the underground paste placement is also provided to the paste plant operator, and is nonstop monitored during filling, to ensure that continuous pour flow is occurring. Video monitoring at stope pour points is vital for observation of pour conditions and for control at the surface plant. In some mining operations, if no video feed is available, no paste backfill placement is allowed to occur unless underground operators are present and in full communication through radio or telephone with the surface backfill plant. Figure 5.19 shows images of paste backfill flowing through a pipeline into underground open stopes.

After the paste backfill materials are placed into the underground mined-out stopes, numerous factors can affect the structural strength and stability of the paste-backfilled stope. These are interactions between the paste backfill and neighbor rock walls, gravity-driven or time-dependent consolidation loadings, paste-backfilled stope geometry, stress distribution in and around the backfilled stope, wall convergences against the paste backfill, chemical shrinkage or volume change, and arching effects developed in the backfill (Aubertin et al. 2003; Fahey et al. 2009; Sivakugan et al. 2013; Festugato et al. 2013; Li 2015).

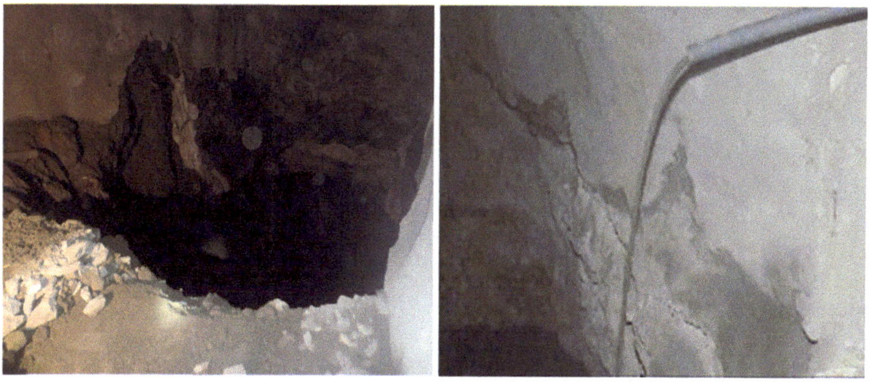

Fig. 5.19 Photos of the typically paste backfill placements to underground open stopes

The preparation, transport, placement, and curing conditions of in situ paste backfill can affect its overall quality and performance. A great deal of investigations have published recently on in situ backfill behavior (Cayouette 2003; Belem et al. 2004; le Roux et al. 2005; Yumlu and Guresci 2008; Grabinsky and Bawden 2007; Thompson et al. 2012). The results show that geo-mechanical designs of in situ paste backfill are conservative in terms of the backfill recipes, barricade design, filling, and curing strategies. Overall, the field backfill observations have shown lower void ratios, higher solids concentration, and higher UCSs near the bottom of the stope, as well as higher void ratios, lower solids concentration, and lower UCSs near the top of the backfilled stope. This is mostly caused by self-weight consolidation, heat transfer in the backfill, and curing conditions (removal of excess water within backfill through cracks at the bottom, or water absorption by soft materials placed at side walls). These processes enhance the CPB's strength development due to pore pressure dissipation and solids skeleton settlement (Belem et al. 2004; Nasir and Fall 2008; Helinski et al. 2010).

Figure 5.20 shows an evolution of mechanical strengths of CPB samples prepared with 4.5 wt% of the binder Portland cement-slag@20:80 wt% for different curing times. It is apparent that the highest compressive strengths were obtained from the capped–drained C–D samples. The C–D condition represents the placement of sample into the mold, which has drainage holes at the bottom and a cover on the top. The drained backfills produce better mechanical strengths than undrained ones for a given CPB recipe and curing time. This is due to the fact that drained backfills have less water-to-cement ratios, which favor the acceleration of hydration reactions with higher strengths. The water loss by drainage gives rise to the settling of backfill and the resultant reduction of the CPB void ratio. Thus, higher strengths are achieved when compared with undrained backfills. In most cases, the water loss during filling and curing of backfill at underground mines can be explained by the consolidation, drainage processes, compaction, and self-desiccation.

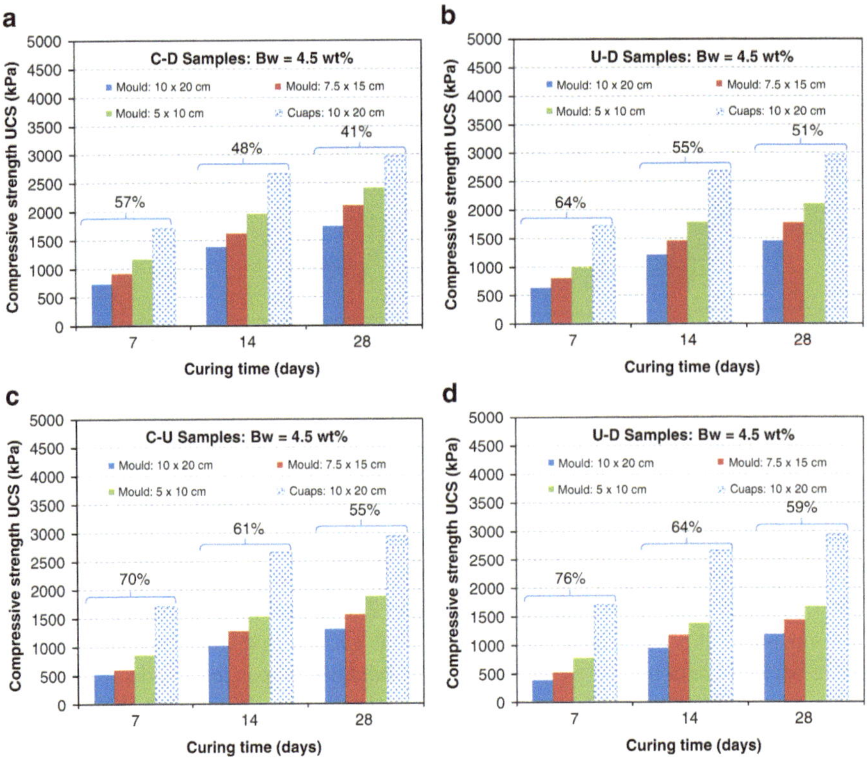

Fig. 5.20 CPB strengths in different placement conditions: (**a**) capped and drained C–D; (**b**) uncapped and drained U–D; (**c**) capped and undrained C–U; (**d**) uncapped and undrained U–U

The direct measurement of volume change or chemical shrinkage at early-age paste backfill materials is vital because it can cause significant reductions in excess pore water pressure (PWP) during hydration (Helinski et al. 2010), playing a role on backfill stiffness and thus strength development. Fourie et al. (2007) stated that the chemical shrinkage is closely related to the multiple-stage cement hydration reactions happened at very early ages of curing and the gradual formation of the stiffness development of paste backfill. It is well known that the volume of the non-hydrated cement is relatively lower than the sum of the hydrated cement and water volumes. The chemical shrinkage is observed within paste backfill through pore water pressure measurements or suction increase due to the capillary depression (Grabinsky and Bawden 2007). Helinski et al. (2007) have also demonstrated experimentally using the dilatometry (or volumetric method) in order to quantify the chemical shrinkage from direct measurements that the higher *w/c* ratios of paste backfills can give rise to less efficiency of the hydration reactions at curing times of less than 30 days. The effectiveness of short-term hydration decreased drastically when the solid (i.e., tailings and sand) is added to the cement and when the *w/c* ratios became higher.

8 Paste Backfill Barricades

After ore has been extracted from the stope through underground mining methods, a barricade is constructed across the undercut to retain the initially fluid paste backfill which will become a solid material after completion of the cement hydration process and later fulfill its ultimate functions for the mine sites. The barricades constructed must be capable of containing the cemented paste backfill to be placed in a single stope typically having a backfill volume of up to 25,000 cubic meters. They also need to be designed to include a drainage system which will allow any seepage water to drain out of the backfilled stope. The following subsections outline in detail the construction and monitoring stages of the barricades designed for the underground stopes.

8.1 Construction Design

The use of a capable barricade design in underground stopes is directly related to the backfill production rates and offers a remarkable safety concern since there are risks associated with barricade failure as it is under-designed as evidenced by observed cracks during stope filling. Minimizing the barricade stresses, paste backfilling process can be enhanced by implementing one of the following two techniques: use higher cement content in the backfill's plug fill, and stop pouring when backfilling reaches over the barricade. The barricade design should reflect suiting almost every stope sizes in the mining areas. The safety and efficiency of the backfill operations can be significantly improved by adopting a stronger barricade designs. This can be achieved by several ways including increasing the effective barricade thickness and using buttress support, or adopting preferably using arched barricades, where there is sufficient space in stope draw points. Arched barricades are much stronger than planar barricades and are widely used in the United States, Canada, and Australia. In addition to planar barricades, arched barricades should be used as the high-capacity barricades which can help to enhance filling rate and lessen stope cycle times (Fig. 5.21).

Barricade construction for paste backfill is largely less complex than for hydraulic fill, due to the absence of transport water drainage requirements. The steel-reinforced shotcrete barricade is constructed across the undercut to contain the initially fluid backfill. The construction of shotcrete barricades includes timber frame with plywood, embedded reinforcement rebars tied to the cemented rebar shear pins drilled along the perimeter, and shotcrete to a nominal depth of up to 40 cm. Planar barricades with a lower working capacity of less than 50 kPa require a slow paste backfilling behind the barricades, which increases the total filling time and hence causes longer stope cycles. If the working capacity of barricade is increased to more than 100 kPa, which reflects arched shotcrete barricade, backfilling can then be conducted continuously until a threshold stress is reached. This provides visibly a sharp drop in stope cycling through shortening of the mid-pour cure period, thereby minimizing

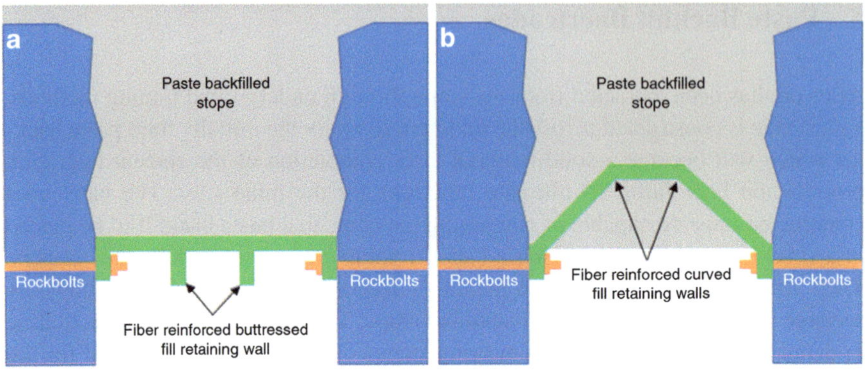

Fig. 5.21 General layouts for (**a**) buttressed planar barricade and (**b**) arched shotcrete barricade

the risk of barricade failures. Some investigators (Yumlu and Guresci 2007; Helinski and Grice 2007; Sainsbury and Revell 2007; Hughes et al. 2010) reported a number of the barricade failures which caused a severe threat to operators and facilities, and resulted in substantial downtimes and cleaning costs. Thus, the design engineer must determine the requirements of CPB strengths and barricade pressures using various types and contents of binders for a given curing time, and collecting some field core samples and characterizing them at laboratory.

8.2 *Video Monitoring*

The barricade pressures depend on numerous parameters including stope dimension, size and type of barricade, type of tailings and binders, binder dosage, filling rate, and fill placement regime. The safety of barricades is provided by real-time CCTV monitoring during filling. Most of the barricades indicate signs of cracking and water seepage. The barricade cracking during filling may be a solid mark of the poor barricade capacity and leads to interruptions in paste backfilling. The water seepage may also state that the density (i.e., solid concentration) of the backfill placed in the stope is low. Note that if there exists high-intensity cracking on the barricades, no paste backfilling is allowed to pour to the stopes. Figure 5.22 provides the captured images of the paste backfill barricades at the beginning and the end of filling.

One can also note that the barricade pressures can be well controlled by the instrumentation of total pressure cells and pore water pressure sensors. The advantage of the instrumentation is that paste can be filled continually to the stopes without waiting for the backfill's hardening process. Even if instrumentation brings a smart solution to the barricade pressure monitoring, there will still be a requirement for the operator to nonstop observe the barricade condition during filling since barricade can fail due to the intensive cracks and/or defects in material and workmanships in the construction of paste backfill barricades.

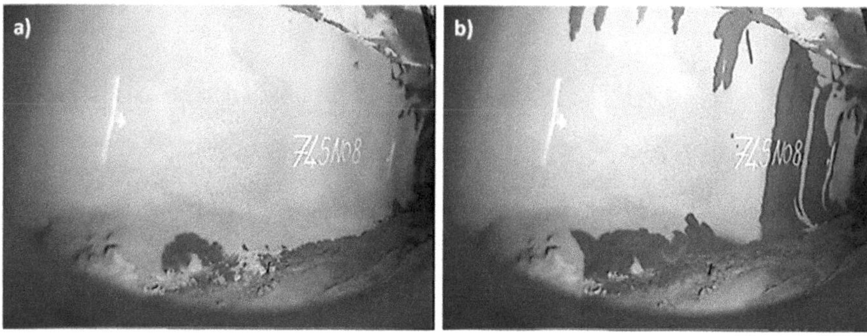

Fig. 5.22 The captured images of paste backfill barricades based on different filling times: (**a**) at the beginning of filling and (**b**) at the end of filling

9 Quality Control Testing

Paste backfill is an integral part of production cycle at most modern mine sites worldwide. An ongoing QA/QC testing program (e.g., slump, compressive strength, and other characteristics) is therefore crucial to ensure a safe mining operation eventually. The hourly cone or cylinder slump tests are conducted on freshly produced backfills and required to check hourly if the CPB's water content is acceptable for its transport and placement characteristics. The paste's flow behavior can be characterized by measuring its rheological parameters (i.e., yield stress, viscosity) and its piping network. The yield stress is most often used at many paste plants to control the quality of produced paste backfills and also has well-defined correlation with cylinder slump tests. The yield stress of paste backfill can be obtained indirectly by measuring its cylinder slump. Figure 5.23 shows photos of both cone and cylinder slump tests.

The filter cake moisture, solids-specific gravity, and grain size distribution of tailings used in backfill production should be measured regularly on weekly composted samples. The density of paste backfill is performed in the paste plant as a primary measure of product quality and consistency. In some plants, the CPB's quality is controlled by comparing measured slump and paste moisture content versus pump power consumption conditions. Mixer power draw is used as an indirect indication for slump of backfill mix and for batching of paste ingredients when plant is in auto mode. The higher power draw means that the paste material in the mixer has higher solids density and yield stress. The mixer power draw can also be considered as an indirect indicator of the slump of the paste material within the mixer. The broken or missing paddles and routinely non-cleaned mixer can alter power draw which may lead to thicker or low-slump paste backfill materials.

The short-, mid-, and long-term strength and durability performance of paste backfill samples are frequently determined by the unconfined compressive strength (UCS) testing. Samples for UCS determination are taken in the surface paste plant during filling production and kept at controlled temperatures under high-relative-humidity

Fig. 5.23 Photos of (**a**) cone slump test and (**b**) cylinder slump test for paste fill rheology

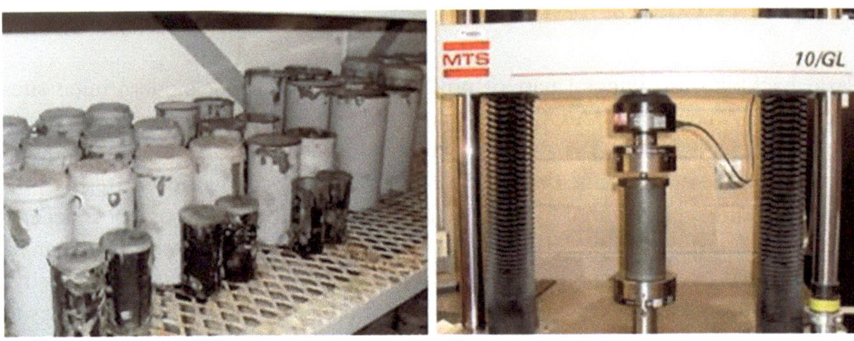

Fig. 5.24 CPB samples cured in a foggy room (*left*) and tested under a UCS test press (*right*)

curing conditions (Fig. 5.24). Small-size cylinder samples (at 5 cm diameter by 10 cm long) are cured for 1, 2, 3, 7, 14, 28, and 56 days and then subjected to UCS tests for a given backfill recipe. These curing times may vary from mine to mine, based on the strength requirements of backfilling. The UCS testing should be always undertaken on hardened backfill samples for several reasons as follows: change in the type of tailings and binders, a possible reduction in binder content and hence the binder-related costs, and the strength needs in underground stope optimization of the CPB recipes.

10 Mine Fill Management

Paste backfill operations are a multidisciplinary approach which encompasses interdependent activities between mine planners, paste plant operators and supervisors, processing engineers, geotechnical design engineers, underground mine operators and supervisors, maintenance employees including paste ingredient suppliers, researchers, technicians, contractors, and consultants. Developing a backfill

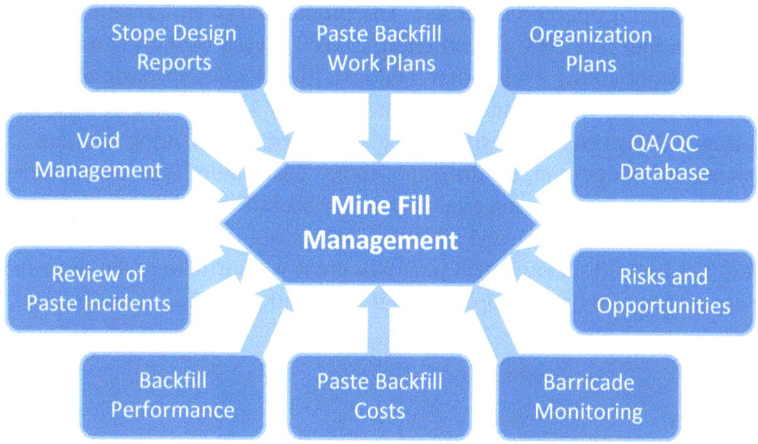

Fig. 5.25 Schematic diagram showing the components of a typical backfill management

management guide is of great significance for establishing the sustainable backfilling operations at optimum conditions and efficiency. This reference document will eventually offer that all key stakeholders have a better understanding of the key aspects of the entire CPB operations from the production stage on surface plant through to the placement and curing stages of the backfills in underground stopes. Figure 5.25 illustrates the main components of a typical fill management guide considered for underground mines.

More specifically, the operational issues on production of the tailings used in paste backfill manufacture and their delivery and placement to underground mined-out stopes in pipelines and/or boreholes through video monitoring belong to the surface processing plant. The paste plant-related issues, such as daily quality control tests (e.g., paste slumps and CPB strengths), optimization of paste backfill recipes, mixer cleanups, filtrate maintenances, cement supplier, operating costs, and housekeeping, will also belong to processing department. The technical issues on the backfill strength requirements, binder dosages required for underground stopes, stope backfilling plans and programs, CPB recipe improvements, quality control data review, stope preparations for paste backfilling, barricade design and construction, etc. will be shared cooperatively between the technical services and mine departments. The three departments (i.e., mining, processing, and technical services) will be sharing equal responsibility for the entire backfill operations. The principal objective is to significantly reduce operating costs, increase production and quality, and enhance safety in sustainable backfill operations.

To better manage a backfill system at mine sites, an engineer should be assigned specifically by the upper management of the mines. This system should be dynamic and updated annually based on the new requirements and implementation of paste backfill. Since backfilling is one of the vital operations in underground hard rock mining, one can say comfortably that a well-managed paste backfill operation will possess a great impact on production optimization (i.e., enhanced productivity and safety, and reduced operational costs). The productivity can be well improved by designing a proper barricade construction, using a higher backfill density, developing a shorter stope filling cycle times, and installing a superior paste backfill delivery and placement techniques through pressure sensor and real-time video monitoring. Moreover, the costs can be improved by ensuring a more correct tailings-cement integration, optimizing a more ideal binder type and content (this will result in less expensive and equal or greater backfill strengths), avoiding a more conservative backfill designs, and adopting a more useful quality control test results. As a result, a backfill management plan will greatly contribute to the mine in every single aspect of paste backfill operations if managed properly.

References

Archibald J, De Gagne D, Nantel J, Hassani F (2003) Underground mine backfill course notes. Edu Mine: Professional Development and Training for Mining, Canada

Aubertin M, Li L, Arnoldi S, Belem T, Bussiere B, Benzaazoua M, Simon R (2003) Interaction between backfill and rock mass in narrow stopes. In: Soil and rock America mechanics symposium, Essen, Germany, p 1157–1164

Awoh AS, Mbonimpa M, Bussière B (2013) Determination of the reaction rate coefficient of sulphide mine tailings deposited under water. J Environ Manage 128:1023–1032

Belem T, Benzaazoua M (2008) Design and application of underground mine paste backfill technology. Geotech Geol Eng 26(2):147–174

Belem T, Harvey A, Simon R, Aubertin M (2004) Measurement and prediction of internal stresses in an underground opening during its filling with cemented backfill. In: The 5th international symposium on ground support in mining and underground construction, Perth, Australia, p 619–630

Benkendorff PN (2006) Potential of lead/zinc slag for use in cemented mine backfill. Mineral Process Extractive Metall: IMM Trans Sect C 115(3):171–173

Benzaazoua M, Fall M, Belem T (2004) A contribution to understanding the hardening process of cemented paste backfill. Miner Eng 17(2):141–152

Bernier RL, Li MG, Moerman A (1999) Effects of tailings and binder geochemistry on physical strength of paste fill. In: The 2nd international conference on mining and the environment, Canada, p 1113–1122

Bertrand V, Monroy M, Lawrence R (2000) Weathering characteristics of paste backfill: mineralogy and solid phase chemistry. In: The 5th international conference on acid rock drainage, Denver, p 863–876

Blowes DW, Lortie L, Gould WD, Jambor JL (1995) Microbiological, chemical, and mineralogical characterization of the Kidd Creek mine tailings impoundment, Timmins area, Ontario. Geomicro-Biology J 13(1):13–31

Bouzahzah H, Benzazoua M, Bussiere B, Plant B (2014) Prediction of acid mine drainage: importance of mineralogy and the protocols for static and kinetic tests. Mine Water Environ 33:54–65

Brackebusch FW (1994) Basics of paste backfill systems. Mining Eng 46:1175–1178

Bussière B (2007) Colloquium 2004: hydrogeotechnical properties of hard rock tailings from metal mines and emerging geo-environmental disposal approaches. Can Geotech J 44(9):1019–1052

Cayouette J (2003) Optimization of the paste fill plant at Louvicourt mine. CIM Bull 96:51–57

Chan BK, Bouzalakos S, Dudeney AW (2008) Integrated waste and water management in mining and metallurgical industries. Trans Nonferrous Met Soc Chin 18:1497–1505

Cihangir F, Ercikdi B, Turan A, Kesimal A, Deveci H, Yazici M,Karaoglu K (2011) Utilisation of sodium silicate activated blast furnace slag as an alternative binder in paste backfill of high-sulphide mill tailings. In: The 14th international seminar on paste and thickened tailings, Perth, p 465–475

Cihangir F, Ercikdi B, Kesimal A, Deveci H (2012) Utilisation of alkali-activated blast furnace slag in paste backfill of sulphide mill tailings: Effect of binder type and dosage. Miner Eng 30:33–43

Cooke R (2006) Thickened and paste tailings pipeline systems: design procedure Part I. In: The ninth international seminar on paste and thickened tailings, Limerick, p 1–10

De Souza E, Archibald JF, Dirige APE (2003) Economics and perspectives of underground backfill practices in Canadian Mining. In: 105th annual CIM general meeting, Montreal, p 1–15

Dirige APE, Archibald JF, Clarke R, Hilkewich T, Frank T (2008) The effect of different mixing techniques on the strength behavior of paste backfill. In: Proceedings of the 42nd U.S. rock mechanics symposium (USRMS), American Rock Mechanics Association, San Francisco, CA, 29 June–2 July, 2008, pp 1–7

Ercikdi B, Yilmaz T, Kulekci G (2014) Strength and ultrasonic properties of cemented paste backfill. Int J Ultrasonics 54(1):195–204

Fahey M, Helinski M, Fourie A (2009) Some aspects of the mechanics of arching in backfilled stopes. Canad Geotech J 46(11):1322–1336

Fall M, Benzaazoua M, Ouellet S (2005) Experimental characterization of the influence of tailings fineness and density on the quality of cemented paste backfill. Miner Eng 18:41–44

Fall M, Benzaazoua M, Saa E (2008) Mix proportioning of underground cemented paste backfill. J Tunnell Underground Space 23:80–90

Festugato L, Fourie A, Consoli NC (2013) Cyclic shear response of fibre-reinforced cemented paste backfill. Géotech Lett 3:5–12

Fourie AB, Fahey H, Helinski M (2007) Using effective stress theory to characterize the behaviour of backfill. CIM Bullet 100(1103):1–9

Fried E, Benzazoua M, Bussière B, Belem T (2007) Study of the leaching behavior and metal fixation within cemented paste backfill. In: The 9th international symposium in mining with backfill, Canada, p 1–13

Ghirian A, Fall M (2013) Coupled thermo-hydro-mechanical-chemical behavior of cemented paste backfill in column experiments. Part I: physical and thermal processes and characteristics. Eng Geol 164:195–207

Ghirian A, Fall M (2014) Coupled thermo-hydro-mechanical chemical behaviour of cemented paste backfill in column experiments. Part II: mechanical and microstructural processes and characteristics. Eng Geol 170:11–23

Godbout J, Bussiere B, Benzaazoua M, Aubertin M (2009) Influence of pyrrhotite content on the physico-chemical behavior of paste backfill. In: The 62nd Canadian geotechnical conference, Canada, p 1–8

Goosen P, Ilgner H, Dumbu S (2011) Settlement in backfill pipelines: its causes and a novel online detection method. In: The 10th international symposium on mining with backfill, South Africa, p 187–195

Grabinsky MW, Bawden WF (2007) In situ measurements for geomechanical design of cemented paste fill systems. CIM Bulletin 100(1103):1–8

Grice AG (2005) Fluid mechanics of mine fill. In: Handbook on mine fill. Australian Centre for Geomechanics, Perth, p 51–63

Hassani FP, Archibald JF (1998) Mine backfill handbook. Canadian Institute of Mining, Metallurgy and Petroleum, Montreal

Hassani FP, Ouellet J, Hossein M (2001) Strength development in underground high sulphate paste backfill operation. CIM Bullet 94(1050):57–62

Helinski M, Grice AG (2007) Water management in hydraulic fill operations. In: Proceedings of the 9th international symposium in mining with backfill, Canadian Institute of Mining, Metallurgy and Petroleum (CIM), Montréal, QC, Canada, 29 April–2 May, 2007, pp 1–11

Helinski M, Fahey M, Fourie A (2007a) Numerical modelling of cemented paste backfill deposition. *J Geotech Geoenviron* 13(10):1308–1319

Helinski M, Fahey M, Fourie AB (2010) Behavior of cemented paste backfill in two mine stopes: Measurements and modeling. J Geotech Geoenviron Eng 13(2):11–182
Henderson A, Revell MB, Landriault D, Coxon J (2005) Paste fill. In: Handbook on mine fill. Australian Centre for Geomechanics, Perth, p 1–179
Hudson-Edwards KA, Jamieson HE, Lottermoser BG (2011) Mine wastes: past, present, future. Elements 7:375–380
Hughes PB, Pakalnis R, Hitch M, Corey G (2010) Composite paste barricade performance at Goldcorp, Inc., Red Lake Mine, Ontario. Canada Int J Min Recl Environ 24(2):138–150
Jakubick A, McKenna G (2003) Stabilization of tailings deposits: international experience. Mining and the Environment III, Sudbury, p 1–9
Johnson JL, Slottee JS (2004) Paste technology: success is in the approach. In: The 11th international conference on tailings and mine waste, Vail, p 305–309
Kesimal A, Yilmaz E, Ercikdi B, Alp I, Yumlu M, Ozdemir B (2003) Paste backfill technology in underground mining—a case study. Turkish J Earth Sci 16(1):45–53
Kesimal A, Yilmaz E, Ercikdi B (2004) Evaluation of paste backfill test results obtained from different size slumps with varying cement contents for mill tailings. Cem Concr Res 34:1817–1822
Kesimal A, Yilmaz E, Ercikdi B, Deveci H, Alp I (2005) Effect of properties of tailings and binder on short- and long-term strength and stability of paste backfill. Mater Lett 59:3703–3709
Klein K, Simon D (2006) Effect of specimen composition on the strength development in cemented paste backfill. Canad Geotech J 43:310–324
Landriault D (2006) They said "It will never work"—25 years of paste backfill 1981–2006. In: The 9th international seminar on paste and thickened tailings, Limerick, p 277–292
Landriault D, Brown R, Counter D (2000) Paste backfill study for deep mining at Kidd Creek. CIM Bullet 93(1036):156–161
le Roux K-A, Bawden WF, Grabinsky MW (2005) Field properties of cemented paste backfill at the Golden Giant mine. Mining Technol 114(2):65–80
Li L (2015) Generalized solution for mining backfill design. Int J Geomech 14:1–13
Lottermoser B (2010) Mine wastes: characterization, treatment and environmental impacts, 3rd edn. Springer, Berlin
McGuinness M, Cooke R (2011) Pipeline wear and the hydraulic performance of pastefill distribution systems: Kidd mine experience. In: 10th international symposium on mining with backfill, South Africa, p 205–212
Moghaddam AS, Hassani FP (2007) Yield stress measurement of cemented paste backfill with the vane method and slump tests. In: The 9th international symposium on mining with backfill, Montreal, p 1–8
Nasir O, Fall M (2008) Shear behaviour of paste fill-rock interfaces. Eng Geol 101:146–153
Nonnen FA (2001) Reduction or elimination of abrasion, wear and corrosion in backfill equipment. In: The 7th international symposium on mining with backfill, Seattle, p 1–12
Ouattara D, Mbonimpa M, Belem T (2010) Rheological properties of thickened tailings and cemented paste tailings and the effects of mixture characteristics on shear behaviour. In: The 63th Canadian geotechnical conference, Calgary, p 118–1185
Ouellet S, Bussière B, Mbonimpa M, Benzazoua M, Aubertin M (2006) Reactivity and mineralogical evolution of an underground mine sulphidic cemented backfill. Miner Eng 19:407–419
Pashias N, Boger DV, Summers J, Glennister DJ (1996) A fifty cent rheomoter for yield stress measurement. J Rheol 40(6):1179–1189
Plante B, Bussiere B, Benzaazoua M (2014) Lab to field scale effects on contaminated neutral drainage prediction from the Tio mine waste rocks. J Geochem Explor 137:37–47
Pokharel M, Fall M (2011) Coupled thermochemical effects on the strength development of slag paste backfill materials. J Mater Civil Eng 23(5):511–525
Potvin Y, Thomas E, Fourie AB (2005) Handbook on mine fill. ACG, Perth
Revell M (2004) Paste: how strong is it? In: 8th international symposium on mining with backfill, China, p 286–294

Sainsbury DP, Revell MB (2007) Advancing paste fill bulkhead design using numerical modelling. CIM Bullet 100(1103):1–10 (Paper 25)
Sivakugan N, Widisinghe S, Wang V (2013) Vertical stress determination in backfilled mine stopes. Int J Geomech 14(5):1–16
Slottee JS, Johnson JL (2009) Paste thickener design and operation selected to achieve downstream requirements. In: The 12th international seminar on paste and thickened tailings, Viña del Mar, Chile, p 69–76
Sofra F, Boger DV (2002) Environmental rheology for waste minimisation in the minerals industry. Chem Eng J 86(3):319–330
Stone D (2007) Factors that affect cemented rock fill quality in mines. CIM Bullet 100:1–6
Tarig A, Nehdi M (2007) Developing durable paste backfill from sulphidic tailings. Waste Resour Manage 160(4):155–166
Thompson BD, Bawden WF, Grabinsky MW (2012) In situ measurements of cemented paste backfill at the Cayeli Mine. Canad Geotech J 49(7):755–772
Ulrich B, Coffin J (2013) Considerations for tailings facility design and operation using filtered tailings. In: The 16th international seminar on paste and thickened tailings, Belo Horizonte, p 201–210
Verburg RBM (2002) Paste technology for disposal of acid-generating tailings. Min Environ Manage 13(7):14–18
Wang X-M, Zhang D-M, Zhang Q-I (2011) Form and mechanism of abrasion in backfill drill hole pipelines in deep mines. In: The 10th international symposium on mining with backfill, South Africa, p 213–220
Weatherwax T, Brosko W, Evans R, Champa J (2010) Role of admixtures in the optimisation of paste backfill systems. In: The 13th international seminar on paste and thickened tailings, Toronto, p 1–11
WesTech (2015) Download paste thickening and backfill process flow sheets, Salt Lake City, USA. http://industries.westech-inc.com/paste-download?submissionGuid=d3d4a9bc-ce96-4b6e-9cfa-588b73421f79
Wilkins M, Gilchrist C, Fehrsen M, Cooke R (2004) Boulby mine fill system: design, commissioning and operation. In: The 8th international symposium on mining with backfill, Beijing, p 43–50
Wilson S, Calverd J (2011) Benefits of paste aggregate backfill. In: The 10th international symposium on mining with backfill, Cape Town, p 1–8
Yilmaz E (2015a) Environmental characterization of surface paste disposal. LAP LAMBERT Academic Publishing, ISBN: 978–3–659-36697-0, Saarbrucken, Deutschland/Germany
Yilmaz E (2015b) Geotechnical characterization of cemented paste backfill. LAP LAMBERT Academic Publishing, ISBN: 978–3–659-60841-4, Saarbrucken, Deutschland/Germany
Yilmaz E, Benzaazoua M, Belem T, Bussière B (2009) Effect of curing under pressure on compressive strength development of cemented paste backfill. Miner Eng 22:772–785
Yilmaz E, Belem T, Benzaazoua M (2014a) Effects of curing and stress conditions on hydromechanical, geotechnical and geochemical properties of cemented paste backfill. Eng Geol 168:23–37
Yilmaz E, Benzaazoua M, Bussière B, Pouliot S (2014b) Influence of disposal configurations on hydrogeological behaviour of sulphidic paste tailings: a field experimental study. Int J Miner Process 131:12–25
Yilmaz E, Belem T, Bussière B, Mbonimpa M, Benzaazoua M (2015a) Curing time effect on consolidation behaviour of cemented paste backfill containing different cement types and contents. Construct Build Mater 75:99–111
Yilmaz E, Belem T, Benzaazoua M (2015b) Specimen size effect on strength behavior of cemented paste backfills subjected to different placement conditions. Eng Geol 185:52–62
Yumlu M, Guresci M (2007) Paste backfill bulkhead monitoring: a case study from Inmet's Cayeli Mine, Turkey. In: The 9th international symposium on mining with backfill, Montreal, p 1–11
Zou DH, Nadarajah N (2006) Optimizing backfill design for ground support and cost saving. In: The 41st U.S. rock mechanics symposium, Madison, p 1–11

Chapter 6
Field Properties and Performance of Surface Paste Disposal

Atac Bascetin, Serkan Tuylu, Deniz Adiguzel, and Orhan Ozdemir

1 Introduction

A significant amount of tailings with a high content of cyanide or sulphur arises from the beneficiation process of metallic mines such as gold, silver, copper, and lead. The influence of these hazardous tailings on the environment may have multiple aspects. The control and configuration of this kind of environmentally dangerous tailings in terms of isolation at disposal site, construction of impermeable layers, transportation from plant, stabilization, safety, their effects on water, and soil quality are the main parameters that could be considered carefully. In this context, tailings management are very important to selection of the optimum disposal method. Also, the some parameters such as physical and chemical characterization of tailings, properties of newly formed material (e.g. acid potential, stabilization, cost and applicability of the paste tailings etc.) should be evaluated. Safely disposal of mine tailings under surface conditions is of great importance in the aspect of environment.

Mining operations have a great amount of tailings after mining milling activities. Technological advancement in mineral processing has increased the feasibility of refining low-grade ores which also has increased the volume of mine tailings (Zou and Sahito 2004). For example, the milling tailings are dumped into the dams of tailings after the thickening process. The process needs large size of dams, and various risks have been faced as failure and leakage due to the stability problem. Additionally, the increased environmental liabilities go along with traditional tailings placement strategies. Actually, this shows that new and existing mines require alternative tailings placement techniques which more effectively meet environmental and regulatory requirements (Bussiere 2007). One of these emerging techniques is called surface paste disposal (SPD) at which each layer of thickened and/or fil-

A. Bascetin (✉) • S. Tuylu • D. Adiguzel • O. Ozdemir
Department of Mining Engineering, Istanbul University, Avcilar, Istanbul 34320, Turkey
e-mail: atac@istanbul.edu.tr

E. Yilmaz, M. Fall (eds.), *Paste Tailings Management*,
DOI 10.1007/978-3-319-39682-8_6

tered tailings is placed and allowed to dry before the next layer is put in a place (Yilmaz 2010). Furthermore, SPD has come into prominence since at last 25 years.

Surface paste disposal (SPD) of tailings is a recent technique for the management of mine tailings. It consists of dewatering (by thickening or filtering) tailings before deposition to obtain a self-supporting and homogenous tailings stack. The main problem in this technique is the cost of filtration. Therefore, recent studies have mainly focused on this subject.

Using paste technology gives great advantages to eliminate the tailings coming from surface and underground mining with respect to economic and environmental issues. Especially, SPD technique allows to stock up great amount of milling tailings on the surface openings by minimizing the environmental and engineering problems. One of the best advantages of the SPD is that there are no need of big dams to store the tailings.

2 Comparison of Conventional Tailings Disposal Methods to Surface Paste Disposal Method

Tailings by mining activities cause an important environmental problem. There are some disposal methods which have been applied in the mining industry for a long time. These methods have some advantages and disadvantages besides environmental considerations. In this part of the chapter, surface paste disposal (SPD) method is compared with conventional disposal/deposition methods which can be classified as

- Tailing dam
- Submarine and riverine tailing disposal
- Dry tailings
- Thickened tailings
- Co-disposal of tailings
- Paste technology

These methods are explained briefly below:

2.1 *Tailing Dam*

Tailing dams are used to store tailings and water together. Tailing slurry is pumped into a pond to allow the sedimentation of solid particles in water. The pond is generally impounded with a dam which is known as tailings dams. In the modern mines, between 40 and 100% of the total tailings slurries are deposited on the mine-site surface (tailings pond) (Benzaazoua et al. 2004a). There are many different subsets of this method. There are two basic types of structures used to retain tailings in impoundments, the raised embankment that can be constructed using upstream, downstream, or centerline methods and the retention dam. Either type of structure can be used to form different types of tailings impoundments (the ring-dike, in-pit, specially dug pit, and variations of the valley design). The most common of them is

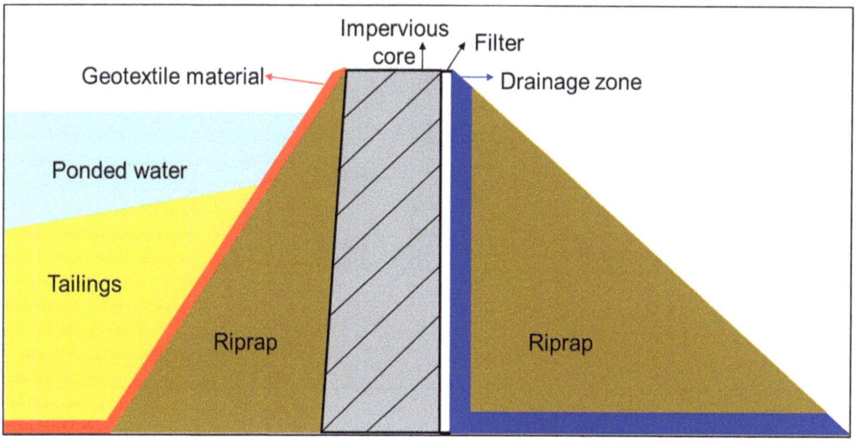

Fig. 6.1 Retention-type dam for tailings disposal

the valley pond. The design choice primarily depends upon natural topography, site conditions, and economic factors. Figure 6.1 shows a typical section of retention-type dam (EPA 1994).

Modern tailings dams are engineered structures for permanent disposing of tailings from mining and milling operations. At some projects, the tailing dams cover several square miles (EPA 1994).

Higher volumes of water and tailings (usually 15% solids and 85% water by weight) are sent to waste dams making it very difficult to control the stability of the dam under conditions such as static and dynamic loads (seismic activities, vibrations caused by blasting, etc.), flooding, and seepage. Meanwhile, 138 mine tailing storage dams have significantly failed so far. Recent examples are given below (Vogt 2012; Bascetin et al. 2013):

- In 1985, 268 people died from the failure of a mine tailings storage dam in Stava, Italy.
- In 1998, the Los Frailes mine tailings dam in Aznalcóllar, Spain, failed releasing five to seven million cubic metres of mine tailings into the Rio Agrio.
- In Hungary in 2010, releasing of 600–700 thousand cubic metres of red mud and water caused huge devastation.
- In 2011, an accident occurred in ETI Silver Company silver production plant in Kutahya, Turkey. A connection wall for four impoundments failed leading to a situation where 25 million cubic metres of water containing cyanide created a great risk to environment.

2.2 *Submarine and Riverine Tailing Disposal*

In this method that is also known as submarine tailings disposal (STD) or deep-sea tailings disposal, tailings are discharged via pipelines into the sea marine environment below euphotic zone. This zone is generally considered to be environmentally

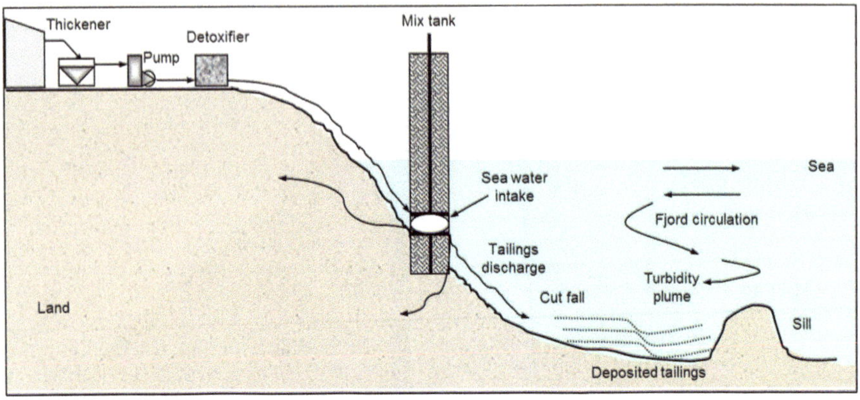

Fig. 6.2 Submarine tailing disposal method

safe (Fig. 6.2). Today's marine disposal discharges are in deep water at final deposition in depths of 30 to 300 m in Norway and over 1000 m in Turkey, Indonesia, and Papua New Guinea. In 2013, submarine tailing disposal was used by 14 mines (Vogt 2013).

Riverine disposal is a very simple method. Tailings are discharged via pipelines into the river in this method. This technique has been practiced throughout mining history. Economics and technical feasibility factors (e.g. mountainous terrain, earthquake prone, extreme rainfall) are considered when considered this method. In 2013, riverine tailing disposal was used by four mines (Vogt 2013).

2.3 Dry Tailings

In this method, the dewatering of tailings is done using vacuums and high-pressure filters which save the water and reduce the impact on the environment (Das and Choudhury 2013). Tailings are filtered to a percent solid greater than about 85%, and transferred using conveyor and truck (Bussiere 2007).

2.4 Thickened Tailings

In conventional tailing disposal method, pulp densities of slurry are low. Increasing the pulp density will eventually result in a "thickened tailings" (Robinsky 1999). In this method, slurry is deposited from a point upstream of the outer retaining wall, and is allowed to flow towards the wall. Because of the reduced water content of the thickened slurry, bleeding water arises during deposition (Blight and Bentel 1983).

2.5 Co-disposal of Tailings

In the co-disposal of tailings method, fine and coarse mine waste is mixed, and it reduces void space in mine waste rock for the disposal of the fine-grained tailings. There are large voids when the waste is placed in a waste dump due to waste rock is commonly contain course grains from blasting of hard rock (Leduc and Smith 2003). The structure of waste rock dumps is typically porous and allowing the flow of oxygen and water. So the structure of waste rock dumps ultimately increases the risk of acid rock drainage. Co-disposal offers an alternative to the current practice of waste and tailings disposal (Wickland et al. 2006).

2.6 Paste Technology

Paste technology is a good alternative for waste management systems because it has many benefits in the aspects of environment, cost, and safety (Yilmaz 2015). In particular, the tailings which include harmful chemical ingredients can be disposed without causing any environmental damage in this method. Paste backfill is a pumpable, flowable, and non-Newtonian fluid consisting generally of mine tailings and cement (Brackebusch 1994). The process tailings are usually used as paste backfill material in some underground minings (Fig. 6.3), but it is not possible to apply this method in open-pit mining. For this reason, tailing wastes must be stored above ground in the open-pit mining as surface paste disposal. Therefore, the behaviour of the paste material is different from the other applications (Bascetin et al. 2012).

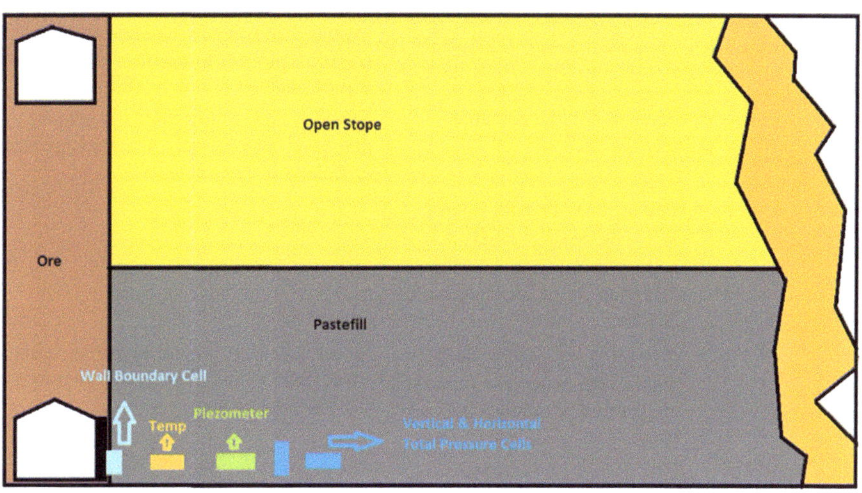

Fig. 6.3 Paste backfill

Paste is defined as a thickened material formed by mixing dewatered fine tailings with water and preferably binder (cement). Addition of cement into paste changes both the chemical and physical characteristics and buffers acid-producing oxidation reactions, resulting in less mobilization of metals (Levens et al. 1996). In order to form a flowable paste, it is required to have particles at least 15% of solid ratio and finer than 20 μm (Brackebusch 1994; Verburg 2001; Kesimal et al. 2003).

Table 6.1 presents some advantages and disadvantages of these methods:

The fundamental problems mentioned below will be eliminated by the paste disposal method (Meggyes and Debreczeni 2006):

Table 6.1 Summary of disposal methods

Method	Advantages	Disadvantages
Tailing dam	Tailings are usually produced by wet processes	Environmental problems and concerns
	Hydraulic transportation by pipeline to the disposal site	Accidents related to the waste dam
Submarine and riverine tailing disposal	Simply the dumping of tailings into the marine environment	Close proximity to off-shelf depths is rare
		Extensive damage to the seafloor can result due to covering by the tailings product
		Control the density and temperature of the tailings product, to prevent it from travelling long distances or even floating to the surface
		Potential direct impacts on fish as well as posing risks to human health (potentially harmful constituents in mine tailings include heavy metals, cyanide and chemical processing agents, sulphide compounds)
Dry tailings	Reduction in the potential seepage rates	Expensive due to increased capital cost to purchase, but necessary if water is needed especially in a dry climate
	Space used	
	Leaves the tailings in a dense	
	Stable arrangement and eliminates the long-term liability that ponds leave after mining is finished	
Thickened tailings	Increasing the storage capacity	Storm-water run-off and any supernatant water collects immediately adjacent to the impounding wall
	Tailings deposited in a dense state are better able to resist imposed loads	
Co-disposal of tailings	Reduced cost according to conventional methods	The deposition strategy must be controlled to optimize the blending of the coarse and fine waste feeds
	Reduced land use according to conventional methods	

(continued)

Table 6.1 (continued)

Method	Advantages	Disadvantages
Paste technology	Reduction of investment costs (construction of large dams is not required)	Paste production may lead to extra costs
	Increased security	Since paste technology is more advanced than conventional methods technological knowledge and infrastructure are required
	Reduction of the negative public perception	
	Smaller disposal site (for the same amount)	
	Protection of water resources (water saving)	
	Reduction of soil and groundwater contamination	
	Reduction of miner's obligation as tuition payments and insurance premiums (decreased responsibilities related to accidents of tailings dam)	
	Paste material can be disposed of with other mine wastes, thus providing additional economic benefits with the ability to accept other waste	
	Very little leachate to reduce the size of water-retaining structures	
	Facilities of reclamation to be better than conventional methods	
	Non-segregation of particles in pipe line and discharging to disposal site	
	Rapid drainage of rainwater due to the sloping surface of the disposal site	

- Ground seismic events (earthquake), vibration, and movement of heavy machinery may cause to the tailings dam accidents.
- Erosion consists of leakage from upstream ponds that would cause failure of dam (bank collapse).
- Embedded drainage and discharging pipes have the risk of leachate.
- An unwanted free waters in tailing dam seepage into the soil and groundwater.
- Acid mine drainage (AMD) high potential of sulphur-rich tailings.
- Reclamation begins after the mine is closed.

The addition of hydraulic binders will improve resistance, stability, and acid-neutralizing potential of the surface paste material, and cement additive in the paste material will also make stable pollutants.

2.6.1 Preventing the Formation of Acid Mine Waters

Discharge of acid-forming tailings in air and classification of these tailings in conventional systems accelerate the oxidation of the tailings; hence this ultimately increases the acid formation (Newman et al. 2001; Newman 2003).

In homogeneous and low-permeability paste applications, only thin layer is oxidized. Therefore, coating of each tailing layer with fresh tailing before the oxidation process prevents the acid formation. Although these types of applications have not been investigated yet, one application in Canada (Kidd Creek Mine) showed that if the layer of the tailings were covered with the new tailings within 12–18 months, the acid production caused no significant problem (Newman et al. 2001). However, some tailings may be more reactive, and since the coverage of top of the old tailings takes longer, the surface of the tailings can be problematic.

In the case of dewatered homogeneous tailings, fine-grained fraction makes water to rise to the surface due to capillary force, and this keeps the tailings saturated and prevents sulphides from contacting with oxygen, and thus prevents the acid formation.

2.6.2 Increased Security

The paste material deposited in tailings dam or disposal area does not flow because of its consolidation is very well. Therefore it can be described that it is more safety than other conventional disposal methods. Since there is not a conventional slime pond, the problem with collapse of the dam remains locally, and it does not trigger ecological disaster.

2.6.3 Improvement in Conditions of Reclamation

The main problem of conventional disposal methods is that the reclamation cannot be done before mining activities are finished. Thus, a flat large area gets rain water that passes through the tailings, and dissolves metals and chemicals which go to groundwater. On the other hand, when these flat areas are allowed to dry, dusting occurs because fine particles are accumulated on the top. In the paste system, rain water at tilted surfaces will immediately drain, and migration of water into the tailings will be prevented, and the dusting will decrease because fine and coarse particles are connected in the matrix. The grade reclamation of the site is another property which prevents both water leaking and dusting.

Implementation of SPD system will allow for progressive reclamation in some of the topography. This will provide an active area of less waste. When the mine is abandoned for any reason, rehabilitation of tailings storage areas will be achieved. Progressive reclamation will ensure decreasing the amount of coverage requested to guarantee the final rehabilitation after the abandoning. The studies done about this subject recently showed that dewatered or conditioned (thickened) tailings worldwide can meet the requirements of disposal (Meggyes and Debreczeni 2006).

2.6.4 Cost

Compared in terms of costs, the paste technology is costly due to the need for advanced technology. In order to prepare a paste material, there may be need for some binding agents such as Portland cement, and classification tools such as disc filter, tank dewatering equipment, and cyclone.

When the comparison is made in terms of investment costs, the investment cost of the paste system must be compared by including structure construction and operating costs in the conventional system and final cost of reclamation (Robinsky 1999; Meggyes and Debreczeni 2006).

In a study conducted by Bascetin et al. (2016), the disposal methods with tailings dam, surface paste disposal (SPD), and tailings disposal using geotextile material methods were compared based on the capital costs, operation costs, and total costs of these methods with a life of 5 years were calculated. The methods with the lowest total capital cost are disposal using the geotextile material, SPD, and tailings dam, respectively. The constructions and land costs of disposal area are the main parts of total capital cost of tailings dam method. However, the methods which have lowest operation cost are the tailings dam and surface paste disposal methods, and geotextile method, respectively. The analysis indicated that the unit costs of the tailings dam, the SPD method, and the geotextile method were calculated as **2.25 $/ton, 2.29 $/ton**, and **7.39 $/ton**, respectively. However, the unit cost of SPD method become **2.79 $/ton** when cement was used. The higher cost of the unit cost of the tailings disposal method using geotextile method was attributed to the high cost of geotextile tubes. The total cost of the SPD and tailings dam methods are very closely to the total cost. However the SPD has many advantages in terms of environmental risks and shear strength. Therefore, it can be concluded which the SPD method is the most appropriate method for the tailings disposal on surface according to economic and environmental consideration.

2.6.5 Costly Traditional Tailing Disposal

First of all, to build up and keep the dams strong in terms of engineering and construction will take a lifetime. While the dam costs a lot, the discharge point of the tailings changes constantly. Meanwhile, cost of reclamation for disposed area will be high. To dispose the paste materials on surface significantly reduces the set and construction, engineering, monitoring, covering, and reclamation costs. Furthermore, the repair and maintenance costs of positive displacement or piston/diaphragm pumps used increase the energy consumption for the paste system.

3 Characteristics of Tailing for Surface Paste Disposal

Tailings are the fine-grained residue of the milling process in which the desired raw materials are extracted from the mined rock, and appear as slurries due to mixing of particles with water during this process. In the case of improper handling and management problem, it will be a serious problem to dispose slurry to waste area using the conventional disposal methods. Several methods such as subaqueous conventional upstream, centerline, and downstream cause some unstable tailings along with liquefiable zones and steep slopes resulting in erosion (Meggyes and Debreczeni 2006). For this reason, there is need for new and economical technologies for waste management in the mining industry such as paste technology which will help safety, human health, and environment (Bascetin et al. 2012).

Disposal of thickened or paste tailings is an attractive option for surface disposal as it eliminates or reduces some of the risks associated with conventional tailings disposal, most notably obviating the need for dams and eliminating the risk of catastrophic failure associated with conventional impoundments. Thickened or paste tailings offer other comparative advantages such as increased water recycling within the mining operation and reduced groundwater seepage out of the tailings impoundment (Henriquez and Simms 2009; Bascetin et al. 2012).

SPD favours water recycling and the control of free water during deposition, and reduces the need for costly retaining of dikes, and also facilitates the site rehabilitation. One of the challenges related to SPD is to prevent the potential reactivity of the tailings (Deschamps et al. 2011; Bascetin et al. 2012).

The main material is fine tailings which are the outcome of mineral processing plants. Physical properties of fine tailings such as particle size distribution, specific surface area, particle shape, and density are of great importance for the methods.

The physical and chemical characteristics of tailings are determined considering the below-mentioned parameters:

- Density
- Particle size distribution
- Particle shape and specific surface area
- Mineralogy
- Acid-sulphate effect (chemical analyses)

In the paste technology, depending on the type of ore, tailings used in the paste mixture can contain sulphide minerals (such as pyrite and pyrrhotite). These minerals are known to be reactive with water and oxygen to produce acidity (sulphuric acid).

In fact, pyrite and other sulphides lost during mineral processing will be oxidized in aqueous media under the effect of oxygen (Table 6.2). Oxidation of sulphide minerals and generation of acid cause leaching of oxidized products. This phenomenon is called acid mine drainage (AMD) which remains one of the critical environmental challenges for the mining industry. AMD can cause acidification and heavy metal release in surface and groundwater (Fig. 6.4). To inhibit the acid formation,

Table 6.2 Some of the sulphide minerals causing AMD

Mineral	Formula	Mineral	Formula
Pyrite	FeS_2	Cinnabar	HgS
Galena	PbS	Realgar	AsS
Pyrrhotite	Fe1-xS	Orpiment	As_2S_3
Sphalerite	ZnS	Cobaltite	(Co,Fe)AsS
Chalcopyrite	$CuFeS_2$	Chalcocite	Cu_2S
Millerite	NiS	Covellite	CuS
Bornite	Cu_5FeS_4	Arsenopyrite	FeAsS

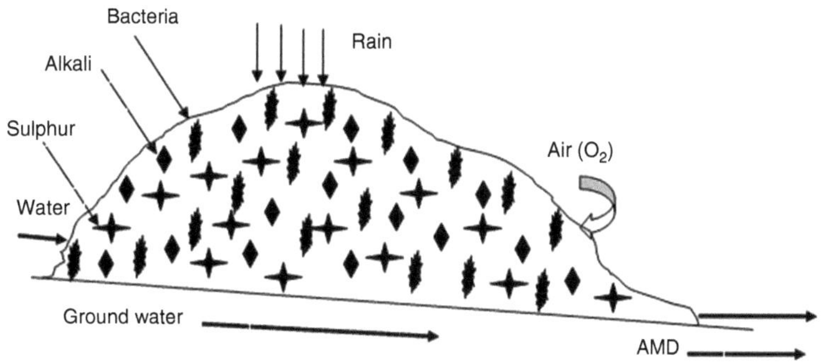

Fig. 6.4 Schematic explanation of AMD

one must exert control on the constitutive elements (i.e. sulphide, water, or oxygen) of the oxidation reaction (Benzaazoua et al. 2000; Dagenais et al. 2005).

The steps involved in the AMD generation process, using pyrite as the example, can be represented by the following reactions. Pyrite can be oxidized in the presence of oxygen and water (Eq. (6.1)). Fe^{2+} produced by (Eq. (6.1)) can oxidize in Fe^{3+} (Eq. (6.2)). When the pH is greater than approximately 4.5, Fe^{3+} tends to precipitate as a hydroxide (Eq. (6.3)). At a lower pH, Fe^{3+} can oxidize pyrite (Eq. (6.4)) (Deschamps et al. 2008):

$$FeS_2 + 7/2O_2 + H_2O \rightarrow Fe^{2+} + 2SO_4^{\ 2-} + 2H^+ \tag{6.1}$$

$$Fe^{2+} + 1/4O_2 + H^+ \rightarrow Fe^{3+} + 1/2H_2O \tag{6.2}$$

$$Fe^{3+} + 3H_2O \rightarrow Fe(OH)_3 + 3H^+ \tag{6.3}$$

$$FeS_2 + 14Fe^{3+} + 8H_2O \rightarrow 15Fe^{2+} + 2SO_4^{\ 2-} + 16H^+ \tag{6.4}$$

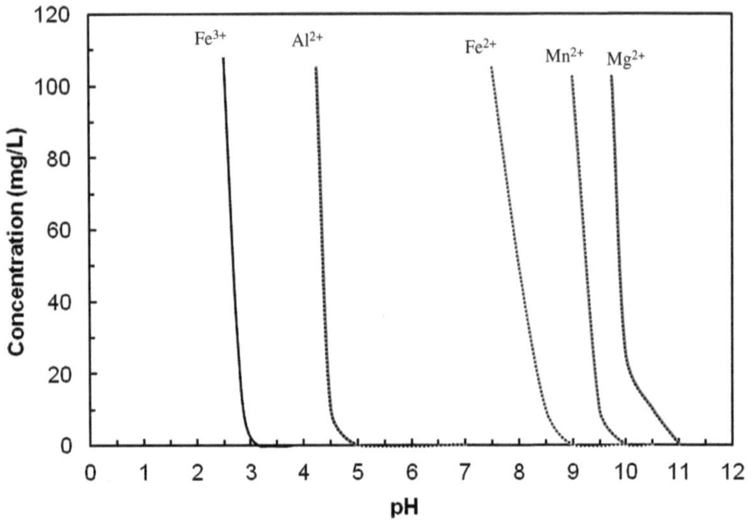

Fig. 6.5 Theoretical solubility of some metal ions (Epa 1983)

The overall sulphide-to-sulphate oxidation is summarized as follows (Eq. (6.5)):

$$FeS_2 + 15/4O_2 + 7/2H_2O \rightarrow Fe(OH)_3 + 2SO_4^{2-} + 4H^+ \tag{6.5}$$

The oxidation of sulphide minerals is an exothermic process. Primary factors of acid generation include sulphide minerals, water, oxygen, ferric iron, bacteria to catalyze the oxidation reaction, and generated heat (Epa 1994). AMD is characterized by high acidity (pH 2–4), high sulphate concentrations (1–20 g/L), and high concentration of heavy metals (Gitari et al. 2008).

There are numerous parameters which affect AMD. In the paste technology, parameters such as pH, AP (acidity potential), NP (neutralization potential), and conductivity are generally considered for determining AMD potential of underground and surface paste disposal.

3.1 *Paste pH*

Paste pH is an important parameter for determining the AMD potential. However, this parameter alone does not provide an indication of the AMD potential of a sample. It is a principal determinant for both mineral reaction rates and solubility. Therefore, it can provide insight into drainage chemistry. Solubility of metals in the paste or solution will increase with decreasing pH. Increasing the solubility of metals in the paste material indicates a risk for AMD. The theoretical solubilities, determined by measurement of each individual ion dissolved in distilled water, are illustrated in Fig. 6.5 (Rescan 2009; Ouellet et al. 2006; Epa 1983). As seen in Fig. 6.5, when the pH of acid mine drainage is varied in the range of 2–4, Fe^{3+} concentration starts increasing sharply.

Iron sulphides must be considered before others because they are the primary minerals in terms of the formation of acid mine drainage. Additionally, many metallic sulphide minerals such as pyrite (FeS_2), alabandite (MnS), galena (PbS), sphalerite (ZnS_2), molybdenite (MoS_2), and chalcopyrite ($CuFe_2S_3$) dissolve after the start of the formation of acidity, and they contribute to the medium pH. After the dissolution, when these minerals are exposed to atmosphere, several factors (oxidation, dilution, mixing, evaporation, and neutralization) lead to mineral depression, and secondary mineral formation can occur. Meanwhile, the development of secondary minerals can be seen in places where tailing stacks, mining waste, and surface water flow. Also, the dissolution of these minerals affects the drainage which cannot be ignored.

Solubility values of minerals in nature depend on many factors. Chemical structures in the rock matrix, particle size and morphology, open spaces on the surfaces, particle surface covered with other minerals, surface areas of the particles exposed to reaction, quality of water, and purities of minerals can be firstly considered. In addition, the physicochemical properties of medium also play an important role. Solution temperature, ionic strength, and pH of solution are important physicochemical parameters in determining the dissolution rate.

In a study done by Morin and Hutt, the pH of the paste samples (a mixture of pulverized sample and water) ranged from pH 3.9 to 8.9 for the 491 rock samples. There was no correlation of paste pH with sulphide (Morin and Hutt 2007). When cement was added into tailings, the pH of the paste increased due to an alkaline medium. There have been several studies that clearly showed the effect of cement addition on pH. For example, after 125 days, the pH value increased from 6.9 to 8 for leachate water for 2% cemented samples while pH value for leachate water of uncemented samples decreased from 7.5 to 6.8 (Yilmaz 2010). The pH value of drainage water collected during the consolidation was 8.1 for uncemented paste sample. With the addition of 0.5% cement, the pH value increased to 11.2, and pH value was obtained as 12.6 with more cement addition (2%) (Deschamp et al. 2011). For uncemented tailings, the pH value decreased from 7.75 to 4, and the pH value was stable at around 8.5 in 30 weeks for cemented tailings (2% cement added into one layer out of three). However, it was also shown that the pH value decreased from approximately 7.6 to 3 almost with a similar trend to uncemented tailings in the case of a misconfiguration of cemented layers (the percentage of cement added was the same in each layer 0.5 and 1 wt.%) regardless of the amount of cement added (Deschamps et al. 2008). The results from another study showed that pH value for tailings with 0.5% cemented addition dropped from 10 to 2 in 16 months (Verburg et al. 2006). In a study by Oulett, it was observed that pH value for tailings with 7% cemented addition increased from 7.56 to 13.05 in 41 days (Ouellet et al. 2006). Table 6.3 presents the changes of pH values obtained from researches which have been conducted by some authors.

Table 6.3 pH values of the paste samples obtained from the literature

Reference	pH	Explanation	Time
Morin and Hutt (2007)	3.9–8.9	Paste (a mixture of pulverized sample and water)	–
Yilmaz (2010)	8	SPD (2% cemented samples)	125 days
	6.8	SPD (uncemented samples)	
Deschamps et al. (2011)	8.1	SPD (uncemented samples)	7 days
	11.2	SPD (0.5% cemented samples)	
	12.6	SPD (2% cemented samples)	
Deschamps et al. (2008)	4	SPD (uncemented samples)	30 weeks
	3	SPD (0.5% cemented samples)	
	8.5	SPD (2% cemented samples)	
Verburg et al. (2006)	2	SPD (0.5% cemented samples)	16 months
Ouellet et al. (2006)	13.05	CPB (7% cemented samples)	41 days

3.2 *Neutralization and Sulphide Acid Potentials*

Neutralization potential (NP) and sulphide acid potentials (AP) are critical aspects for prediction of AMD. While AP is calculated using the amount of sulphur which produces acidity, NP is a measure of the carbonate material available to neutralize acid, and it is derived from acid digestion. NP and AP are usually reported in units of kg $CaCO_3$ equivalent/ton of solid sample. The net neutralizing potential (NNP) is also determined by subtracting AP from NP, and is a measure of the difference between the neutralizing and acid-forming potentials. The value for NNP may be either positive or negative. Interpretation of NP values also requires consideration of the mineralogical composition of the waste. Different minerals can neutralize acid drainage at different rates and pH ranges (Lawrence and Scheske 1997; Morin and Hutt 2006; Epa 1994; Ritcey 2005).

Deschamps et al. (2011) reported that NP was 64.4 kg $CaCO_3$/t and 80.7 kg $CaCO_3$/t for the uncemented paste and the cemented paste, respectively. In this case, the addition of 2 wt% of cement increased the NP of the paste by about 16.3 kg $CaCO_3$/t. In a study conducted by Benzaazoua et al. (2004b), the tailing showed low sulphide content (2.78%), but it was also acid generating due to the absence of neutralizing potential (21.8 kg $CaCO_3$/t).

3.3 *Electrical Conductivity*

Electrical conductivity (EC) is a measure of how well a material accommodates the movement of an electric charge. Electrical conductivity is a very useful property since values are affected by such things as a substance's chemical composition and

Table 6.4 Electrical conductivity and pH values of different type of waters

Water type	Electrical conductivity, EC (μ℧/cm)	pH
Ultrapure water	0.055	7
Drinking water	50–500	6–8.5
Domestic tap water	600	6.5–8.5
Seawater	50,000	7.5–8.4
Laronde Mine, tailings pore water (Yilmaz 2010)	8240	9.05
Laronde Mine, cemented paste tailings leachate water (Yilmaz 2010)	4500–6400	6.8–8
Doyon Mine, tailings pore water (Benzaazoua et al. 2004b)	5390	8.23
Doyon Mine, cemented paste tailings leachate water (Benzaazoua et al. 2004b)	2410	8.13
Bulyanhulu mine, cemented paste tailings leachate water (Deschamps et al. 2011)	4000–7000	10–13

the stress state of crystalline structures. A higher EC value indicates a strong ion charge caused by heavy metal mobilization or ion exchange in paste material. Paste material which has strong ion charge may have AMD potential. In general, time domain reflectometry (TDR) is used to successfully measure EC value of the paste (Sparks et al. 2001; Snyder et al. 2003).

In a study conducted by Benzaazoua et al. (2004b), the low hydraulic conductivity of paste material allowed it to retain metal ions. Despite the tailings pore water was very ion-charged as suggested by the high value of the electrical conductivity (5390 μ℧/cm) at a pH 8.23. After the moistening test on cemented tailing sample, the electrical conductivity value decreased to 2410 μ℧/cm for the leachate water at a pH 8.13.

Laboratory physical model was created by Deschamps et al. (2011) where sulphidic paste tailings were deposited in nine layers. The electrical conductivity of the leachates collected after the deposition of the cemented bottom paste layers was higher (between 6000 and 7000 μ℧/cm at a pH between 12 and 13) than for the subsequent uncemented layers (around 4000 μ℧/cm at a pH between 10 and 12) due to longer contact time with water.

In a study conducted by Yilmaz (2010), the tailings pore water had a high value of the electrical conductivity of 8240 μ℧/cm at a pH 9.05. After the moistening test on cemented tailing sample, electrical conductivity value decreased to 4500–6400 μ℧/cm at a pH 6.8–8.

As seen in Table 6.4, the EC value of seawater is greater than the EC values of tailings pore water. Although the tailings in Table 6.4 contain heavy metal ions at relatively high concentrations, total dissolved ion concentration is lower than the seawater.

Furthermore, electrical conductivity and pH may give an insight about the AMD potential when considered along with the other parameters such as chemical composition.

4 Properties of the Paste Material for Surface Disposal

Nowadays, paste technology has been safely used to dispose of metallic mine tailings having sulphur or cyanide content under surface conditions. Meanwhile, these waste materials have already been disposed using backfill operations in Turkey. Recently, studies about disposal of paste material with hazardous tailings have been increasing (Cincilla et al. 1997; Landriault et al. 1997, 2001; Crowder et al. 2000, 2002; Grabinsky et al. 2002; Verburg 2002; Theriault et al. 2003; Benzaazoua et al. 2004b; Yilmaz 2010, 2015; Yilmaz et al. 2014). With the use of this method, the amount of free water on the surface will be relatively low. Additionally, more homogeneous material having less segregation will be formed. At the same time, the improvement of hydro-geotechnical properties of the material will result in better stability. Finally, the addition of hydraulic binder into the paste mixture will increase the strength, durability, and acid-neutralizing potential of the paste material.

A big amount of the problems caused by the chemicals that are being used in mineral processing is related to the transportation and on-site accidents in which most emerge from the transportation of the hazardous materials, as well as seepage and flood failures. A huge amount of mud having hazardous chemicals and heavy metals is spread due to these accidents, and this seriously affects the potable and irrigation water sources at the dam site. Additionally, hydrogen cyanide gas resulted from processing of free cyanide in the mud and water is another danger. When these problems are considered, to use paste technology for transportation and disposal of tailings such as cyanide will be more feasible compared to other methods.

Paste material is generally formed by mixing dewatered fine-size tailings (70–85%wt of total mixture) with the desired amount of water (approximately 15–30%wt) in the absence or presence of binder, mostly cement, about 2% (Cincilla et al. 1997; Grabinsky et al. 2002; Verburg 2002; Theriault et al. 2003; Benzaazoua et al. 2004b; Deschamps et al. 2008, 2011; Yilmaz 2010; Bascetin et al. 2013, 2014).

Paste is defined as a thickened material formed by mixing dewatered fine tailings with water. In case of need, binder (cement) may be a component for the paste. Dewatered solids which are being used in the mixture with a portion in between 70 and 85% by weight are tailings obtained from the plant. A binder portion can be utmost 2% considering the economy and the strength of waste material. Water addition is usually in between approximately 15 and 30% of total mixture by weight (Cincilla et al. 1997; Grabinsky et al. 2002; Verburg 2002; Theriault et al. 2003; Benzaazoua et al. 2004b; Deschamps et al. 2008, 2011; Yilmaz 2010; Bascetin et al. 2013, 2014; Yilmaz et al. 2014).

In order to form a flowable paste, it is required to have at least 15% of particles finer than 20 μm in diameter (625 mesh). In the paste system, colloidal electrical particle charge which gathers up solid particles and water molecules provides the containment of water between the particles of fine-grained material. Therefore, the paste can easily be transported inside a pipe. Additionally, mineralogy and particle shape affect the amount of fine particles necessary (Brackebusch 1997; Tenbergen 2000; Meggyes and Debreczeni 2006).

The water content or density of a paste for a given slump consistency depends on the size distribution of the particles. The fine particles must be wetted more for their surface areas. Surface disposal of paste would be possible with a slump of 25 cm (Tenbergen 2000; Pullum 2003; Benzaazoua et al. 2004a; Meggyes and Debreczeni 2006; Yilmaz 2010). Piston/diaphragm pumps (higher pipeline pressures) can be considered for high-slump paste transport.

Paste mixtures are non-Newtonian fluids, and classified as Bingham plastic fluids that have a significant yield stress but have a relatively constant viscosity as flow rate increases. Depending on the characteristics of the paste material, viscosity of paste material can either decrease or increase. Viscosity of paste material decreases with great pumping rates because most of the paste materials show pseudoplastic behaviour (Brackebusch 1997). When the paste material has low water contents, resistance to flow will be larger. On the other hand, if the solids content is too low, it will flow too far, and will not form the desired stack. Also, there is relationship between water content and yield stress of paste material (Henriquez and Simms 2009). The rheological properties of pastes (viscosity, yield stress, etc.) have been investigated for optimizing the surface disposal parameters and determining pumpability of paste material, as well as understanding how these properties and parameters can be modified (Kwak et al. 2005).

All of the properties of the process tailings affect the properties of the paste either positively or negatively. Most significant parameters that affect the behaviours of the paste are settlement of the solids, consolidation, and desiccation. Those three factors are significantly influenced by the porosity of the paste, void content, and density of solids. In situ stability of the paste after full saturation also affects these factors. Geotechnical and mechanical properties that affect the behaviours of cemented paste material depend on the physical, chemical, and mineralogical properties of tailings, binder types, and their proportions. Conducted studies on the paste material showed that there is a possibility of reduction in the strength of the material during the curing time as a result of chemical reactions caused by the aggressive environment (Amaratunga and Yaschyshyn 1997; Ouellet et al. 1998; Benzaazoua et al. 2002; Kesimal et al. 2002). Cohesion, density, and solid content (%) are deterministic parameters in the paste applications. Cohesion of the paste depends on the particle size distribution, specific surface area, particle shape, binder quality, and chemical resistivity.

Due to the geotechnical characteristics of the pastes, they are very similar to soil materials. Therefore, to determine the geotechnical properties of the paste material is essential for durability of the paste. Some important geotechnical properties are compressive strength, porosity, shear strength, hydraulic conductivity, water content, and consolidation.

Porosity has a strong influence on the quality of paste mixtures as it is a parameter which directly affects the mechanical and geochemical properties of the paste. It is an important parameter which has a substantial influence on water-solid relations and interactions in paste mixtures (Bascetin et al. 2012).

The total porosity and the pore size distribution are usually determined by mercury intrusion porosimetry (MIP) which is particularly useful to assess pore size variation in paste tailings materials. Two parameters are often used to characterize the MIP pore size distribution curve: threshold diameter and critical diameter. The threshold diameter (TD) represents the narrowest path in interconnected pores. More specifically, this diameter represents the pore size above which there is comparatively little mercury intrusion, and below which occurs the main intrusion; it could also be defined as the largest pore diameter for which significant intruded mercury volume is detected. The critical pore diameter is the most frequently occurring diameter in the interconnected pores as it corresponds to the onset of the steepest slope on the cumulative porosity curve or to the peak value on the differential porosity curve (Deschamps et al. 2011).

Cement may also affect the porosity of the paste. MIP tests showed that the threshold diameter (consequently the pore size) of the 2 wt% cemented paste collected during the core sampling was slightly higher than that of the unleached samples (Deschamps et al. 2011). The porosity of cemented tailings decreased relative to the surrounding uncemented tailings ranging from an 8 to 18% decrease (McGregor and Blowes 2002; Bascetin et al. 2012).

In a study conducted by Mcgregor and Blowes (2002), the porosity calculations determined that the porosity within the Fault Lake cemented layer decreased from an average value of 0.51 for the underlying uncemented tailings to a minimum of 0.44 within the cemented layer. The decreasing in the total porosity relative to the surrounding uncemented tailings may also result in the cemented layers acting as a hydraulic and diffusive barrier towards the migration of infiltrating precipitation and atmospheric gases such as oxygen (McGregor and Blowes 2002). Furthermore, the addition of fly ash to paste mixture will improve the flocculation of the tailings particles' arrangement and will reduce its hydraulic conductivity and porosity. Therefore, the overall result of the proposed treatment procedure is lowering of the magnitude of the hydraulic conductivity and prevention or reduction of the penetration of moisture into the treated mortar (Mohamed et al. 2002).

The shear strength of paste fill is a result of friction and interlocking of particles, and possibly cementation or bonding at particle contacts. Shear strength is a term used in paste mechanics to describe the magnitude of the shear stress. At failure, shear stress along the failure surface (mobilized shear resistance) reaches the shear strength. The paste grains slide over each other along the failure surface.

For surface disposal proposes or in aspects of pumpability, shear stress also gains importance. As mentioned above, paste mixture is a fine-grained material which has also some water in mixture. Thus, this soil-alike material adopts flow behaviours under stress. Recent advances in technology have permitted economic dewatering and pumping of thicker tailings to the point where they behave as a non-Newtonian fluid. It was assumed that the material is a Bingham fluid which means that the paste acts like a rigid body until a specific value of shear stress is achieved (Henriquez and Simms 2009). However, under bigger shear stresses material loses its rigidness and starts to flow (Bascetin et al. 2012).

Water content is generally defined as the ratio of the weight of the water to the solids for a given unit volume of material. If material has water content less than porosity, it is the definition of unsaturated conditions. When a material is unsaturated, the water is subjected to a negative pressure relative to atmospheric pressure. This negative pressure, called suction, is caused by the surface tension which exists between air and water in contact with the soil matrix (matrix suction) as well as the affinity between water and solid (osmotic suction) (Belem et al. 2001).

In an SPD study conducted by Benzaazoua et al. (2004a, b), a laboratory physical testing model was developed to generate in situ conditions using multiple paste layers disposed on top of each other. It was found that the changes in water content ratio were smaller for bottom layers than upper layers. The reason for this can be that bottom layer was deposited on geo-textile while upper layers were deposited upon already dry surface paste fill. The main reason of the water flow from upper layers to lower layers is matric suction which is lower for bottom layers than upper layers (Benzaazoua et al. 2004b).

The water content of sulphide-containing paste tailings is also one of the most important control parameters on the rate of sulphide oxidation. Diffusion of oxygen in water is slower than in air. The presence of even small amounts of moisture in the pore space can have significant effects on oxidation rates. For a determined evaporation, the degree of saturation of a paste is higher than that of an equivalent tailings material. The water in the paste sample is largely retained regardless of disposal conditions (Verburg 2002). It should be noted that these observations related with the bulk of the paste mass; larger changes in water contents should be anticipated in the outer layers of paste deposits. Verburg (2002) mentioned that based on these considerations, the rate of sulphide oxidation in paste should be reduced relative to the rate of sulphide oxidation in equivalent tailings (Bascetin et al. 2012).

There are a few detailed studies about the water content of sulphidic paste tailings. In one of these studies, sulphidic paste tailings were deposited in a laboratory-scale physical model in nine layers. The proportion of binder (cement) was added only in the two bottom layers (this configuration should improve the environmental behaviour). The measurements showed that the evolution of the water content during the drying phases was different for cemented and uncemented layers (Deschamps et al. 2011).

In a study conducted by Henriquez and co-workers (2009), a dynamic imaging and modelling method was used to determine the effect of water content on flow behaviour of the paste. Relationship between water content and yield stress was determined in this study. While there is some variation in the slumps obtained using different size cylinders, there is a consistent relationship between gravimetric water content and yield stress for all the tests. It is also known that the yield stress and viscosity increase with the decrease of water content. Variations in flow behaviour with water content are predominantly caused by physical conditions, and not chemical interactions (Kwak et al. 2005). Water content is strongly correlated with mechanical properties of the paste.

Hydraulic conductivity values were determined from samples of bentonite-paste tailings mixtures at various water contents. It can be noticed that the hydraulic conductivity decreases in all samples as the water content is increased (Fall et al. 2009).

Hydraulic conductivity, commonly known as permeability coefficient or permeability (m/s) and symbolically represented as K, is described as the ease with which water can move through pore spaces or fractures. Permeability is an important parameter for determining the hydromechanical behaviours (i.e. rheological properties, static stability) of the paste material (Bascetin et al. 2012).

Permeability is the velocity of fluid flow through a porous medium. The tests showed that the permeability of paste is generally half an order of magnitude lower than ordinary tailings due to the fact that paste is produced from run-of-mill tailings which have not undergone the grain size segregation observed during tailings deposition. The paste maintains full distribution of particle sizes. This resulted in reduced permeability characteristic of poorly sorted materials. Tensile stresses in the tailings are also responsible for reduced infiltration. This occurs because the gravitational downward pull on the liquid surrounding the tailings particles is countered by upward capillary suction (Verburg 2002).

Admixture of small amounts (for example 1–2% by weight) of binder materials with pozzolanic and/or cementitious properties may further decrease the permeability of paste, thereby providing additional environmental protection. Disposal of tailings in paste form removes the direct cause for infiltration and seepage (Verburg 2001).

5 Application of Surface Paste Disposal Method for Industrial Scale

Mining companies are reconsidering high risk of conventional tailings impoundments retaining saturated tailings and contaminated supernatant water. Disposal of process wastes in the paste form is an attractive option for waste management as it eliminates or reduces the risks associated with conventional methods.

The surface paste disposal (SPD) method can be used to minimize environmental risks and prevent such disasters. The environmental benefits associated with SPD are both the physical characteristics of the paste itself and the operational advantages associated with its placement. The advantage of this is that there is very little free water available to generate leachate. In addition, the relatively low permeability of the poorly sorted full plant tailings limits infiltration or paste thickener resulting in a reduced seepage volume present in the deposited paste. Dewatered solids are mixed with water to form a mixture of 70–75 wt% solid ratio. A binder portion (approximately 2%) can be added considering the economy and the strength of

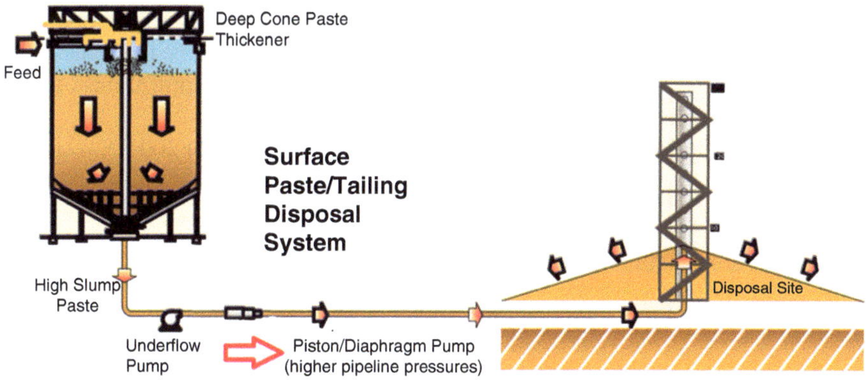

Fig. 6.6 Deep-cone paste thickener system for SPD (FLSmidth 2010)

Table 6.5 Consistency, transporting, and slope of the tailing in surface deposition (Brzezinski 2001; Bussiere 2007; Yilmaz 2010)

Consistency	Solid (by weight) content of tailing	Pumping system	Deposition slopes (%)
Slurry (slump 0 inc)	25–45% or <50%	Centrifugal pump (lower pipeline pressures)	0–1
High density slurry (slump 0 inc)	50–65% or <70%	Centrifugal or piston/ diaphragm pump (high pipeline pressures)	1–4
High slump (approx 10 inc) paste	70–%75	Piston/diaphragm pump (higher pipeline pressures)	4–8
Low slump (7–7.7 inc) paste	75–85%	Dual-piston positive-displacement pump (high-pressure pipelines)	It is a generally used backfill method in underground mining

waste material. Water addition is usually in about 25–30 wt% of total mixture. The mixture then can be transferred to a surface disposal field through pipelines, and forward-displacement pumps (Fig. 6.6). The paste is placed layer by layer with a suitable time interval between the deposition of each layer to provide the necessary conditions for the mixture to settle desiccate and consolidate. Therefore, a stable and durable deposit with desired slope angles can be achieved. Saturated and well-consolidated stack will have desired permeability to prevent the diffusion of oxygen and seepage of pore water (Bascetin et al. 2013).

A yield stress range of the order of say 200±25 Pa is proposed at the point of discharge as marking the transition between slurry and high-slump paste (MMSD 2002). For the sake of clarity, slurries can be subclassified according to the extent of thickening into slurry, high-density slurry, high-slump paste, and low-slump paste. Paste has low viscosity which is non-segregating channel flow in thick layers (>100 mm) (Table 6.5).

Paste is mainly prepared for using cemented backfill in the underground mine. However, a practicable deposition system can be designed to suit the flow characteristics of the paste; surface disposal operations may increasingly utilize this consistency of material in the future.

The system components for surface paste disposal (Brzezinski 2001):

- Appropriate selection thickener units to produce paste of the required characteristics to match the disposal site design objectives
- Choice of the right pump for the required duty to pipeline transport the tailings to the disposal site
- Disposal site is designed taking into account the following parameters:
  - Deposition method and discharge locations
  - Toe containment earthworks
  - Drainage and sealing
  - Properties of the tailings
  - Site topography
  - Climatic conditions
  - Storage life required
  - Environmental regulatory issues, etc.

Optimum surface disposal sites must have topography with ground slopes less than 5%. Thus, maximum storage is provided using thin-layer deposition from a limited number of discharge points. Discharge locations are not excessive distance and elevation from plant. Regional climate has evaporation equal to or greater than precipitation. Climatic conditions (summer and winter) are very important to maximize evaporation, drying, and consolidation of each layer. Major seismic events should not be at the disposal area. Base sealing of disposal area should be good to protect groundwater aquifers against pore water and run-off seepage.

6 Case Studies

6.1 Laboratory-Scale Designs

Recently, the detailed study about utilization of flotation tailings for the application of SPD has been carried out at Mining Engineering Department, Istanbul University, Turkey. In this study, firstly, in order to simulate the field disposal conditions and testing of the layer configurations, a unique laboratory-scale test cabin seen in Fig. 6.7a was used for the experiments. The length, width, and height of the test cabin are 200 cm, 70 cm, and 50 cm, respectively. The sides of the cabin are made of a transparent material of plexiglas to provide sufficient visibility. The bottom of the test cabin was covered with a geotextile filter to prevent material loss and to allow seepage for the sampling (Fig. 6.7b).

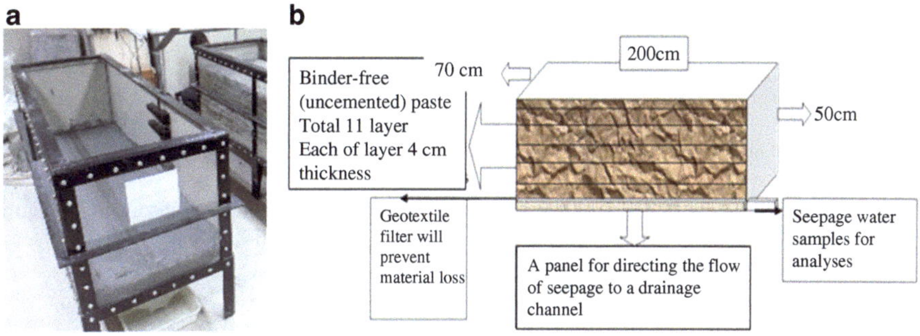

Fig. 6.7 Test cabin used for the experiments

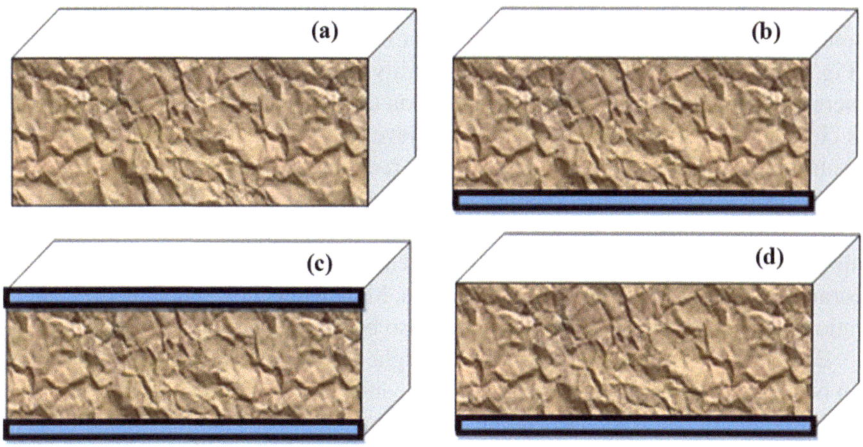

Fig. 6.8 Details of different test cabins with different set-ups

It was planned to apply different set-ups in the test cabins to determine the optimum design layout by testing several SPD configurations in terms of the important parameters which affect surface paste disposal method (Fig. 6.8).

Set-up 1 was the test cabin consisting of completely uncemented paste tailings which poured layer by layer (Figure 6.8a). This set-up was used to compare other test cabins where different configurations were tried.

Set-up 2 was the test cabin where first layer consisted of cement by the weight of 2% of the amount of solid, and remaining layers consisted of the paste tailings with uncemented (Fig. 6.8b). The solid-water ratio of the sample for each set-up was adjusted as the slump value of 10″ (250 mm). With this test configuration, it was planned to make the alkaline level of the first layer increase, and hence to prevent the mobilization of heavy metals into groundwater. Also, the cement used as a

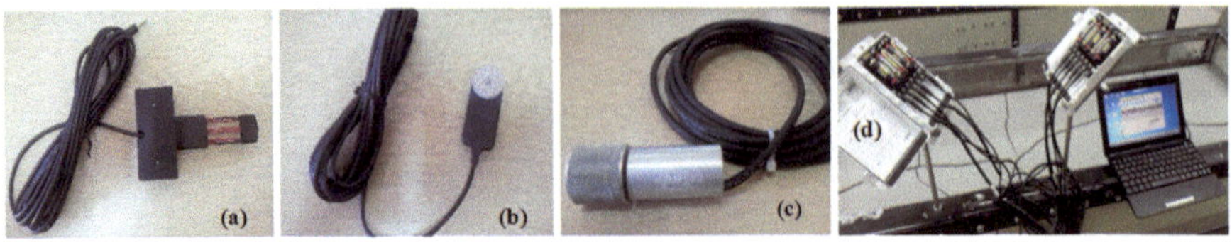

Fig. 6.9 Sensors and data logger: (**a**) Decagon 5TM moisture and temperature sensor; (**b**) Decagon MPS-1 dielectric water potential (suction) sensor; (**c**) Apogee SO-100 & 200 Series oxygen sensor; (**d**) Decagon Em50 Data logger Decagon DataTrac3 Software

binder in this set-up increased stability of the paste materials while strengthening the bonds between the particles.

Set-ups 3 and 4 will be used in future research. Briefly, in the case of set-up 3, first and tenth layers will consist of cement by the weight of 2% of the amount of solid, and then remaining layers will consist of the paste tailings with uncemented (Fig. 6.8c). Thus, it is thought that the stability of the top layer of the field will be increased further, and this will help recreation and reclamation work after the mine is closed. In addition, the high alkaline of level in the top layer which comes into contact with surface water will contribute to reducing the risk of AMD.

Set-up 4 will have similar configuration as set-up 2. But, in this set-up, first layer will consist of cement by the weight of 1.5% of the amount of solid in order to minimize the cement ratio (Fig. 6.8d). The cement as a binder is one of the important parameters increasing the cost of this process. Meanwhile, how reducing the cement ratio in the binder affects the process will also be investigated.

Each set-up has 11 layers with 4 cm in height of the each layer. During the setting up of the layers, next layer is cast on the previous layer after the completion of the drying period of that layer. Depending on the temperature and humidity conditions in the laboratory, drying period was chosen as approximately 7 or 8 days. Also, all of the layers must have an equal drying period. Oxygen, matric suction, volumetric water content, and temperature sensors were placed on the first, fifth, and tenth layers. The sensors and their ancillary equipment are seen in Fig. 6.9. After the completion of the drying period for the last layer, wetting and drying cycle was conducted considering the climate conditions of the region where the samples were provided from.

The planned configurations for set-ups 1 and 2 are seen step by step in Fig. 6.10.

The experiments were carried out at 23 °C±2 and humidity levels of 55%±5, and graphs for volumetric water content and oxygen consumption obtained in experiments carried out with uncemented paste are shown in Fig. 6.11 (Tuylu 2016).

As seen in Fig. 6.11a, the volumetric water content of the first layer decreased to 70% from 100% after the first pouring. Additionally, Fig. 6.11b shows the oxygen content of the layers affected due to cracks occurred. In addition, after the pouring of the first two layers, low values of the volumetric water and oxygen contents of the layers were obtained. This can be attributed to decrease in the inter-particle space of the layer and hence consolidated. In this study, the data of the sensors in other layers and the results for analysis of the paste materials is not presented here.

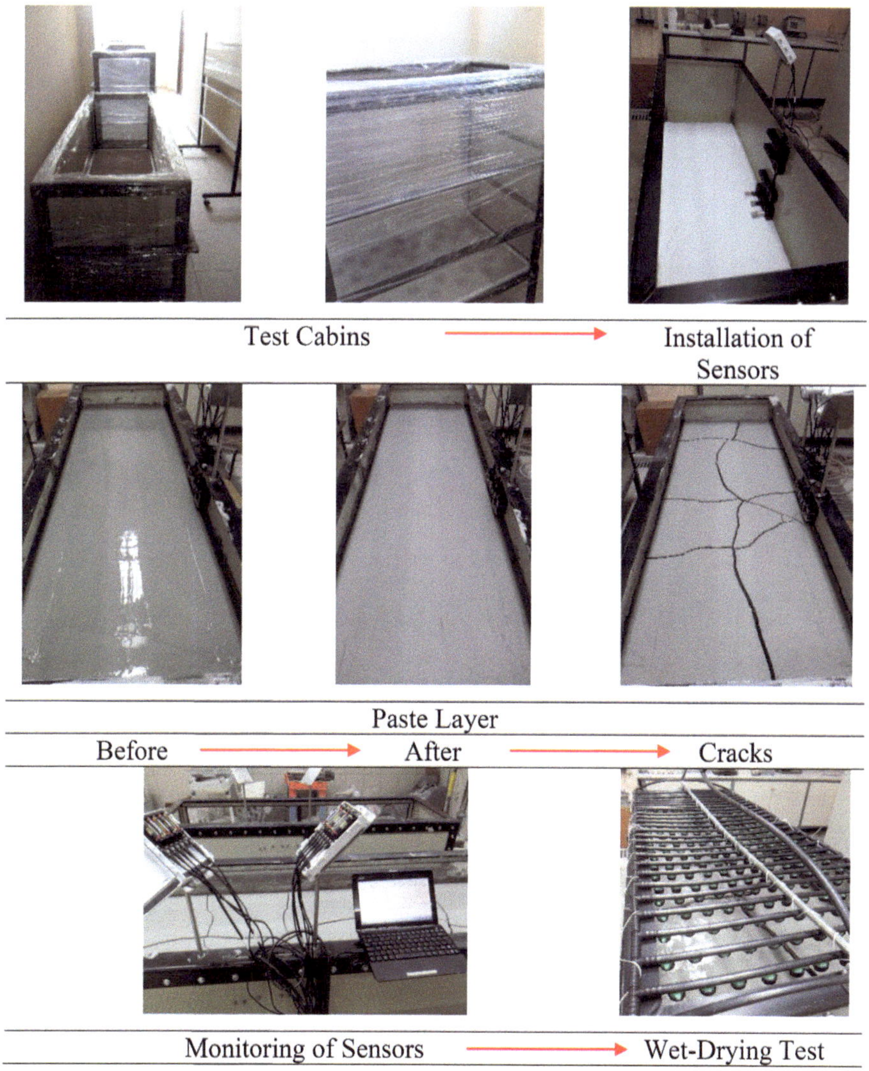

Fig. 6.10 Experimental studies about SPD in mining engineering laboratory

6.2 *Case Studies: Industrial Practice and Researches*

Theriault et al. (2003), Martin et al. (2006), and Simms et al. (2007) reported that Bulyanhulu mine in Tanzania is the first gold mine to employ surface paste disposal (SPD) worldwide (Fig. 6.12). Here, tailings are deposited alternately via several

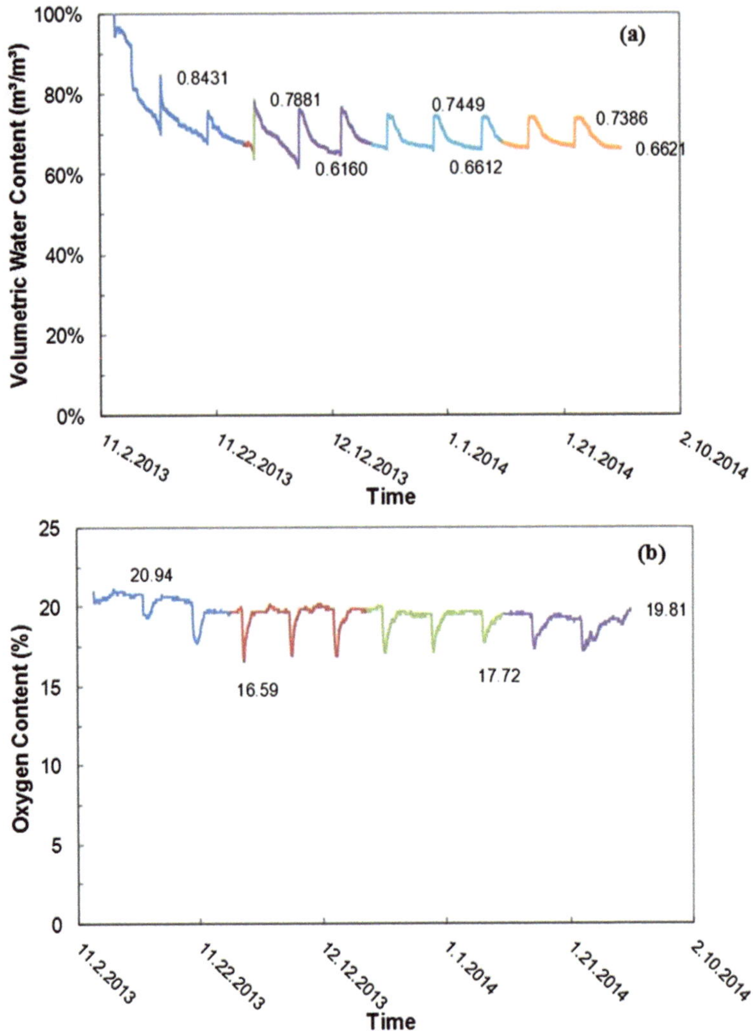

Fig. 6.11 Sensor measurement of uncemented paste first layer: (**a**) volumetric water content, (**b**) oxygen content

towers. The tailings flow away from each tower to form a conical deposit. Each layer had a lower thickness than 30 cm. The fresh tailings were poured on top of drying and hardening layer after a waiting period of 5–30 days. No cement was added into the tailings to prepare paste material. Uncemented paste was deposited by a solid content of 73% which corresponded to a 250 mm slump cone height (Yilmaz 2010).

Fig. 6.12 Industrial practice of SPD (Theriault et al. 2003)

Fig. 6.12 (continued)

In a study conducted by Yilmaz (2010), Fig. 6.13 illustrates the preliminary results of a field cell testing undertaken to investigate the effect of in situ climate conditions and material properties on the performance and quality of fine-grained, sulphide-rich paste tailings at the LaRonde mine in Quebec, Canada. A total of two field cells are constructed in the surface disposal site of the mine and categorized as cemented cell CC and uncemented cell UC as seen in Fig. 6.13.

7 Summary and Conclusions

Recently, several methods, namely, dry tailings, thickened tailings, co-disposal of tailings, and paste technology, have been developed for safe disposal of mill plant tailings. Among these methods, paste technologies offer many advantages for tailings management systems in both environmental and economic aspects. In this method, there are two different applications which are surface paste disposal method and underground paste backfill method.

Meanwhile, behaviours of paste material in field conditions should be examined in many ways. In situ properties of paste material are geochemical (as AMD risk, heavy metal mobilization) and geotechnical such as hydraulic conductivity, consolidation, crack propagation, and solidification behaviours in all kind of weather conditions. Understanding and successfully determining the effects of those parameters are significantly important for quality of the paste and paste applications.

In summary, the surface paste disposal (SPD) method can be more economically and environmentally friendly rather than the classic deposition/disposal methods.

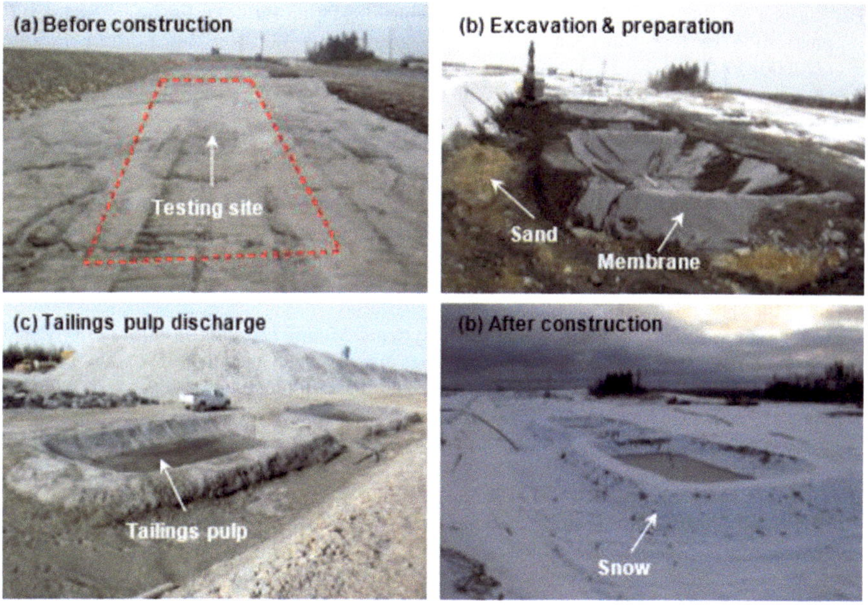

Fig. 6.13 Experiments for SPD at the LaRonde mine in Quebec (Yilmaz et al. 2014)

Acknowledgements This work was supported the Scientific and Technological Research Council of Turkey (Project Numbers: 116M721) and also by the Executive Secretariat of Scientific Research Projects of Istanbul University Project Numbers: 40376, 19062, 21527, and 27845. The authors are grateful to the Scientific and Technological Research Council of Turkey and the Executive Secretariat of Scientific Research Projects of Istanbul University.

References

Amaratunga LM, Yaschyshyn DN (1997) Development of a high modulus paste fill using fine gold mill tailings. Geotech Geol Eng 15(3):205–219

Bascetin A, Tuylu S, Sertabipoglu Z, Adiguzel D, Binen IS (2012) The study of geotechnical parameters for paste technologies: a general review. In: SME annual meeting, Preprint 12–055, Feb. 19–22, 2012, Seattle

Bascetin A, Tuylu S, Adiguzel D, Akkaya U, Binen IS (2013) Investigation of suitability of Pb-Zn mine tailings for surface paste disposal. 23rd World Mining Congress, Montreal, CANADA, 11–15 August 2013, Paper 301

Bascetin A, Tuylu S, Adiguzel D, Akkaya UG (2014) The study of surface paste disposal technology for Pb-Zn mine tailings. In: SME annual meeting, Preprint 14–020, Feb. 23–26, 2014, Salt Lake City

Bascetin A, Tuylu S, Adıguzel D, Ozdemir O (2016) New technologies on mine process tailing disposal. J Geol Resour Eng 2(2016):63–72

Belem T, Bussière B, Benzaazoua M (2001) The effect of microstructural evolution on the physical properties of paste backfill. Tailings and Mine Waste '01 2001 Balkema, Rotterdam, ISBN 90 5809 182, p 365–374

Benzaazoua M, Bussiere B, Kongolo M, McLaughlin J, Marion P (2000) Environmental desulphurization of four Canadian mine tailings using froth flotation. Int J Miner Process 60(2000):57–74
Benzaazoua M, Belem T, ve Bussiere B (2002) Chemical factors that influence the performance of mine sulphidic paste backfill. Cem Concr Res 32:1133
Benzaazoua M, Fall M, Belem T (2004a) A contribution to understanding the hardening process of cemented pastefill. Miner Eng 17(2004):141–152
Benzaazoua M, Perez P, Belem T, Fall M (2004b) A laboratory study of the behaviour of surface paste disposal. In: Proceedings of the 8th international symposium on mining with backfill, Beijing, China, September 19–21, p 180–192
Blight GE, Bentel GM (1983) The behaviour of mine tailings during hydraulic deposition. J S Afr Inst Min Metall 83:73–86
Brackebusch FW (1994) Basics of paste backfill systems. Mining Eng 46(10):1175–1178
Brackebusch FW (1997) Mineral processing tailings disposal. US Patent 5,636,942. Jun. 10, 1997
Brzezinski LS (2001) Surface disposal of thickened tailings. Mining Environ Manage 9(5):7–12
Bussiere B (2007) Colloquium 2004: hydro-geotechnical properties of hard rock tailings from metal mines and emerging geo-environmental disposal approaches. Canadian Geotech J 44:1019–1052
Cincilla W, Landriault D, Verburg R (1997) Application of paste technology to surface tailings disposal of mineral wastes. In: Proceedings of the 4th international symposium on tailings and mine waste. A.A. Balkema, Vail, Fort Collins, p 343–356
Crowder, J., Grabinsky, M., Landriault, D., 2000. Consolidation testing and SEM images of tailings pastes for surface disposal. In: Proceedings of the 53th Canadian geotechnical conference, Montreal, p 625–632
Crowder J, Grabinsky M, Klein K (2002) Laboratory characterization of tailings paste for surface disposal. In: Proceedings of the 55th Canadian geotechnical conference, Niagara Falls, Ontario, p 401–409
Dagenais A, Aubertin M, Bussière B, Martin V (2005) Large scale applications of covers with capillary barrier effects to control the production of acid mine drainage. Post-Mining 2005, November 16–17, Nancy, France
Das R, Choudhury I (2013) Waste management in mining industry. Indian J Sci Res 4(2):139–142
Deschamps T, Benzaazoua M, Bussière B, Aubertin M, Belem T (2008) Microstructural and geochemical evolution of paste tailings in surface disposal conditions. Miner Eng 21:341–353
Deschamps T, Benzaazoua M, Bussière B, Aubertin M (2011) Laboratory study of surface paste disposal for sulfidic tailings: Physical model testing. Miner Eng 24(2011):794–806
EPA (1983) Design manual: neutralization of acid mine drainage. Environmental Protection Agency Office of Research and Development Industrial Environmental Research Laboratory, US, EPA-600/2–83-001, January 1983
EPA (1994) Technical document, acid mine drainage prediction. Environmental Protection Agency Office of Solid Waste Special Waste Branch, US, EPA530-R94–036, NTIS PB94–201829, December 1994
Fall M, Adrien D, Célestin JC, Pokharel M, Touré M (2009) Saturated hydraulic conductivity of cemented paste backfill. Miner Eng 22(2009):1307–1317
FLSmidth Brochure (2010) EIMCO® Deep Cone® Paste thickeners. www.flsmidth.com, Revision 10/05/2010, Minerals Processing Technology Center FLSmidth Salt Lake City, Inc. 7158 S. FLSmidth Drive Midvale, UT
Gitari WM, Petrik LF, Etchebers O, Key DL, Iwuoha E, Okujeni C (2008) Passive neutralization of acid mine drainage by fly ash and its derivatives: a column leaching study. Fuel 87:1637–1650
Grabinsky, M.W., Theriault, J., Welch, D., 2002. An overview of paste and thickened tailings disposal on surface. In: Proceedings of the first symposium on mine waste and environment, Rouyn-Noranda, Quebec, p 5–12
Henriquez J, Simms P (2009) Dynamic imaging and modelling of multilayer deposition of gold paste tailings. Miner Eng 22(2009):128–139

Kesimal A, Yılmaz E, Erçıkdı B, Alp İ, Yumlu M, ve Özdemir B (2002) Laboratory testing of cemented paste backfill. Madencilik 41(4):11–20
Kesimal A, Erçikdi B, Yilmaz E (2003) The effect of desliming by sedimentation on paste backfill performance. Miner Eng 16(2003):1009–1011
Kwak M, James DF, Klein KA (2005) Flow behaviour of tailings paste for surface disposal. Int J Miner Process 77(2005):139–153
Landriault DA, Cincilla WA, Gowan MJ, Verburg RBM (1997) Paste disposal—the future of tailings management practice. In: Proceedings of the second international conference on mining and industrial waste management, Johannesburg, South Africa, p 1–15
Landriault DA, Welch D, Frostiak J, Evans D (2001) Bulyanhulu mine: blended paste backfill and surface paste deposition: the state of the art in paste technology. In: Proceedings of the seventh international symposium on mining with backfill, Seattle, p 1–14
Lawrence RW, Scheske M (1997) A method to calculate the neutralization potential of mining wastes. Environ Geol 32(2):100–106
Leduc M, Smith ME (2003). Tailings co-disposal: innovations for cost savings and liability reduction. Vector Engineering, Inc.: 9
Levens RL, Marcy AD, Boldt CMK (1996) Environmental impacts of cemented mine waste backfill. Report of Investigations 9599, United States Department of the Interior and Bureau of Mines, p 1–22, ISSN 1066–5552
Martin V, Aubertin M, McMullen J (2006) Surface disposal of paste tailings. In: Proceedings of the 5th ICEG environmental geotechnics: opportunities, challenges and responsibilities for environmental geotechnics, vol. 2. Cardiff, p 1471–1478
McGregor RG, Blowes DW (2002) The physical, chemical and mineralogical properties of three cemented layers within sulfide-bearing mine tailings. J Geochem Explor 76(2002):195–207
Meggyes T, Debreczeni A (2006) Paste technology for tailings management. Land Contamin Reclamation 14(4):815–827
MMSD (2002) (Mining, Minerals and Sustainable Development) Mining for the future Appendix A: large volume waste working paper. April 2002, No:31, p 18–19
Mohamed AMO, Hossein M, Hassani FP (2002) Hydro-mechanical evaluation of stabilized mine tailings. Environ Geol 41:749–759
Morin KA, Hutt NM (2006) Conversion of minerals into neutralization potentials with units of CaCO3 equivalent, MDAG.com. Internet Case Study #20
Morin KA, Hutt NM (2007) Morrison project—prediction of metal leaching and acid rock drainage, Phase 1. Report by the Minesite Drainage Assessment Group (MDAG), March 23, 2007
Newman P (2003) Paste, the answer to dam problems. Int J Mater World 11(3):24–26
Newman P, Cadden A, White R (2001) Paste—the future of tailings disposal? Securing the future. In: International conference on mining and the environment. June 25–July 1, Skelleftea, Sweden, p 594–603
Ouellet J, Benzaazoua M, Servant S (1998) Mechanical, mineralogical and chemical characterization of a paste backfill. Tailings and Mine Waste'98, Colorado, p 139–146
Ouellet S, Bussiere B, Mbonimpa M, Benzaazoua M, Aubertin M (2006) Reactivity and mineralogical evolution of an underground mine sulphidic cemented paste backfill. Miner Eng 19(2006):407–419
Pullum L (2003) Pipeline performance. 2003 international seminar on paste and thickened tailings, 14–16 May 2003, Melbourne, Section 8, p 1–13
Rescan™ (2009) Environmental Services Ltd. (Proj. #0793–001-02). Appendix 15, Prediction of metal leaching and acid rock drainage, Phase 1 (Part 1). Metal leaching and acid rock drainage characterization report. Pacific Booker Minerals, Inc., Report Version D.1, September 2009
Ritcey GM (2005) Tailings management in gold plants. Hydrometallurgy 78(2005):3–20
Robinsky EI (1999) Thickened tailings disposal in the mining industry. E.I. Robinsky Associates Ltd., Toronto, 210p
Simms P, Grabinsky MW, Zhan G (2007) Modelling evaporation of paste tailings from the Bulyanhulu mine. Can Geotech J 44(12):1417–1432
Snyder KA, Feng X, Keen BD, Mason TO (2003) Estimating the electrical conductivity of cement paste pore solutions from OH−, K+ and Na+ concentrations. Cem Concr Res 33(6):793–798

Sparks JP, Campbell GS, Black RA (2001) Water content, hydraulic conductivity, and ice formation in winter stems of Pinus contorta: a TDR case study. Oecologia 127:468–475. doi:10.1007/s004420000587

Tenbergen RA (2000) Paste dewatering techniques and paste plant circuit design. Tailings and Mine Waste 2000, Balkema, Rotterdam, p 75–84

Theriault J, Frostiak J, Welch D (2003) Surface disposal of paste tailings at the Bulyanhulu gold mine. In: Proceedings of the 2nd mining environment conference, Sudbury, Ontario, p 1–8

Tuylu S (2016) Determination of optimum design specifications for surface paste tailings disposal. Ph.D. Thesis, Advisor Prof. Dr. Atac Bascetin and Co-Advisor Prof. Dr. Mostafa Benzaazoua, Istanbul University Institute of Science

Verburg RB (2001) Use of paste technology for tailings disposal: potential environmental benefits and requirements for geochemical characterization. IMWA Symposium

Verburg RBM (2002) Paste technology for disposal of acid-generating tailings. Mining Environ Manage 13(7):14–18

Verburg R, Newman B, Fordham PM (2006) Surface paste disposal of high-sulfide tailings—field cell monitoring and pilot plant testing. In: 7th international conference on acid rock drainage (ICARD), March 26–30, Published by the American Society of Mining and Reclamation (ASMR), 3134 Montavesta Road, Lexington, KY 40502, 2170–2187

Vogt C (2012) International assessment of marine and riverine disposal of mine tailings. Secretariat, London Convention/London Protocol International Maritime Organization (IMO) London, England &United Nations Environment Programme-Global Program of Action, November 30, 2012

Vogt C (2013) International assessment of marine and riverine disposal of mine tailings (Full Report), November 2013

Wickland BE, Wilson GW, Wijewickreme D, Klein B (2006) Desing and evaluation of mixtures of mine waste rock and tailings. Can Geotech J 43:928–945

Yilmaz E (2010) A field investigation of the behaviour of fine-grained, sulphide-rich paste tailings under a surface disposal condition. Post-Doc Project Final Report, Université du Québec en Abitibi-Témiscamingue (UQAT), 2010, Canada

Yilmaz E (2015) Environmental characterization of surface paste disposal. LAP LAMBERT Academic Publishing, ISBN: 978–3–659-36697-0, Saarbrucken, Deutschland/Germany

Yilmaz E, Benzaazoua M, Bussière B, Pouliot S (2014) Influence of disposal configurations on hydrogeological behaviour of sulphidic paste tailings: a field experimental study. Int J Miner Process 131:12–25

Zou DH, Sahito W (2004) Suitability of mine tailings for shotcrete as a ground support. Canad J Civil Eng 31(4):632–636

Chapter 7
Instrumentation of Underground Backfill and Surface Disposal

Isaac Ahmed and Reza Moghaddam

1 Introduction

Instrumentation is widely used in paste or thickened tailings systems from controlling the production process to diversion of material to monitoring of deposited backfill. The key uses of instrumentation at the plant level, transportation system and as placed in the paste backfill are discussed. The types of instruments used and conceptual control strategies are introduced. From an instrumentation and controls perspective, at the plant level there are two particular areas of interest, the tailings thickener and, in backfill systems, the paste mixer. Specific parameters are measured to optimise the performance of the thickener. The thickener contents (bed level and bed pressure) and thickened tailings solids content in the underflow are observed, providing a running synopsis of the equipment. Likewise, the heart of the paste backfill plant is the paste mixer, primary process parameters are monitored, trending adjustments to binder (commonly cement, fly ash, blast furnace slag or some combination) and water addition with feedback on paste backfill consistency.

Unlike underground applications, the mechanical behaviour of thickened tailings or paste for surface disposal is fairly understood (e.g. Crowder 2004; Li et al. 2009; Al-Tarhouni et al. 2009; Been and Li 2009; Simms et al. 2010). This includes (a) the stability of thickened tailings stacks in both the short term during operation and long term in the closure mode; (b) dynamic response of thickened tailings; and (c) post-deposition behaviour. Similar to conventional tailings surface disposal, the thickened tailings deposition practices are in a pragmatic way, with no particular

I. Ahmed (✉) • R. Moghaddam
Paste Engineering & Design, Golder Associates Ltd., Sudbury, ON, Canada
e-mail: iahmed@golder.com

E. Yilmaz, M. Fall (eds.), *Paste Tailings Management*,
DOI 10.1007/978-3-319-39682-8_7

instrumentation installation during deposition. Due to the accessibility of the sites, the strength parameters and in situ properties of the deposited thickened tailings can be obtained after deposition using in situ techniques. Therefore, in addition to instrumentation use at the paste backfill plant and underground distribution system, this chapter addresses the instrumentation of underground backfill, which is less understood.

Modern bulk underground mining methods typically create underground openings by extracting stopes (units of the ore body) with volumes of the order 10^2 to 10^4 m^3 each. The stability of underground openings is affected by several factors including in situ stress, rock mass properties, opening size and time since excavation. Ground conditions are subject to deterioration by losing confining forces on the rock mass after mining. Backfilling provides regional ground support, helping to slow down this time-dependent deterioration by maintaining confinement for the mine-wide rock mass during continued mining.

The rate of backfilling and the quality of the backfill have a significant impact on production rates and therefore mining economics, both on the individual stope level (e.g. the cycle time per stope and as-delivered cost of the fill) and on the larger mine planning level—in particular, the stope extraction sequence and associated development costs. Therefore, systems are required to quickly deliver high-quality backfill to support the rock mass surrounding underground openings.

Hard rock precious metal and other mines produce tailings with solids that can be characterised as non-plastic silts. The tailings that contain sufficient fine particles (i.e. at least 15% less than 20 μm) develop a paste-like consistency when sufficiently dewatered and thickened. And when binding material is added, the resulting mixture is referred to as paste backfill. Advantages of paste backfill over other types of backfill include relatively rapid rates of delivery to and filling of the stope; non-segregation of the fill within the stope; minimal water management (as compared to hydraulic fills); and the ability to fill tightly in confined spaces (as compared to rock fill).

The mining industry was initially slow to adopt paste backfill technology due to initial problems in paste production and transport. However, these issues are now routinely resolved and the engineering focus has shifted to the engineering design of paste backfill systems and to monitoring pate backfill performance in situ. Despite the now widespread application of paste backfill, the design of paste backfill systems is still largely empirically supported by hands-on experience. It is therefore imperative that practical methods be developed for the design of paste backfill systems which requires, in part, a fundamental understanding of paste backfill's material properties. This is true not only during the production phase at the plant level, but while filling the stope as well, especially during the first few hours and days as the fill hydrates and gains stiffness and strength.

Information gathered with the instrumentation may be stored to provide historical data trending. This allows the operations to monitor any changes in the paste backfill during production, as placed, and provide the means to troubleshoot.

2 Instrumentation at the Plant

2.1 *Tailings Thickener*

The thickener is used to consolidate the tailings from a dilute slurry to thickened tailings. For example, a dilute slurry of 15–30 wt% solids may be received by the thickener and discharged at 60–70 wt% solids. The inherent settling characteristics of the tailings ultimately dictate the achievable underflow solids content. In some, more demanding surface disposal applications where a higher level of dewatering is required a deep bed (alternatively known as deep cone or paste) thickeners are employed to produce even higher solids underflow. Typically, with these types of thickeners, a recirculation system is used to shear thin the thickener underflow to improve its pumpability.

Flocculant is used to enhance the settleability of the tailings. The overflow or water extracted from the dilute slurry is often reused at the mill or may be further treated as necessary for discharge off site. The settling rate of the tailings solids is tied to the flocculant dosage. This is commonly expressed in grams of flocculant used per tonne of tailings. In many metal mine applications, a flocculant dosage rate of 20–80 g/tonne is commonly seen. Periodic test work to evaluate the flocculant dosing requirements will enhance the thickener operation by updating the dosage rate.

2.2 *Thickener Instrumentation and Controls*

Instrumentation employed on the tailings feed stream to the thickener feedbox is used to measure both the slurry volumetric flow rate and slurry density. The tailings mass flow rate can then be determined. In a standard application a rubber-lined magnetic flow tube is used to measure the slurry flow rate, while a nuclear density gauge assesses the solids content. In applications where the slurry content affects the generated magnetic field, other types of flow measuring instruments such as Coriolis or sonar flow meters may be employed. With these two parameters identified, on an instantaneous basis the tailings solids feed rate can be determined. Consequently, flocculant at a predetermined ratio may be dosed to the thickener. The concentration of the flocculant in the prepared solution is a known value. As the flocculant dosing pumps are installed with a variable frequency drive the tailings solids feed modulates the pump speed to maintain ratio of flocculant to tailings solids. As a result, changes to the tailings mass flow rate result in proportional changes to the flocculant dosing rate.

To monitor the operation of the thickener, oftentimes an ultrasonic bed-level monitor and pressure transmitter are employed. The ultrasonic bed-level monitor is mounted just below the liquid level of the thickener. It emits ultrasonic energy towards the bottom of the tank and with the rebounding signals interprets the thickener bed profile. A series of readings are taken and an averaged profile is obtained. The different fluid layers within thickener cross section can be determined, such as the interface between the supernatant and the tailings settling zone, low-density mud layer, high-density mud layer and tank bottom. The pressure transmitter is mounted on the bottom conical part of the thickener. It is used primarily to monitor the thickener bed pressure. With some installations, as the thickener underflow pumps operate with a variable speed drive (VSD), the bed pressure is tied to the underflow pump speed to maintain the bed pressure set point. The bed pressure can be correlated to an underflow solids content (by weight) which is normally the set operational target. Alternatively, if the thickener underflow piping is equipped with a nuclear density gauge to measure the thickened tailings density (and hence solids content by weight), the thickener underflow pumps can be modulated by the density of the thickened tailings directly. The underflow pumps may slow down to build the bed depth within the thickener to enhance solids consolidation. Conversely, if the solids content is above the set point, the underflow pump will speed up. Under either control strategy, consideration is placed on the interface level of the supernatant and the tailings settling zone to maintain the thickener overflow clarity. In addition, a minimum underflow discharge pipeline velocity should be kept to avoid the settling or deposition of the tailings particles in the pipeline. Hence, a flow meter is normally installed in the underflow pipeline as well. This will allow the minimum thickened tailings flow rate set point to override the modulating underflow pump speed.

A simplified typical thickener instrumentation package with control parameters is shown in Fig. 7.1. The mill tailings mass flow rate and solids content dictate the mixed flocculant feed rate which in turn is tied to the dilution water feed at a

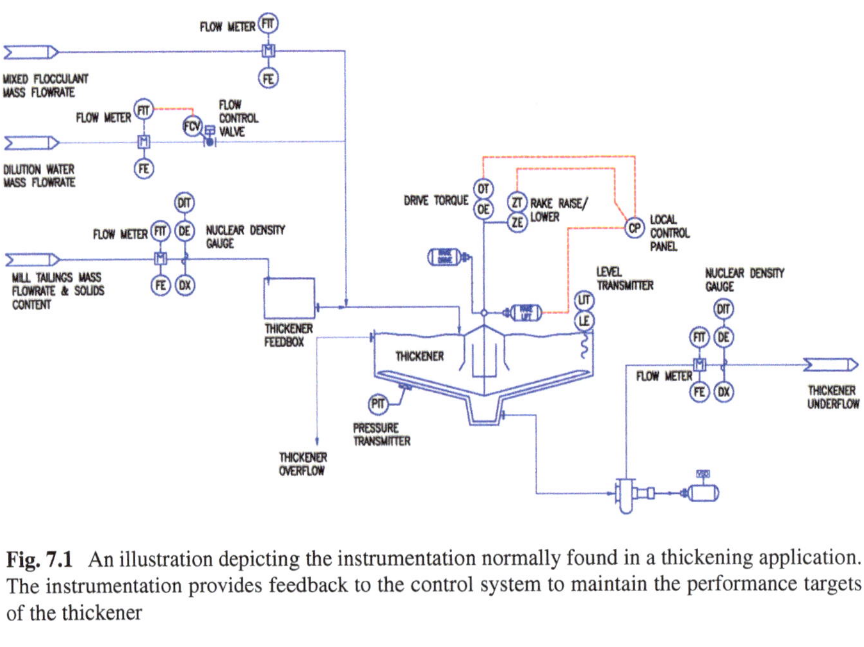

Fig. 7.1 An illustration depicting the instrumentation normally found in a thickening application. The instrumentation provides feedback to the control system to maintain the performance targets of the thickener

predetermined ratio. The rake mechanism torque is monitored against set points to determine if there is a need to raise the rakes. Lowering of the rakes is usually completed manually by an operator at the thickener bridge who can observe the performance of the thickener. The level and pressure transmitters as previously described are also illustrated, as are the instrumentation normally found on the discharge of the thickener underflow pipeline.

2.3 *Recirculation Loop Instrumentation*

Each supplier of high-compression thickeners to deep bed-style thickeners has its own proprietary thickener underflow recirculation, or also commonly known as shear thinning systems. The objective is the same: it is to reduce the viscosity of the thickened tailings and improve its pumpability. As most thickened base metal, precious metal and oil sand tailings exhibit shear thinning behaviour this technology has a wide application. A separate recirculation pump, on a variable speed drive, with a higher flow capacity compared to the thickener underflow pump, is employed. Auxiliary proprietary equipment is installed in addition to instrumentation used to monitor the flow rate and the pipeline pressure of the recirculation system. Depending on the application, the recirculation loop may be operated full time or on an as-needed basis. For instance, one application calls for running the recirculation pump continuously at a base flow rate of 130% of the underflow pump. This set point is determined based on the thickened tailings characteristics and can be changed to suit the specific material. The operation of the recirculation pump may be set and modulated by the underflow pump by this ratio. In another application, the recirculation pump speed is operated proportional to the thickener rake torque. In this example, when the rake torque reaches 20% of the maximum available to the thickener drive, the recirculation pump starts at 20% speed and increases in 20% increments with the rake torque. Hence, at 24% of the maximum rake torque, the recirculation pump will reach 100% pump speed.

2.4 *Paste Backfill Mixer*

The paste backfill mixer is at the heart of the paste backfill plant. Each component entering the mixer is measured to produce a paste backfill material of a target slump with the resulting strength requirement. By correlating the mixer power draw with the paste backfill slump, the consistency of the material can be monitored continually. In a continuous paste backfill plant, the tailings solids enter the mixer from the filter cake conveyor and the slurry bypass line. It is commonly named the slurry bypass line, as the slurry bypasses the disc filters. The filter cake conveyor is equipped with a belt scale that measures the mass flow rate of the filter cake. Coupled with grab sample values of the filter cake moisture content, the tailings, on a dry basis, reporting to the mixer from the disc filter can be determined. With a target paste backfill slump, a certain quantity of slurry is added to the mixer. The slurry bypass pipeline is equipped with a rubber-lined magnetic flow tube and a nuclear density gauge. The different types of instruments that can evaluate the tailings mass flow were discussed previously in Sect. 2.2. The slurry is added in proportion to the filter cake solids mass flow rate to provide the initial dilution necessary to produce paste backfill close to the target slump. In most cases, the filter cake produced from the filtration step has a high solids content beyond the typical 178 mm slump paste backfill. Hence dilution is necessary to reduce the solids content to the target slump. In addition, this controlled dilution process enables the mixer to produce a consistent-quality paste backfill. With feedback from the mixer power draw, dilution water may be added to the mixer to fine-tune the paste backfill consistency. The standard approach is to provide a slightly reduced amount of required dilution slurry such that a small amount of dilution water is added to the mixer. The dilution water line is outfitted with a magnetic flow tube (typically) such that the amount of water introduced to the paste backfill mixer is captured. If an admixture system (rheology-modifying agent) is installed at the paste backfill plant, the diluted solution may be introduced to paste backfill mixer in lieu of dilution water. Similar to dilution water the admixture discharge line is equipped with a magnetic flow tube tracking the quantity of admixture use. For a batch paste backfill plant, one of the key differences is that the filter cake is dispensed from a weigh hopper. The weigh hopper is mounted on load cells allowing for a predetermined amount of re-pulped filter cake to dispense into the batch mixer. Dilution slurry and water are added in the same manner as the continuous process in accordance to the tailings solids and the mixer power draw.

Binders, commonly cement, fly ash and blast furnace slag or some combination, are introduced into the paste backfill mixer in proportion to the mass flow rate of tailings solids. Since the total mass flow rate of tailings solids entering the paste backfill mixer is known, binder is dispensed into the mixer at a ratio that would give the target binder content. For a continuous paste backfill plant a weigh belt conveyor is often used to measure the quantity of binder delivery. The weigh belt conveyor is mounted on load cells and operates on a variable speed drive (VSD). The weigh belt conveyor works in tandem with the rotary valve, also equipped with a VSD to provide a consistent bed of binder material. Maintaining a constant bed of material will allow for a more repeatable and accurate means of measuring the binder mass flow rate. Since the binder mass flow rate is a function of the tailings solids addition, the weigh belt conveyor and rotary valve will speed up and slow down to match the target binder content. In a batch paste backfill plant, binder from the silo is dispensed into a weigh hopper. Load cells on the weigh hopper measure the quantity of binder release into the batch mixer.

Figure 7.2 provides an illustration of the measured parameters (binder, slurry, dilution water) that enter the mixer. The addition rate of these materials is governed by the filter cake mass flow rate which is tied to a target paste backfill mix design. The mixer motor torque provides further feedback to the control system which would override the dilution water flow rate set point. In a standard gravity discharge

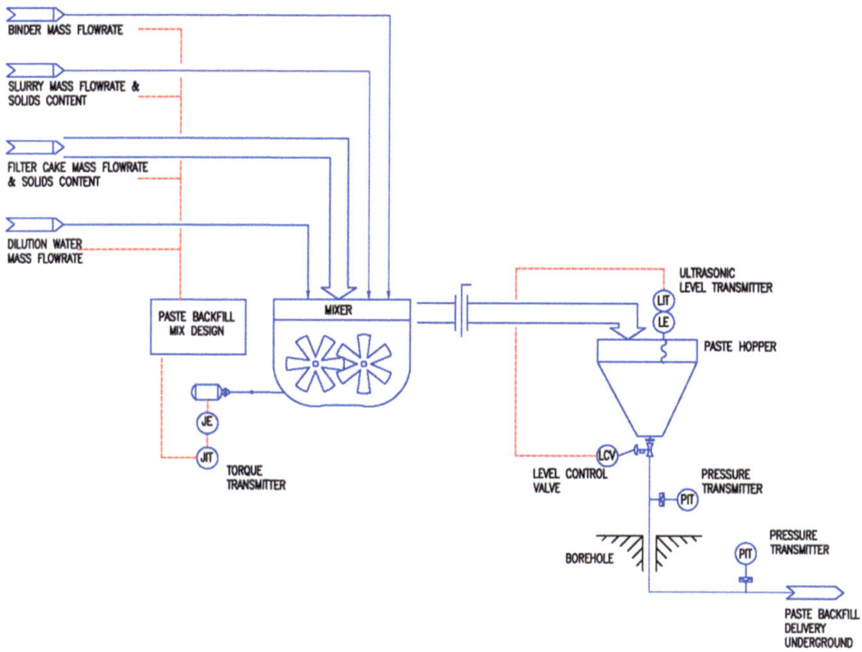

Fig. 7.2 A depiction of a typical paste backfill mixer set-up, where measured inputs are assessed against the primary filter cake mass flow rate and solids content parameter. Instrumentation is used to control the production of paste backfill to a target mix design

system as shown, a level control valve, normally a vacuum-rated pinch valve, modulates to maintain a certain level in the paste hopper. Should the paste hopper level become too high, the upstream equipment will either slow down or shut down.

There are certainly different control strategies, each aimed to provide a consistent, repeatable mix design.

Grab samples from the mixer discharge can be used to correlate and fine-tune the measured slump to the trending on the mixer power draw. Continuous feedback on the target power draw against slump allows for a tighter control on the paste backfill consistency with the changing operating conditions.

2.5 *Paste Backfill Underground Distribution System*

Through an interconnected series of piping, paste backfill is distributed to different areas underground. The underground distribution system (UDS) which includes the piping, pipe supports, bracing and boreholes is designed specific to each application. Throughout the distribution network, the pipeline pressure is measured, to monitor the paste backfill routing, actual pressure versus design as well as troubleshooting. Figure 7.3 is an illustration of a standard pressure transmitter installation

Fig. 7.3 A pressure transmitter installed on an underground distribution system to monitor pipeline pressure at specific locations. A rupture spool can also be seen in the background

on a UDS. These are installed strategically, typically at borehole to level distribution transition sites, where high pipeline pressures are often experienced. The pressure transmitters can be connected to the mine underground infrastructure such that the data can be displayed and captured in real time. A rupture spool is normally placed upstream of the pressure transmitter. The idea of the rupture spool is to allow for failure to occur at a localised and contained area should a pipeline overpressure condition exist. These rupture spools are specially designed and fabricated with consideration of the pipeline failure pressure. In some cases, they are designed to rupture at 200–250% of the normally rated pipeline pressure (Archibald et al. 2009).

Based on the elevation changes, pipeline lengths and estimated pipeline friction losses, a flow model is often constructed to examine the extent of the paste backfill distribution against the maximum allowable operating pressure (MAOP) of the system (Fig. 7.4). The MAOP is defined by the system piping selection, its pressure rating. Interfaced against the actual operating pressure, it may be readily compared to the system design pressure (hydraulic grade line) and MAOP. This allows the operation to understand the distribution envelope with a given paste backfill slump and optimise the piping selection within the system.

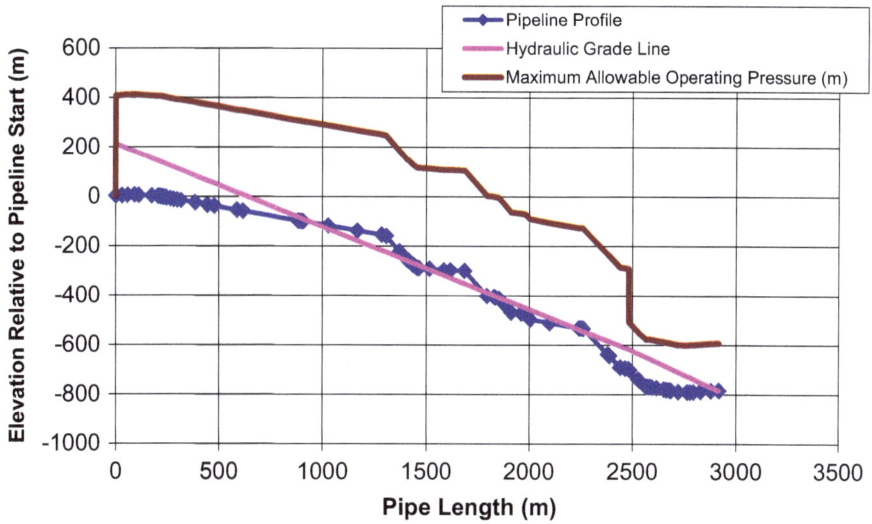

Fig. 7.4 Typical flow model results showing the extent of paste backfill delivery for a given material. The maximum allowable operating pressure is referenced along the pipeline profile

3 Paste Backfill Filling Process

A typical paste backfill filling process for bulk mining operations is described in the following section.

Overcut and undercut accesses are developed to vertically delineate the planned stope. The plan dimensions of the stope depend on the ore body geometry and the open spans that can be supported by the host rock, but the vertical dimensions vary from about 25 m (long hole methods) to 200 m or more (Alimak raise methods). Once the ore has been blasted and extracted, the open stope must be backfilled as quickly as possible; otherwise the unsupported rock starts to fail into the open stope.

Backfilling involves creating a suitable barricade at the bottom access and delivering the backfill into the stope through the top access. Paste backfill barricade design varies considerably, ranging from "rammer jam" piles constructed of waste rock forced into a plug in the undercut using heavy equipment to reinforced concrete slabs that are notionally engineered according to accepted civil engineering-reinforced concrete design methods.

For paste backfill, the delivery of paste is invariably by end-of-pipe deposition where the delivery pipe is run into the overcut location, and the paste is prepared in a surface or underground backfill plant and delivered through an underground system to the overcut. The paste backfill plants, in general, have good control of the delivery rates; hence the rate of fill rise within the stope can be regulated; however it is a function of the stope geometry as well. At many mines the initial rise rate is a few decimetres per hour, at least in the initial stages of filling adjacent to the undercut barricade where the geometry is often wider to accommodate the extractive mining methods.

Where stability of the barricade is a concern, some mines choose to fill to a limited extent (e.g. a few metres) above the undercut brow and then wait for the resulting "fill plug" to cure with the resulting stiffness and strength gains providing a degree of protection for the barricade. The costs associated with this technique include longer stope cycle time and increased labour, both of which can be considerable and hence continuous pours are desired. Another technique used by some mines is to increase binder content in the equivalent "fill plug" location, the design assumption being that higher binder content will result in faster development of stiffness and strength albeit at an increased binder cost. Alternatively, in situ backfill monitoring is another technique to assess the evolution of the paste backfill material properties during the filling process.

4 Backfill Monitoring Process

In situ backfill monitoring involves three stages including preparation, instrumentation installation and data acquisition. These stages might be different depending on the instrumentation location. The key areas of consideration for the instrumentation in a backfill system are presented in this section. Thompson et al. (2009) illustrate different installation configurations, methods in application and monitoring results.

4.1 Instrumentation Within the Main Stope

Several individual instruments are usually assembled as a cluster within a wire cage, which are sequentially mounted within a steel protective frame. There might be several cable-connected instrument clusters required to be installed at different locations into the stope. In this case, the instruments are required to be suspended within the stope using a cable system and the stope preparation usually begins before the stope is mined. This includes installation of a pulley system in overcut prior to blasting. The instrument clusters will be pulled up from the undercut or lowered down from overcut using a combination of winches and tuggers after blasting. Figure 7.5 shows suspended instrument clusters installed within the main stope.

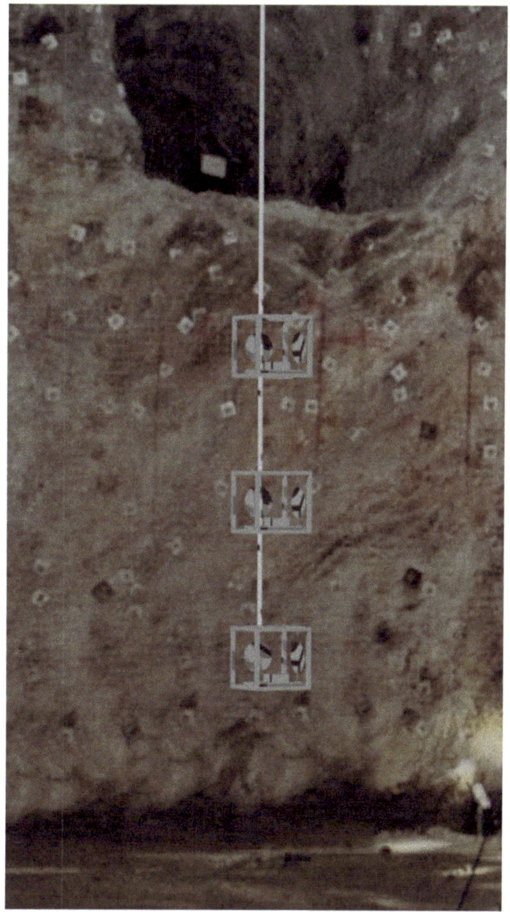

Fig. 7.5 Cable-connected suspended instrument cluster concept within a stope

4.2 Instrumentation Behind the Stope Barricade in the Undercut

Individual or a cluster of instruments can be installed behind the stope barricade after all instruments within the stope are installed. The individual instruments should be mounted on a frame right behind the barricade. However, an instrument cluster in a cage must be suspended by mounting the cage on a support structure (Fig. 7.6). The support structure will be moved to the desired location in the undercut by a remote control LHD vehicle.

In the case of an instrument cluster, selection of individual transducers and their integration into a system must be done carefully prior to installation as the individual transducers may be sourced from different manufacturers. It would be more practical to have one manufacturer to build the entire system to ensure that the integrated system would work properly. A single multi-strand wiring cable should be

Fig. 7.6 Instrument clusters mounted on T-support structure concept

used to connect all instruments within a cluster with a single connector. This helps avoid mistaken connections between different instrument clusters and amongst transducers. In addition, the wiring cable must be protected to avoid abrasion and excessive tension during installation.

Upon completion of installation within the stope and behind the stope barricade, the stope barricades will be constructed. The paste filling process would be the same as the process described in the previous section. Once paste reaches the instruments, the data will be acquired using dataloggers located in the underground. The type of dataloggers and number of channels depend on the type and number of instruments installed in the stope. The interval time for data acquisition varies depending on the stage of hydration and/or filling process. Since the paste backfill's material properties may change drastically before the initial set of hydration, the time intervals should be minimised for this period. To monitor the backfill behaviour in real time, the dataloggers could be connected to the surface via Ethernet.

4.3 Instrumentation Design

Some general guidelines will be discussed in the section for selection of the individual instruments that can be used to better understand the behaviour of paste backfill. However, discussion with instrument suppliers and manufacturers is always

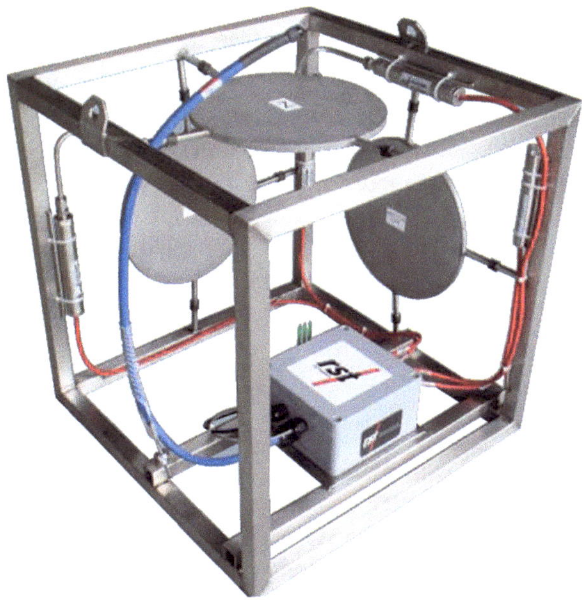

Fig. 7.7 TEPCs (RST Instruments Inc.)

warranted when considering a particular mine application. The typical instruments for backfill monitoring are further discussed.

4.3.1 Total Earth Pressure Cells (TEPC)

To measure the total vertical and horizontal stresses, one vertical and two horizontal TEPCs are installed in orthogonal directions, as presented in Fig. 7.7. This arrangement allows the assessment of the transition from hydrostatic loading to non-hydrostatic loading, which occurs when paste backfill gains shear strength. For a paste backfill application, it has been recommended (Grabinsky 2010) that the TEPCs should consist of a high-aspect-ratio circular plate (usually with sensing face on one side) and a vibrating wire (VW) sensor. The sensing plate with a relatively large diameter, compared to its thickness (i.e. high aspect ratio), minimises the stiffness contrast effect and creates a uniform contact pressure on the sensing face of the plate. The advantages of using VW sensors are their electrical stability, improved signal processing technology incorporated in the data acquisition systems and their precision. The selection of the VW sensors is also related to their relative robustness under the expected static and dynamic loading conditions induced by mining activities. The TEPC range should probably be specified for the upper limit of anticipated pressures while ensuring that the precision remains within acceptable tolerance for the intended application.

The effect of temperature on the response (i.e. stresses) of the TEPCs needs to be calibrated. Temperature can affect the registration of the entire cell assembly (i.e. VW and sensing plate) or VW sensors. Some of the VW sensors contain a thermal data channel for subsequent temperature corrections. This temperature channel can also be used to evaluate temperature changes in the backfill due to hydration of cementing materials. The TEPCs should be calibrated based on the actual field stress and temperature conditions, as described by Grabinsky and Thompson (2009) and Daigle and Zhao (2004).

4.3.2 Piezometers

Piezometers are installed within a backfill system to measure pore water pressure evolution during filling and hydration processes. The effective stresses can be estimated using the total stresses (obtained from TEPCs) and pore pressure values. The zero effective stress condition can be observed during the filling process when the backfill system has not developed shear strength. This condition might be the most critical stress state of a backfill system particularly when the barricade design is of concern. In geotechnical engineering, zero effective stress may also refer to the state when a loose saturated soil undergoes liquefaction.

For the paste backfill application, VW-based piezometers have shown a prominent performance, as described by Grabinsky (2010). The tips of the piezometers in this case must be filled with de-aerated water at the stope location. This helps eliminate the initial air pressure effect due to differential pressures at the time of manufacturing and the underground condition.

4.3.3 Matric Suction Sensors

To measure negative pore pressure due to generation of matric suction, matric suction sensors are recommended for backfill monitoring. Matric suction may be generated in paste backfill through self-desiccation because of the hydration of cementitious materials, as identified in various laboratory studies (Grabinsky and Simms 2006; Helinski et al. 2007; Simms and Grabinsky 2009). Self-desiccation implicates the determination of water retention curve and mechanical properties during curing, which are important in the stope design.

The air entry value (AEV) is the key feature of the matric suction sensors that needs to be specified for the intended application. In unsaturated soil mechanics, the AEV is defined as the matric suction value that must be exceeded before air recedes into the soil pores. In other words, the measurements are only valid when the pore pressure exceeds the AEV. For the paste backfill application, suctions of several hundred kPa may occur as the pore sizes are small. Therefore, a sensor with an AEV of a couple of kPa would be sufficient.

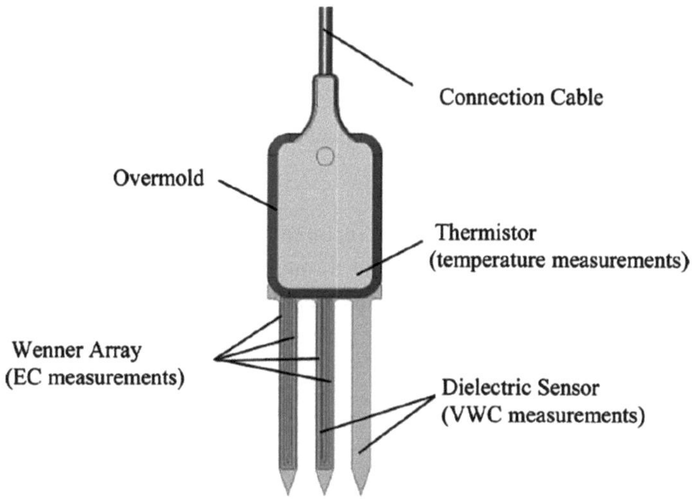

Fig. 7.8 A three-prong capacitance probe (adapted from Decagon Devices Inc.)

4.3.4 Electromagnetic (EM) Probes

Electrical conductivity (EC) and dielectric permittivity (DP) are two most important EM parameters that can be obtained using EM probes. Since most conduction takes place in the pore space of the paste backfill during hydration, monitoring the bulk EC changes can provide useful information regarding the in situ hydration process. EC is sensitive to temperature, increasing with increasing temperature. Therefore, EC data must be correlated with changes in temperature to better understand this process.

On the other hand, DP data can be used to measure the volumetric water content (VWC) of soils including uncemented tailings. However, the interpretation of VWC obtained from DP data might not be straightforward (if not impossible) for paste backfill applications as DP are influenced by changes in EC, porosity and pore space tortuosity which occur during hydration (Simon and Grabinsky 2012). Therefore, it would be useful to measure EC, DP and temperature at the same location within the media. Figure 7.8 shows a schematic of a commercially available three-prong capacitance probe for measuring the EM parameters. Commercially available EM probes may use the time domain reflectometry (TDR) or the capacitance technique to measure the DP of materials.

In addition to composition and structural properties, the frequency of the applied EM field influences the material's EM parameters (Klein and Simon 2006). Therefore, it is important to verify the frequency of the probe for the intended application because low frequencies (<10 MHz) are highly susceptible to changes in salinity and temperature (Decagon Devices Inc. 2009). It was shown that commercially available single-frequency (i.e. 70 MHz) capacitance probes are a useful tool to measure the VWC of soils but not for hydrating materials such as paste backfill

(Simon and Grabinsky 2012). Although monitoring EM parameters are of interest, the application of EM probes in paste backfill designs has been considered as a research tool only (Grabinsky 2010).

4.3.5 Thermistors

Temperature affects several parameters including hydration process, EC and shear strength gain. Temperature measurement throughout the fill cycle is an important parameter to monitor. With the temperature known the data may be more readily interpreted.

4.3.6 Other Transducers

In addition to above-mentioned transducers, other transducers such as accelerometers and dynamic pore water pressure transducers can be used to monitor the properties of the paste backfill system under dynamic loading due to adjacent mining activities. The displacement transducers may also be used on the free face of the barricades to understand its performance. The displacement data can be used in an advanced structural analysis and to optimise the design of future such barricades.

5 Summary

In modern mines, instrumentation is widely used, from controlling plant operational performance to the monitoring of paste backfill pipeline pressure and its in situ as-deposited characteristics. The paste production process is tightly governed to produce a material that meets the target design specifications. It must be flowable or pumpable to reach the designated stope location and develop the necessary strength as a structural fill, when required. Data gathered with commercially available instruments placed prior to filling provides an in situ look at the behaviour of paste backfill while it is curing. This information can be interpreted and fed back into the mine design model to further improve operational efficiencies. Key parameters such as fill sequences, stoping sequences and binder usage in paste backfill may be optimised with the data collected.

References

Al-Tarhouni M, Simms P, Sivathayalan S (2009) Seismic behaviour of thickened gold mine tailings. In: Jewell RJ, Fourie AB, Barrera, Wiertz J (eds) Proceedings twelfth international seminar on paste and thickened tailings. Australian Centre for Geomechanics, Perth, Australia, pp 291–300

Archibald J, De Souza E, Beauchamp L (2009) Compilation of industry practices for control of hazards associated with backfill in underground mines—part II underground transport and stope placement. In: Diederichs M, Grasselli G (eds) ROCKENG09: proceedings of the 3rd CANUS Rock Mechanics Symposium, Toronto, Canada

Been K, Li AL (2009). Soil liquefaction and paste tailings, Paste 2009. The 12th international seminar on paste and thickened tailings, 21–24 April 2009, Viña del Mar, Chile

Crowder JJ 2004. Deposition, consolidation, and strength of a non-plastic tailings paste for surface disposal. Ph.D. thesis, University of Toronto, Toronto, Ontario, Canada

Daigle L, Zhao JQ (2004) The influence of temperature on earth pressure cell readings. Can Geotech J 41(3):551–559. doi:10.1139/t04-004

Decagon Devices Inc. (2009) ECH2O dielectric probes vs. time domain reflectometers (TDR), Application note: 13397-02

Grabinsky MW (2010) In situ monitoring for ground truthing paste backfill designs. In: Proceedings of the 13th international seminar on paste and thickened tailings, 3–6 May 2010. Australian Centre for Geomechanics, Toronto, ON, pp 85–98

Grabinsky MW, Simms P (2006) Self-desiccation of cemented paste backfill and implications for mine design. In: Jewell R, Lawson S, Newman P (eds) Proceedings ninth international seminar on paste and thickened tailings. Australian Centre for Geomechanics, Perth, Australia, pp 323–332

Grabinsky MW, Thompson BD (2009) Thermally induced stresses in cemented paste backfill. Geotech News 27(3):36–40

Helinski M, Fourie A, Fahey M, Ismail M (2007) Assessment of the self-desiccation process in cemented mine backfills. Can Geotech J 44(10):1148–1156. doi:10.1139/T07-051

Klein K, Simon D (2006) Electromagnetic properties of cemented paste backfill. J Environ Eng Geophys 11(1):27–41

Li AL, Been K, Ritchie D, Welch D (2009) Stability of large thickened, non-segregated tailings slopes. In: Jewell RJ, Fourie AB, Barrera, Wiertz J (eds) Proceedings twelfth international seminar on paste and thickened tailings. Australian Centre for Geomechanics, Perth, Australia, pp 301–312

Simms P, Grabinsky M (2009) Direct measurement of matric suction in triaxial tests on early-age cemented paste backfill. Can Geotech J 46(1):93–101. doi:10.1139/T08-098

Simms P, Dunmola A, Fisseha B, Bryan R (2010) Generic Modelling of desiccation for cyclic deposition of thickened tailings to maximize density to minimize oxidation. In: Proceedings of the 13th international seminar on paste and thickened tailings, 3–6 May 2010. Australian Centre for Geomechanics, Toronto, Ont, pp 293–301

Simon D, Grabinsky MW (2012) Electromagnetic wave-based measurement techniques to study the role of Portland cement hydration in cemented paste backfill material. Int J Min Reclam Environ 26(1):3–28

Thompson BD, Grabinsky MW, Counter DB, Bawden WF (2009) In situ measurements of cemented paste backfill in longhole stopes. In: Diederichs M, Grasselli G (eds) ROCKENG09: proceedings of the 3rd CANUS rock mechanics symposium, Toronto, May 2009

Chapter 8
Cemented Paste Backfill Pressure Monitoring and Field Testing

Kemal Karaoglu and Erol Yilmaz

1 Introduction

Mine backfill, classified primarily as cemented rock fill (CRF), cemented hydraulic fill (CHF), and cemented paste backfill (CPB), is lengthily employed in most modern underground mine operations, reducing volumes of surface-disposed mine wastes and costs relating to mine backfilling operations (Hassani et al. 1998; Yilmaz 2011). This environmental benefit as well as cost-effectiveness have spurred the acceptance of CPB as an economical and efficient alternative to both CRFs and CHFs. CPB also offers a number of advantages over other backfill materials in operational and financial terms, including pipeline delivery that aids fast filling to open or blind stopes (Grice 2013; Yilmaz 2015). A stope is defined as the site of ore production in an orebody. After extraction of ore, the stope is ready for backfilling. A steel-reinforced shotcrete barricade is constructed across the undercut to contain the initially fluid paste materials (Karaoglu et al. 2013). CPB is then poured into the stope through a borehole or pipeline system from the surface paste plant to the overcut access. Generally, a plug fill of few meters high (up to 7 m) is poured into the stope and then the residual fill with 2–5% cement is placed. The plug fill holds relatively high cement (5–7%) and is usually left 2–7 days for curing prior to the residual fill to avoid excess pressure on the barricade (Ouellet et al. 1998; Belem et al. 2004; Ouellet and Hassani 2005; Yilmaz et al. 2014).

The internal pressure induced by semifluid CPB loading is initially hydrostatic. The backfill then gains shear strength due to effects of cement hydration and consolidation (le Roux et al. 2005; Yilmaz et al. 2009; Huang et al. 2011). Therefore, the resultant horizontal stresses are pointedly reduced and hence barricade pressures become lower than that of the hydrostatic condition. There is also a phenomenon

K. Karaoglu (✉) • E. Yilmaz
Cayeli Bakir Isletmeleri A.S., PO Box 42, Madenli Beldesi, Cayeli, Rize 53200, Turkey
e-mail: karaogluk@fqml.com

E. Yilmaz, M. Fall (eds.), *Paste Tailings Management*,
DOI 10.1007/978-3-319-39682-8_8

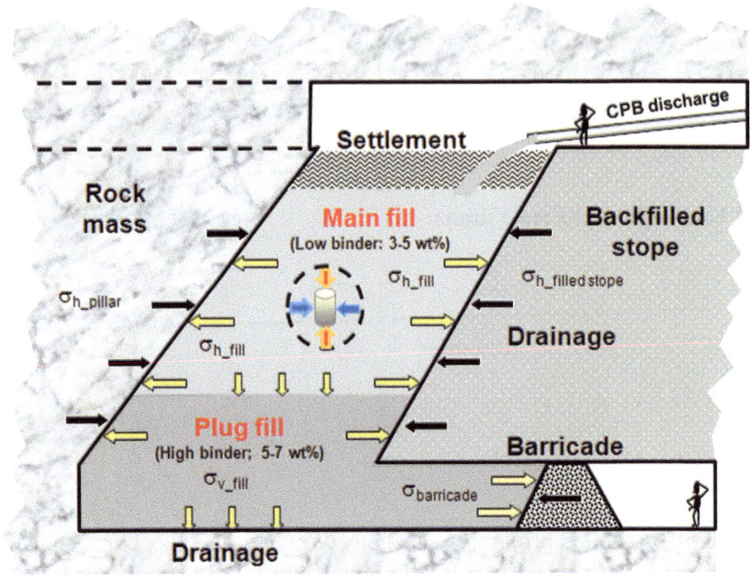

Fig. 8.1 Schematic view of the events occurred in an underground paste-filled stope

known as "arching effect" in a CPB stope (Potvin et al. 2005; Pirapakaran and Sivakugan 2007; Fahey et al. 2009; Thompson et al. 2011). Once CPB is placed in the stope, it tends to settle under its own weight, and the stiffer rock walls tend to embrace CPB in place. This causes the generation of shear stresses along the CPB-wall contact interfaces and reduction of the vertical and horizontal stresses in CPB (Li et al. 2005; Nasir and Fall 2008). The safer and more efficient CPB strategies (it may lead to the reduction of stope cycle times) are designed in mines if pressures in CPB and pressures at barricades are known measurably by operators. Eventually, CPB pressures can fairly change as a result of cement type and rate, material characteristics, stope geometry, and filling rate (Fig. 8.1).

To assess the mechanisms linked with the behavior of early-age CPB during and after filling, in-stope measurements and monitoring of effective stress, pore water pressure, temperature, suction, and electrical conductivity are required to improve the design of CPB-filled stopes. Recently, investigators have published field works at a number of different CPB operations worldwide (Landriault et al. 2000; Cayouette 2003; Yumlu and Guresci 2007; Grabinsky et al. 2007; Grabinsky and Thompson 2009; El Aatar 2010; Fall and Nasir 2010; Hughes et al. 2010; Thompson et al. 2010b, 2012; Fahey et al. 2011; Helinski et al. 2011; Wu et al. 2014; Doherty et al. 2015). Results show that geo-mechanical designs of in situ CPB are really conservative in terms of CPB recipes, barricade design, and filling and curing strategies. Generally, field observations have shown lower void ratios, higher solids concentration, and higher strength near the bottom of the stope, as well as higher void ratios, lower solids concentration, and lower strength near the top of the CPB-filled stope. This is principally caused by self-weight or time-dependent consolidation, interaction between CPB and rock, and effective curing conditions (curing under pressure and removal of excess water within CPB through cracks at the bottom, or water absorption by soft rocks or materials placed inside walls, curing temperature).

The objective of this chapter was to assess the evolution of internal stresses in CPB during and after stope filling. As part of the CPB instrumentation and pressure monitoring program at the Cayeli mine, two stopes were instrumented using total pressure cells, piezometers, and temperature and conductivity sensors. The pressures induced in the paste backfill (in lateral and vertical directions) and on the barricade (in lateral direction) were recorded during and after placement of paste backfill. It is believed that this work will contribute to knowledge on the understanding of internal pressures developed within the backfill over time.

2 The Cayeli Copper-Zinc Mine

The Cayeli mine is an underground copper and zinc operation located on the Black Sea coast of north-eastern Turkey and operated by Cayeli Bakir Isletmeleri A.S., a wholly owned subsidiary of First Quantum Minerals Ltd. About 1.2 million tons of ore are extracted each year to produce 220,000 tons of concentrate. Cayeli is the largest underground base metal mine in Turkey. It produced its first concentrate in 1994 and is expected to operate until at least 2019. Cayeli is an example of mining and milling experience translating to achieving throughput that exceeds design capacity, as its initial design capacity of 600,000 tons per year has grown to 1,200,000 tons per year.

2.1 *Geology and Ore Types*

The orebody is a steeply dipping volcanogenic massive sulfide deposit. The ore body's two parts, the upper main zone and lower deep zone, are separated by a major discontinuity which is acknowledged as a fault of synvolcanic growth (Fig. 8.2). This feature strikes subparallel to the ore body and dips to the east. The two zones become closer to each other gradually at the southern of the deposit. Mainly there are six ore types. Stockwork Ore which underlies the main massive sulfide lens is the thickest ore layer. It is in close proximity to synvolcanic fault which offsets the main and deep ore lenses. Above the Stockwork zone occurs yellow ore (Cu rich).

In order to identify black and yellow ore, structure properties and zinc content features can be used. Zinc content of the black ore is higher than 4.5% and the amount of copper is less than zinc. Another type of the ores is clastic ore which can be generally found in small zones and at the top of the ore deposit. In clastic ore structure there are small sphalerite pieces sizes of which range from 1 to 40 cm and it has a feature of quite complex mineralogy.

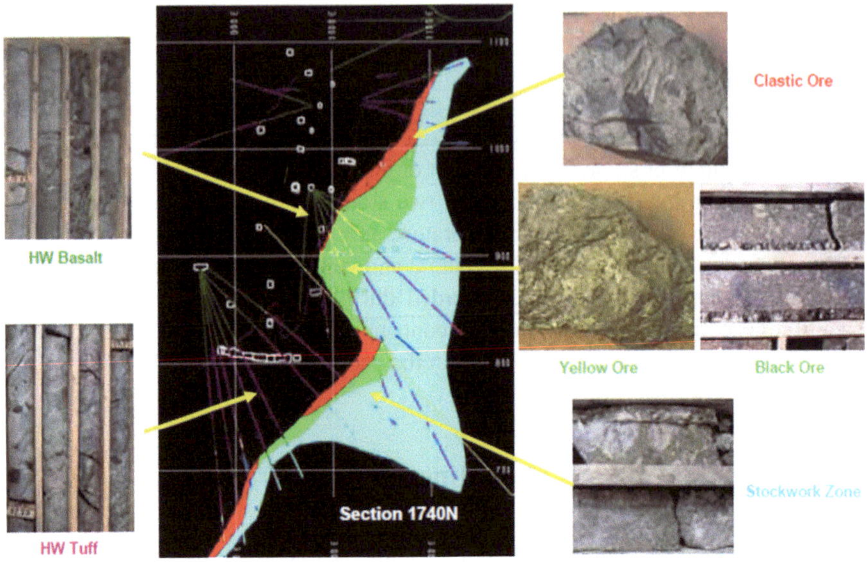

Fig. 8.2 Cross section of the Cayeli mine geological structures, including main ore types

The small massive sphalerite pieces contain a large amount of chalcopyrite and these pieces can negatively affect plant performance. The deposit consists of 52% yellow ore (copper rich), 28% black ore (zinc rich), and 20% clastic ore. Black ore occurs above or lateral to the yellow ore. Clastic ore characterized by sphalerite fragments occurs at the top or around the edges of the orebody. Stock ore is characterized by veins of pyrite and chalcopyrite and occurs in the footwall rhyolite. Yellow ore consists of pyrite and chalcopyrite clasts, up to 20 cm in size, in a sulfide matrix with less than 10% sphalerite. Black ore consists of pyrite and chalcopyrite clasts from 2 mm to 20 cm in diameter in a grained matrix of pyrite and sphalerite containing more than 10% sphalerite. Clastic ore consists of pyrite, chalcopyrite, and sphalerite clasts, from 2 mm to more than 20 cm in diameter in a sulfide matrix. The diagnostic of this ore type is the sphalerite clasts. This ore type is the most difficult ore type to be processed metallurgically owing to the fine inclusions of chalcopyrite in the sphalerite—the so-called chalcopyrite disease.

2.2 *Mining Method Description*

Cayeli's mine design is based on underground bulk mining methods with the use of delayed backfill to extract ore in a sequential manner. The primary mining method selected for mining the orebody (Fig. 8.3a) is transverse and longitudinal long-hole,

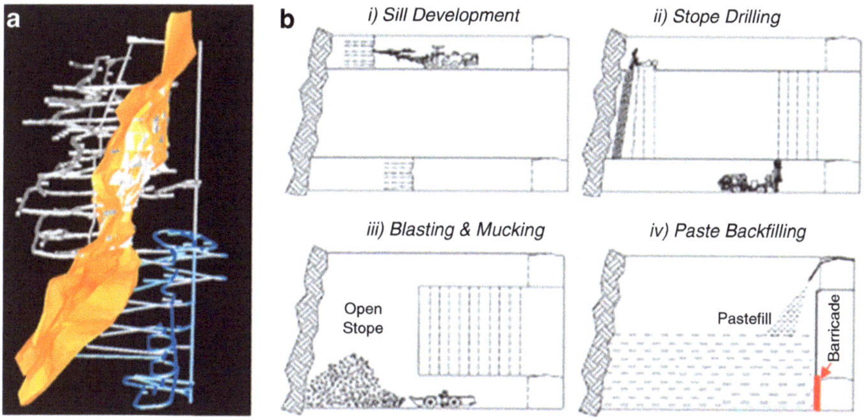

Fig. 8.3 Orebody (**a**) and mining method (**b**) at Cayeli copper-zinc mine

open, and blind stoping with cemented paste backfill and loose or consolidated waste rock backfill. The stopes are mined in primary, secondary, and tertiary sequencing. Both primary and secondary stopes are mined transversely but tertiary stopes are mined longitudinally. The Cayeli mining method includes (1) driving sills for stope development, (2) slotting and ring blasting, (3) remote mucking, and (4) barricade construction and filling (Fig. 8.3b).

In comparison with many bulk mining operations in the world, weak ground conditions at Cayeli mine dictate relatively small stopes. The stopes are generally 7–10 m wide with stope lengths varying between 15 and 30 m. The stopes are mined in single lifts between 15 and 20 m high sublevels. Each year, more than 100 stopes are paste-backfilled at the mine (1450 tons of paste backfill are placed per day). The backfill is an integral part of the mining method at Cayeli. There are currently three types of backfilling in use at Cayeli mine site. These are waste fill using development waste rock, cemented waste fill, and cemented paste backfill. On average Cayeli places about 250,000 m^3 of paste backfill annually. The stope filling method consists of pouring first the plug (with a binder proportion varying from 6.5 to 8.5 wt%) up to 7 m behind the draw point followed by a curing time of 2 days. After that the rest of the stope, also known as the residual fill, is backfilled with a binder proportion varying from 4.5 to 6.5 wt%.

3 Instrumentation and Installation of Underground Stopes

To monitor the evolution of stresses in CPB during and after stope filling and to determine the evolution of barricade loads, a fieldwork program including two stopes has been carried out at Cayeli (Thompson et al. 2010a; Karaoglu et al. 2013). These stopes were instrumented using total pressure cells, piezometers, and

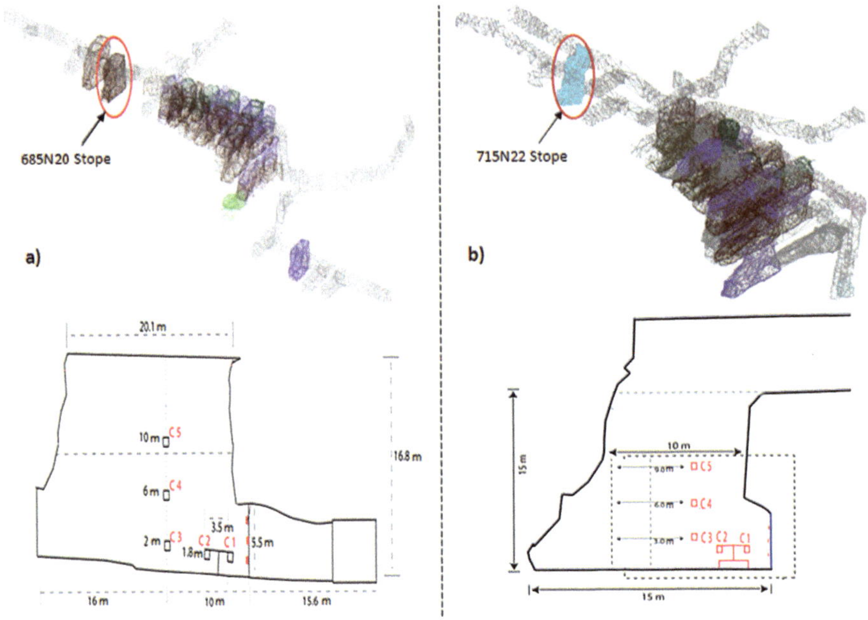

Fig. 8.4 Positions of paste stopes tested and in-stope instruments: (**a**) 685N20 and (**b**) 715N22

temperature and conductivity sensors. The instruments used during the field tests and their installations and operations are provided in this section.

3.1 *Stope Geometry*

The two monitored stopes were 685N20 and 715N22, as shown in Fig. 8.4. These stopes were located at different underground level and filled roughly 15 days. The 685N20 stope had a height of ~17 m while the 715N22 stope had a height of 15 m. The 685N20 stope dimension is 25 m × 10 m in the undercut, and 23 m × 11 m in the overcut. This is a stage 1 stope, being the first stope to be mined from a panel containing two stopes. For this reason, the access drift to the stope is moderately long and the backfill barricade hinders from a stope by 3–4 m. The 715N22 stope dimension is 20 m × 8 m × 10 m in the undercut, and 15 m × 8 m × 5.5 m in the overcut. At the time of paste backfilling, no other stopes had been mined in the local area and the 715 level was the deepest active mining level in the mine.

3.2 *In-Stope Instruments*

Instrumentation aiming at investigating total and pore pressure, temperature, and electrical conductivity was installed in two test stopes, namely 685N20 and 715N22.

Each cage (total five in a single stope) featured three total pressure cells (RST Instruments, capable of reading up to 500 kPa with a resolution of ±0.1% full-scale

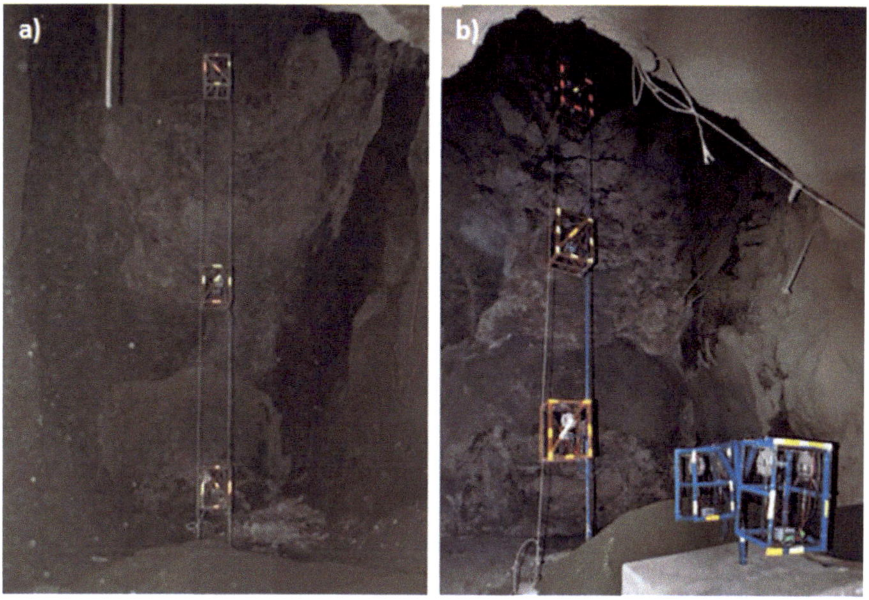

Fig. 8.5 Photos of the cage instruments being placed in the stopes: (**a**) 685N20 and (**b**) 715N22

range), one piezometer (RST instruments, capable of reading up to 700 kPa with a resolution of ±0.025% full-scale range), one electrical conductivity sensor (Hoskins Scientific, Decagon ECH20 TE probe), and one negative pore water pressure transducer (Campbell Scientific, heat dissipation matric potential sensor 229-L). Figure 8.5 shows photos of the instrumented cages in the stopes.

Two trial stopes were selected and instrumented each with three pressure cells to monitor the evolution of the pressures developed within paste backfill over time. A special care was taken to adjust the total pressure cells either vertically or horizontally. Two instrumented cages, designed in the *T* structure, were inserted very near the backfill barricade while the others (two cages in the plug fill and one cage in the residual fill) were inserted in CPB material. The obtained total pressures were corrected using a 0.5 kPa/°C temperature correction factor. The pore water pressures were measured using vibrating wire piezometers, 19 mm in diameter and 136 mm in length. A thermistor is built into both the piezometer and pressure cell bodies to permit temperature measurement. In order to observe any changes in cage orientation during paste backfilling, compasses and tilt meters were also attached into the cages.

3.3 *Barricade Instruments*

In addition to the CPB strengths, another critical factor in designing an efficient mine fill system for underground support is nonstop backfilling process where the barricade constructed should meet pressures developed in paste backfill, without

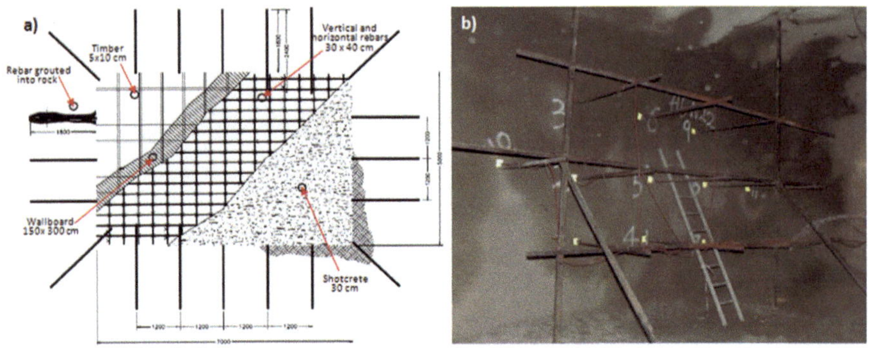

Fig. 8.6 (**a**) Schematic view of the CPB barricade design and (**b**) instrumented fill barricade

failure. The unwanted events or consequences, such as fatality, property damage, production loss, or stopping, can take place if the barricade fails due to overfilling, fast filling rate, poor plug fill curing age, insufficient fill monitoring, and barricade design. Thus, mine operators want to make sure about their barricade design by monitoring lateral loads on the barricades. In this regard, three pressure cells and piezometers were attached to the central axis of the barricade, at heights of 1.4, 2.8, and 4.2 m. The barricade constructed in the mine was flat (without curvature), having 5.6 m high and 8.5 m wide.

For both stopes tested, the backfill barricades were erected in a standard way as follows: (1) rebar grouted into rock, (2) wooden frame installed to contain and provide a backing for shotcrete, (3) vertical and horizontal rebar extended from the rebar grouted in rock, and (4 shotcreteing of the constructed barricade or fence. Figure 8.6 shows the design plan of cemented paste backfill barricade construction and attached instruments for monitoring the internal pressures developed during and after backfilling.

4 Paste Backfill Characteristics

Cemented paste backfill (CPB) was prepared by using the following three main ingredients: full tailings, binding, and mixing water. The materials used for CPB preparation were supplied internally, except for the cement. Two types of tailings, namely spec ore tailings (SOT) and clastic ore tailings (COT) manufactured in ore processing plant, were used to make CPB mixes while the cement was Portland composite cement. In this subsection, all the ingredients were subjected to evaluate in terms of grain size, specific gravity, and chemical composition.

4.1 Characterization of Processing Tailings

Both SOT and COT materials were thoroughly sized by using a Malvern Mastersizer particle size analyzer and the gained results are shown in Fig. 8.7. The grain size distribution of tailings was closely similar due to identical ore geology.

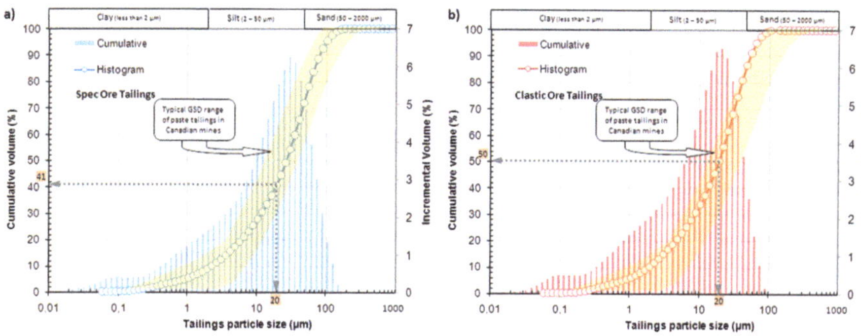

Fig. 8.7 Grain size distribution curves of (**a**) spec ore tailings and (**b**) clastic ore tailings

The SOT were found to have nearly 41 wt% finer than 20 μm and the COT were found to have nearly 50 wt% finer than 20 μm, which indicates that both tailings can be classified as a medium tailings material according to Landriault (2006). The specific gravity of tailings was also measured using a pycnometer. The results indicated that spec and clastic ore tailings had a specific gravity of 4.53 and 4.38, respectively. From the chemical analyses, one can speak that the SOT was dominated by iron oxide (65.3%), detecting minor quantities of silicon dioxide (4.5%) and aluminum oxide (2%) as well as trace amounts of magnesium, calcium, potassium, sodium, chromium, manganese, and phosphorous oxides (all less than 2%). Also, COT had similar phases dominated by iron oxide (48%) and minor amounts of silicon dioxide (13%) and aluminum oxide (4.2%), in common with trace amounts of similar minerals (all less than 2%).

4.2 Portland Composite Cement and Mixing Water

The Askale Portland composite cement PCC [CEM II/A-M (P-LL) 42.5 R] was used in the tests. The main properties of PCC are listed in Table 8.1. The fine filling material it contains tightens the micropores in CPB and increases the imperviousness. In its own class, PCC is in the group with high early and final strength. PCC increases quantity in building chemical productions due to its low density and contributes in processability in ready-mixed concrete productions. PCC also ensures long-term strength after adding trass to the mix.

As mixing water, a combination of process water and/or tap water was used for CPB mixes. The process water, listed in Table 8.2, was relatively aggressive, with sulfate (SO_4^{2-}) content of 2512 ppm. The relatively high concentration of calcium Ca (334 ppm) is due to the lime added during ore processing.

The pH, Eh (redox potential), and EC (electrical conductivity) parameters of the mixing waters were analyzed using a Benchtop pH/ISE Meter, Orion Model 920A with a Thermo Orion Triode combination electrode, and were determined at 7.83, 0.16 V, and 8.53 mS/cm for process water and 6.44, 0.39 V, and 0.25 mS/cm for tap water, respectively.

Table 8.1. The chemical, physical, and mechanical requirements of Portland composite cement

Chemical requirements	Value (%)	Physical requirements	Value
Total SiO_2	20.76	Residue on 45 μm sieve	1.2%
Insoluble residue	6.05	Residue on 90 μm sieve	–
Al_2O_3	4.96	Blaine's specific surface	4748 cm^2/g
Fe_2O_3	4.13	Specific gravity	2.99
CaO	56.73	Setting time—initial (min 60)	155 min
MgO	1.72	Setting time—final	220 min
SO_3 (max 4%)	2.96	Water requirement	34.8%
Loss on ignition	5.59	Le Chatelier's soundness	1 mm
Na_2O	0.38	Mechanical requirements	Strength (MPa)
K_2O	0.77		
Cl^- (max 0.1%)	0.0218	1-day cured strength	15.3
Total	98.01	2-day cured strength (min 20)	27.7
Free lime	1.51	7-day cured strength	47.8
Total additives (12–20%)	14.98	28-day cured strength (min 42.5)	48.2

Table 8.2. Chemical and geochemical analysis of pore water and tap water

Geo-chemical parameter	Process water	Tap water
Magnesium Mg (ppm)	2.13	2.07
Calcium Ca (wt%)	334	52.4
Sulfate content SO_4^{2-} (ppm)	2512	142
Electrical conductivity EC (mS cm^{-1})	8.53	0.25
Redox potential Eh (V)	0.16	0.39
pH	7.83	6.44

4.3 Stope Filling Design and Pouring Strategy

Using a computerized control system, the CPB mixes were produced in a surface paste plant first and then transported to the stopes via reticulation pipes by a combination of pumping and gravity systems. The stopes are typically backfilled in two stages. The bottom 8 m of the all stopes in the first sublevel of every stoping block (typical stope height is 15 m) is filled with 8.5 wt% cement (Stage 1) while the remaining part is filled with 6.5 wt% cement (Stage 2) in order to create a cemented sill pillar to enable recovery of ore from the block below.

In all other sublevels, the first 10 m of every stope, representing the initial plug fill behind the barricade, is filled with 6.5% cement. The remaining top parts of primary, secondary, and tertiary stopes are filled with 6.5%, 4.5%, and 4.5% cement content, respectively. Depending on the used cement and stope geometry, the average rise rates are maintained below 35 cm/h in the undercut, and below 42 cm/h in the main stope volume. The slump consistency varies between 175 and 190 mm and the target design strength for vertical exposure is 1.0 MPa after curing time of 28 days using a safety factor of 2.

Note that the first part of the 685N20 stope was filled with SOT containing 8.5 wt% cement content while the remaining part of the stope (second part) was filled with 6.5 wt% cement content. The corresponding CPB rise rate was 22 cm/h. The volume of the 685N20 stope was 3260 m^3 while its filling time lasted 68 h. Both first and second parts of the 715N22 stope were filled with COT containing 6.5 wt% cement. The corresponding CPB rise rate was 37 cm/h. The volume of the 715N22 stope was 1215 m^3 while its filling time lasted 41 h. The 685N20 stope was poured continuously while the 715N22 stope was poured sequentially after a curing waiting time of 3 days.

5 Field Testing Results and Discussion

5.1 In-Stope and Barricade Pressure Monitoring

Figures 8.8 and 8.9 show the evolution of total pressure, pore pressure, and temperature recorded on both barricade and cages 3 and 5 up to 7 days of paste backfilling in the 685N20 and 715N22 stopes. On the barricade, total pressures rise consistently with increasing paste levels in the stope, reach a peak value after the completion of paste backfill pouring, and then tend to reduce once the hydration of cement commences up. The reduction in barricade pressures takes place after a curing time of 3 days for the 685N20 stope and of 4.5 days for the 715N22 stope, respectively. Pore water pressures on barricade constructed for the 685N20 stope are relatively lower (~50 kPa) than those obtained from the 715N22 stope. The latter provides a relatively high pore pressures (up to 100 kPa) mainly because of the presence of water-retention clayey minerals within clastic ore tailings. The former stope (685N20) is filled with spec ore tailings which contain less clayey minerals, such as quartz.

Cages 3 and 5 demonstrate similar pressure trends for both stopes. Vertical pressures were always higher than horizontal pressure in short and long axes. Cages 3 and 5 in the 685N20 stope provide a highest vertical stress of ~200 and ~100 kPa while the peak horizontal pressures remain around 100 kPa. The corresponding vertical and horizontal pressures for the 715N22 stope are, respectively, ~250 and 100 kPa for Cage 3 and ~120 and ~80 kPa for Cage 5. One can say that pressures increase during the second pour. This is because the second pour acts as a layer and gives extra hydrostatic loading on the first layer of paste filling.

Figures 8.8 and 8.9 also provide the evolution of temperatures induced by the cement hydration for the stope filling of different tailings types, namely spec and clastic tailings. For the 685N20 stope, the temperature of the barricade varies between 40 and 43 °C while remaining at a constant temperature of 50 °C for both Cage 3 and Cage 5. The corresponding temperature values for the 715N22 stope are 40–46 °C, 51 °C, and 39 °C for barricade, Cage 3, and Cage 5, respectively.

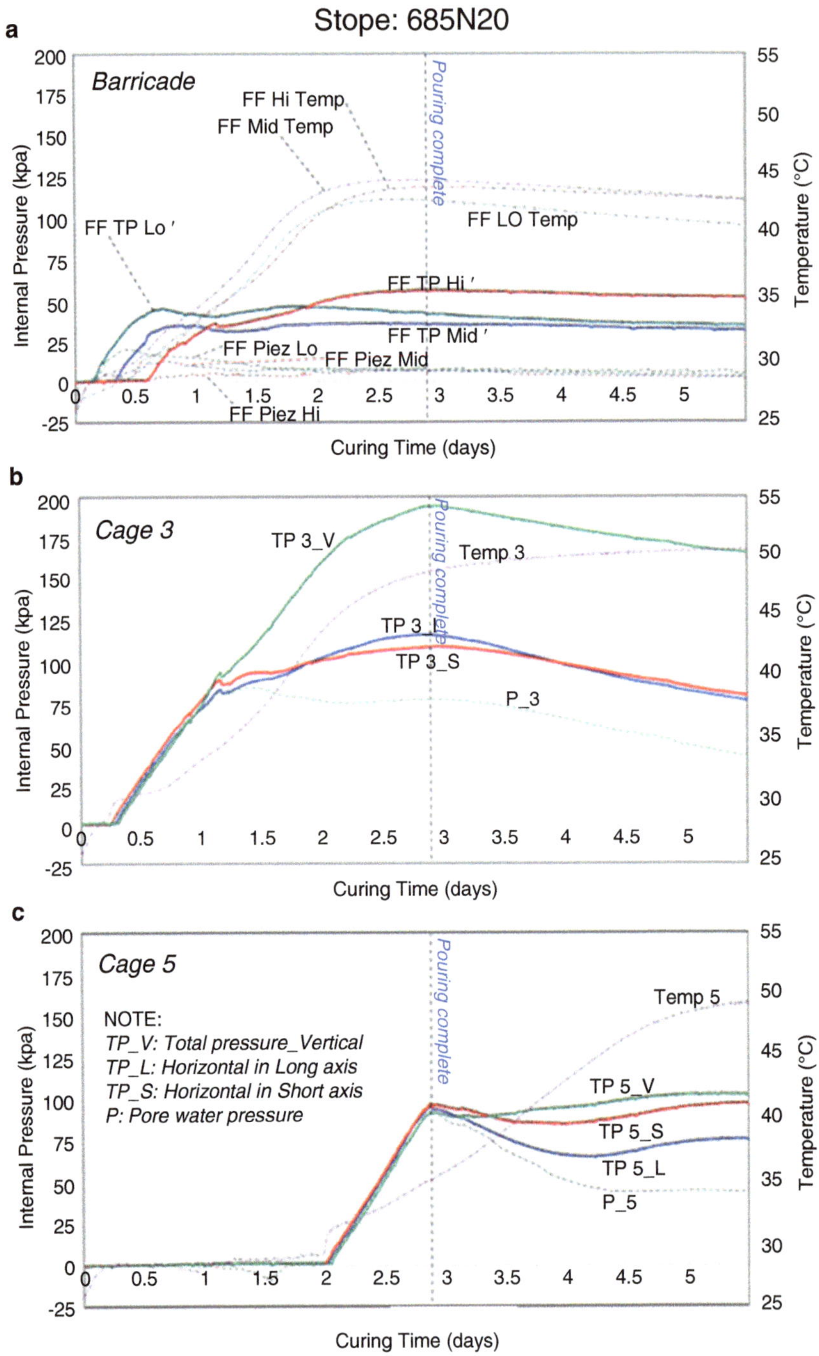

Fig. 8.8 Variation in total pressure, pore pressure, and temperature with time for the 685N20 stope: (**a**) Barricade, (**b**) Cage 1, and (**c**) Cage 3

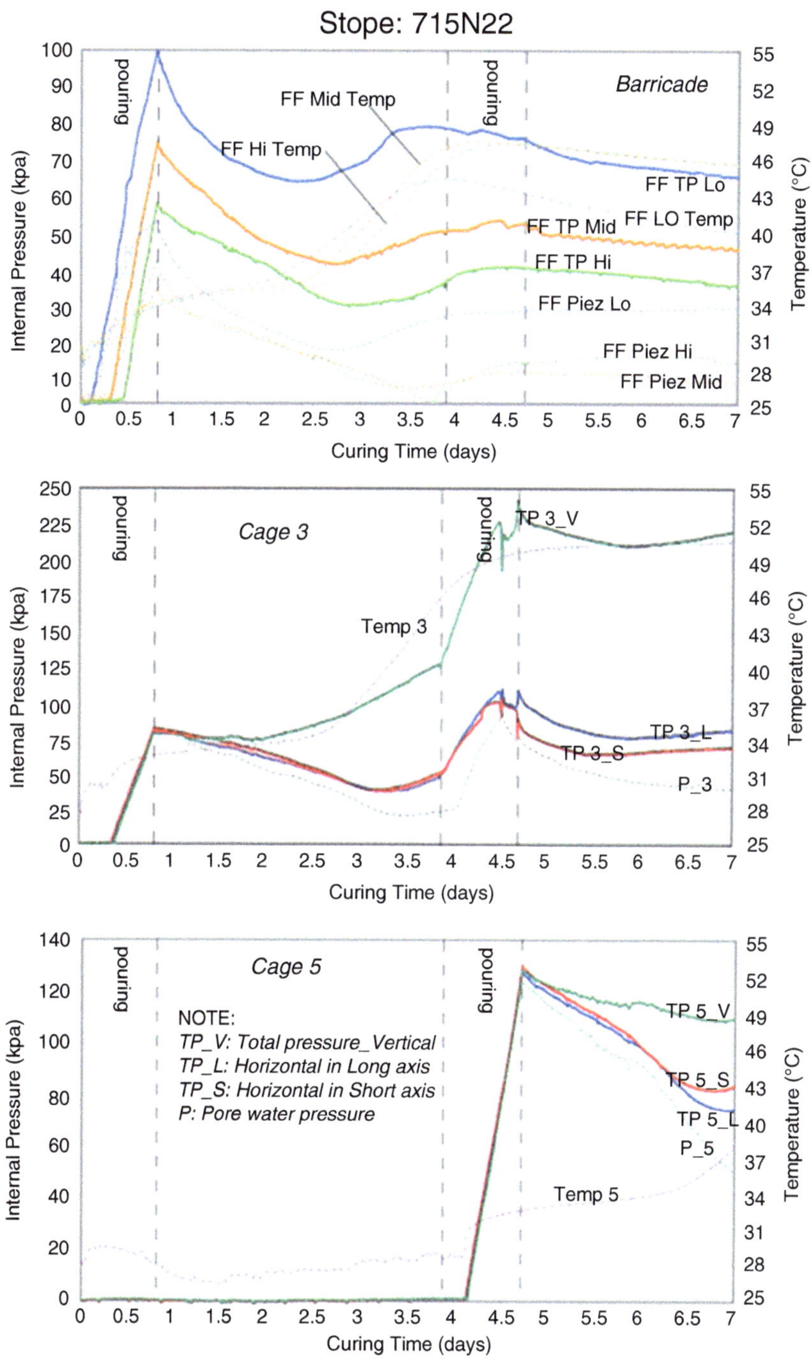

Fig. 8.9 Variation in total pressure, pore pressure, and temperature with time for the 715N22 stope: (**a**) Barricade, (**b**) Cage 1, and (**c**) Cage 3

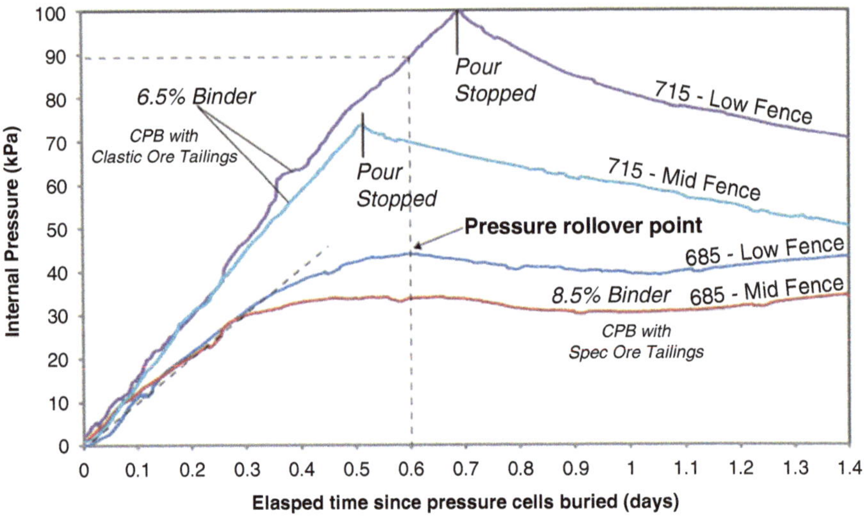

Fig. 8.10 Variation in pressure with time for CPB-filled stopes having spec and clastic tailings

5.2 *Effect of Tailings Type on Pressure Development and Temperature*

Figure 8.10 illustrates the variation in internal pressure with curing time for paste-filled stopes containing spec and clastic ore tailings at different cement rates. As expected, the pressures developed in the stope are dependent on paste fill rise rate, stope geometry, CPB type, and backfill cement contents.

The 715N22 stope barricade pressures rise constantly up to 0.7 days while the 685N20 stope barricade pressures rise constantly up to 0.5 days. Note that the 685N20 stope was filled with spec ore tailings at a binder content of 8.5 wt% while the 715N22 stope was filled with clastic ore tailings at a binder content of 6.5 wt%. The latter stope was filled continuously since its volume is relatively small (1200 m^3), allowing high backfilling rates. Clastic CPBs with 6.5 wt% binder generate higher internal pressures (70–100 kPa) than spec CPBs with 8.5 wt% binder.

Figure 8.11 shows the variation in temperature with time for paste-filled stopes containing spec and clastic ore tailings at different cement rates. Due to the fact that cement hydration generates heat (higher temperatures), a direct comparison between hydration characteristics for CPBs with spec and clastic tailings can be made in terms of the binder content used. The 685N20 temperatures increase rapidly with the rate of temperature increase and become a plateau after 3 days. The higher the binder content (8.5 wt%), the higher the acceleration of cement hydration becomes for a given tailings type. For example, the temperature recorded in Cage 5 of the 685N20 stope filled with 6.5 wt% binder becomes 40 °C while the corresponding temperature data become, respectively, 45 °C and 47 °C for the Cages 3 and 1 of the 685N20 stope filled with 8.5 wt% binder. The 715N22 stope temperature data shows that, up to 2.5 days, temperature does not change significantly and then tends to

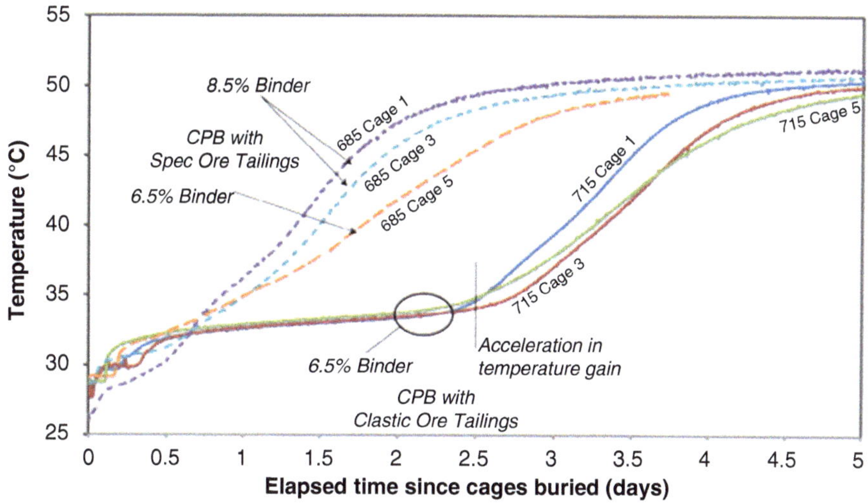

Fig. 8.11 Variation in temperature with time for CPB-filled stopes having spec and clastic tailings

increase to a peak of 50 °C. Results clearly indicate that the temperature of spec CPBs accelerates 2 days earlier than that of clastic CPBs, leading to faster cement hydration and strength gains.

5.3 *Effect of Stope Depth on Mechanical Strength Development*

Various CPB pieces were collected from different stope depths and then brought to the laboratory for strength testing. Figure 8.12 shows the evolution of compressive strength with curing time for paste-filled stopes with different tailings (spec ore tailings SOT and clastic ore tailings COT) and binder contents (6.5 and 8.5 wt%), obtained at different depths from the top. As expected, the strength increases with increasing stope depth after curing time of 28 days, following similar trend as overburden pressure ($\sigma_v = \gamma_{wet} h$). Note that an average unit weight of the backfill was taken as 23.5 kN/m^3. For 6.5 wt% binder, the compressive strengths of 962 kPa, 1050 kPa, 1098 kPa, 1185 kPa, 1264 kPa, and 1342 kPa were obtained on spec CPB samples at stope depths of 0 m, 4 m, 8 m, 12 m, 16 m, and 20 m, respectively. The corresponding compressive strengths for clastic CPB samples were, respectively, 820 kPa, 870 kPa, 920 kPa, 976 kPa, 1065 kPa, and 1142 kPa. This confirms that the strength gain of the studied CPB-filled stopes increases from top to bottom due to the self-weight consolidation effect and enhanced hydration and curing conditions. For a binder content of 8.5 wt%, this increase was, respectively, calculated as 33%, 30%, 30%, 27%, 25%, and 23% for spec CPB samples, and 34%, 33%, 31%, 31%, 30%, and 28% for clastic CPB samples. One can say that spec CPBs offer higher strengths than clastic ones due to their mineralogical composition which allows less water in their structure.

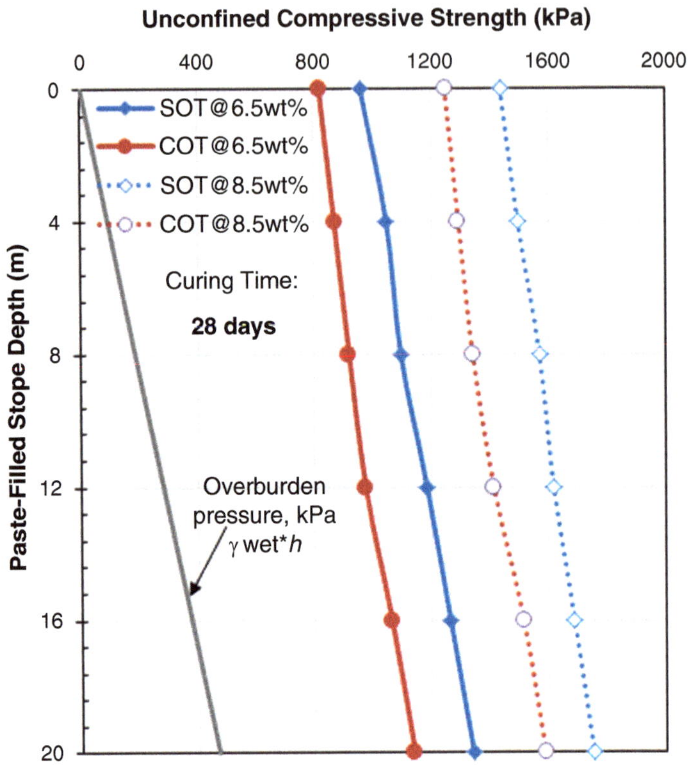

Fig. 8.12 Variation in compressive strength with stope depths for different CPB samples

One can also conclude that the density increases with increasing stope depth, decreasing the corresponding void ratio or porosity as a result of consolidation pressure during curing. This is because the lower the void ratio *e*, the closer the CPB particle packing, and hence, the stronger the material (high cohesion). Eventually, the compressive strength of laboratory-prepared CPB samples without effective stress during curing is inappropriately unsuitable for use in backfilling design, as it is representative of only the top of backfilled stopes where no pressure is applied to the samples. Instead, taking the cored samples from underground stopes, as practiced in this study, is beneficial for an efficient CPB design. However, it should be kept in mind that this approach may not always be practical or may be costly, risky, and time consuming for mix optimization.

5.4 Effect of Stope Depth on Geotechnical Index Parameters

Figure 8.13 shows the evolution of geotechnical index parameters (gravimetric water content, degree of saturation, void ratio, and specific surface area) with stope depth for different CPB samples after curing time of 28 days. For 6.5 wt%

binder, the gravimetric water content of spec backfill samples (17.2–24.5%) is quite lower than that of clastic backfill ones (19.3–26.3%) when stope depth is increased from 0 to 20 m. As the binder content is increased to 8.5 wt%, the obtained increase rates are on average 17% and 19% for spec and clastic CPBs, respectively (Fig. 8.13a). The void ratio of CPB is strongly influenced by the drainage ability of early-age paste backfills (up to 3–5 days). The spec paste fills show less porosity and smaller degree of saturation than clastic paste ones for a given binder content (Fig. 8.13b, c). During binder hydration, the capillary voids available in the CPB sample will be filled with hydrated products, thus increasing the internal cohesion. Accordingly, the volume of water available for drainage will significantly be reduced. The gravimetric water content of the CPB samples is proportional to corresponding void ratio. The higher the water content, the greater the CPB porosity becomes for a given tailings type and binder content. It can be inferred from Fig. 8.13d that, as the binder increases, the specific surface area of spec CPB samples is higher than that of clastic ones because of the more efficient cement hydration under effective stress. Similarly, void ratio and specific surface area are strongly influenced by the drainage ability of CPB samples. The strength of spec CPB samples is remarkably consistent with water content. The highest binder content of CPB exhibits the lowest water content and degree of saturation, regardless of binder content and tailings type. This observation is readily explained by the different initial water/cement ratios and amounts of water required for cement hydration.

Similar observations are also made for clastic CPB samples with different binder contents. It is observed that the gravimetric water content of clastic CPBs was higher than that of spec ones for a given binder content. It is also observed that the clastic CPBs provide higher degree of saturation and void ratio than spec ones. Clastic tailings keep more water than spec tailings. Thus, less water for spec CPB samples was required to fill the relatively small amounts of voids between particles, which increased the solids concentration. This phenomenon is related to the lower water/cement ratio and higher strength for a given binder content.

6 Summary and Conclusions

The objective of the present work is to better understand the behavior of CPB and improve its backfilling strategy. To reach this goal, two stopes at the Cayeli mine were instrumented at different locations. The total pore pressures and temperature of paste backfill during and after stope filling were monitored for both in-stope and barricade. The 685N20 stope being filled with spec tailings had a largest stope volume with slow rise rate (23 cm/h) while the 715N22 stope being filled with clastic tailings had a small stope volume with relatively high rise rate of 35 cm/h. Hydrostatic loading shows that CPB is hydrating and gains its shear strength. For the 6.5% binder, no break in hydrostatic loading occurred within 20-h loading period. The peak pressure was measured as ~200 kPa at the 685N20 stope height of 6 m.

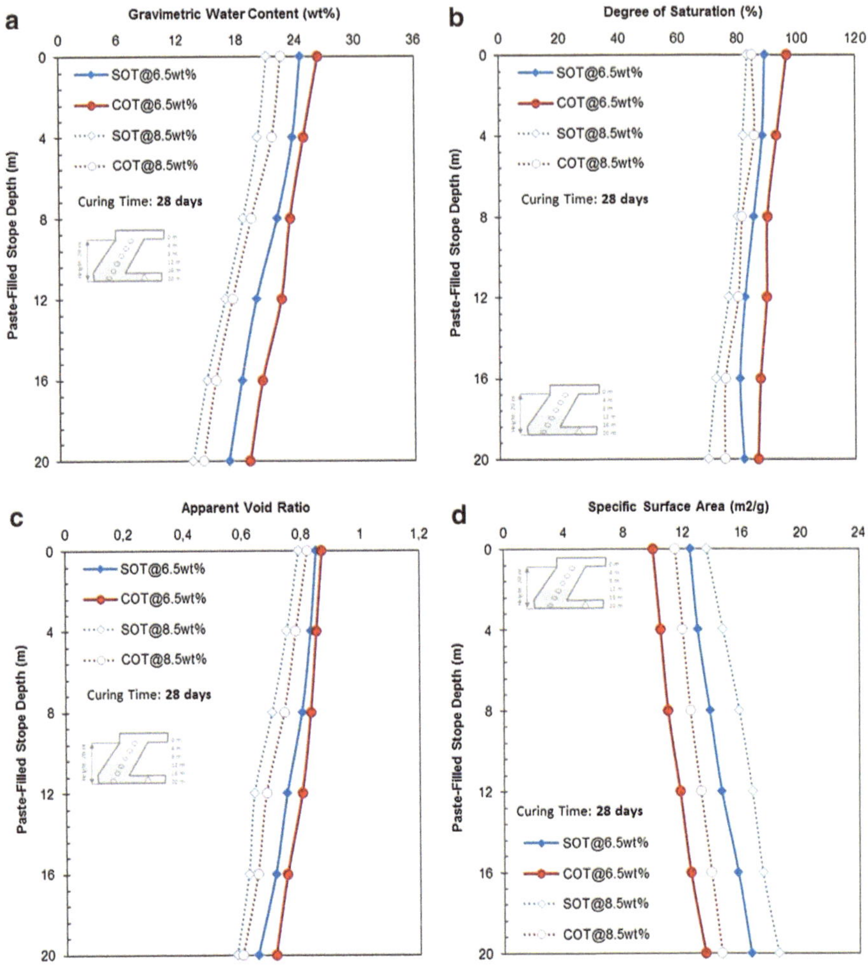

Fig. 8.13 Variation in geotechnical index parameters with stope depth for different CPB samples: (**a**) water content; (**b**) degree of saturation; (**c**) void ratio; and (**d**) specific surface area

The horizontal pressure was decreased with distance from the center of the stope. For the 715N22 stope, there was a major variation between hydration responses for specs and clastic CPBs with 6.5 wt% binder. The temperature of spec CPB increases 2.5 days earlier than that of clastic CPB due to improved material characteristic of spec tailings.

Furthermore, measuring the strength gain developing in CPB-filled stopes is of vital value to assess the overall structural stability of CPB as regional and local support part. To do this, a great deal number of the CPB samples were prepared first and then subjected to unconfined compressive strength and geotechnical index testing. Results show that the stope depth acts a layer of paste filling and provides that the highest the stope depth, the highest the material properties evolves over curing time.

Results are in good agreement with already applied backfill design data at the Cayeli mine, but may not be appropriate for other metal mines since their characteristics become quite different.

References

Belem T, Harvey A Simon R, Aubertin M (2004) Measurement and prediction of internal stresses in an underground opening during its filling with cemented backfill. In: The 5th International Symposium on Ground Support in Mining and Underground Construction, Perth, Australia, p 619–630

Cayouette J (2003) Optimization of paste backfill plant at Louvicourt mine. Can Inst Min Bull 96(1075):51–57

Doherty JP, Hasan A, Suazo GH, Fourie A (2015) Investigation of some controllable factors that impact the stress state in cemented paste backfill. Can Geotech J 52:1–12

El Aatar O (2010) Consolidation behavior of cemented paste backfill material. M.Sc. thesis, Université du Québec en Abitibi-Témiscamingue (UQAT), Rouyn-Noranda, Québec, Canada

Fahey M, Helinski M, Fourie A (2011) Development of specimen curing procedures that account for the influence of effective stress during curing onthe strength of cemented mine backfill. Geotech Geol Eng 29:709–723

Fahey M, Helinski M, Fourie A (2009) Some aspects of the mechanics of arching in backfilled stopes. Can Geotech J 46:1322–1336

Fall M, Nasir O (2010) Mechanical behavior of the interface between cemented tailings backfill and retaining structures under shear loads. Geotech Geol Eng 26:779–790

Grabinsky M, Bawden W, Thompson B (2007) In situ monitoring of cemented paste backfill in an Alimakstope. In: The 60th Canadian Geotechnical Conference, Ottawa, Canada, Paper 386

Grabinsky M, Thompson BD (2009) Thermally induced stresses in cemented paste backfill. Geotech News 27(3):36–40

Grice T (2013) Mine backfill- a cost centre or an optimisation opportunity? Aust Centre Geomech Newslett 41:1–2

Hassani F, Fotoohi K, Doucet C (1998) Instrumentation and backfill performance in a narrow vein gold mine. J Rock Mech Min Sci 35(4–5):392–396

Helinski M, Fahey M, Fourie A (2011) Behavior of cemented paste fill in two stopes: measurements and modeling. Geotech Geoenviron Eng 137(2):171–182

Huang S, Xia K, Qiao L (2011) Dynamic tests of cemented paste fill: effects of strain rate, curing time, and cement content on strength. J Mater Sci 46:5165–5170

Hughes PB, Pakalnis R, Hitch M, Corey G (2010) Composite paste barricade performance at Goldcorp Inc. Red Lake Mine, Ontario. J Min Reclam Environ 24:138–150

Karaoglu K, Kucukates K, Thompson BD (2013) Paste backfill pressure monitoring at Inmet's Cayeli underground copper and zinc mine. In: 23rd World Mining Congress, Montreal, Canada

Landriault D (2006) Keynote address: they said "It will never work"—25 years of paste backfill 1981–2006. In: The Ninth international seminar on paste and thickened tailings, Ireland, p 277–292

Landriault D, Brown R, Counter D (2000) Paste backfill study for deep mining at Kidd Creek. Can Inst Min Bull 93:156–161

le Roux K, Bawden WF, Grabinsky M (2005) Field properties of cemented paste backfill at the Golden Giant Mine. In: The 8th international symposium on mining with backfill, Beijing, p 233–241

Li L, Aubertin M, Belem T (2005) Formulation of a three dimensional analytical solution to evaluate stress in filled vertical narrow openings. Can Geotech J 42:1705–1717

Nasir O, Fall M (2008) Shear behavior of cemented paste fill-rock interfaces. Eng Geol 101(3–4):146–153

Ouellet J, Bidwell TJ, Servant S (1998) Physical and mechanical characterization of paste backfill by laboratory and in-situ testing. In: The 5th international symposium on mining with backfill, Brisbane, Australia, p 249–254

Ouellet J, Hassani F (2005) Study of cemented paste fill and rock mass interaction and behavior in filled stopes. In: The 40th US symposium on rock mechanics, Anchorage, Alaska, Paper 801

Pirapakaran K, Sivakugan N (2007) A laboratory model to study arching within a hydraulic fill stope. Geotech Test J 30(6):1–8

Potvin Y, Thomas E, Fourie A (2005) Handbook of mine fill. ACG Publishing, Perth, Australia

Thompson BD, Bawden WF, Grabinsky MW (2011) In-situ monitoring of cemented paste fill pressure to increase backfilling efficiency. Can Inst Min J 2(4):1–10

Thompson BD, Bawden WF, Grabinsky MW (2012) In situ measurements of cemented paste backfill at the Cayeli Mine. Can Geotech J 49(7):755–772

Thompson BD, Bawden WF, Grabinsky MW (2010a) Fieldwork report for paste backfill project test Stopes 685N20 and 715N22 at Cayeli Bakir Mine, Turkey. Cayeli fieldwork report, University of Toronto, p 1–82

Thompson BD, Bawden W, Grabinsky MW, Karaoglu K (2010b) Monitoring barricade performance in a cemented paste backfill operation. In: The 13th International Seminar on Paste and Thickened Tailings, Toronto, Ontario, Canada, p 85–98

Wu D, Fall F, Cai SJ (2014) Numerical modeling of thermally and hydraulically coupled processes in hydrating cemented tailings backfill columns. Int J Min Reclam Environ 28(3):173–185

Yilmaz E, 2015. Geotechnical characterization of cemented paste backfill. LAP LAMBERT Academic Publishing, ISBN: 978-3-659-60841-4, Saarbrucken, Deutschland/Germany

Yilmaz E (2011) Advances in reducing large volumes of environmentally harmful mine waste rocks and tailings. Miner Resour Manag 27(2):89–112

Yilmaz E, Belem T, Benzaazoua M (2014) Effects of curing and stress conditions on hydromechanical, geotechnical and geochemical properties of cemented paste backfill. Eng Geol 168:23–37

Yilmaz E, Benzaazoua M, Belem T, Bussière B (2009) Effect of curing under pressure on compressive strength development of cemented paste backfill. Miner Eng 22:772–785

Yumlu M, Guresci M (2007) Paste backfill bulkhead monitoring: A case study from Inmet's Cayeli Mine, Turkey. In: The 9th Int. symposium on mining with backfill, Montreal, Quebec, p 1–11

Chapter 9
Risk Assessment for Paste Tailings Projects

Enrique (Ike) Isagon

1 Introduction

1.1 Preamble

All mining operations and unit processes have varying degrees of risk associated with them. As new technologies are developed, so the risk profile changes. Advances in paste fill and paste tailings deposition have changed the risk profile of underground operations and surface tailings management. This chapter considers factors that need to be included in risk assessments of high solids concentration tailings applications in mining, such as underground paste backfill or surface paste tailings deposition.

As for all processes, these have the potential to be associated with risks that have varying degrees of consequence to the mining operation, ranging from minor dilution of ore to major losses in production. In some extreme cases these risks could result in injuries or fatalities, suspension of operations and damage to corporate reputation. If not carefully assessed and effectively mitigated at the appropriate time, these risks could impose a major impact on day-to-day operations, the owner and ultimately the shareholders.

In practice, there are several methods used to conduct a risk assessment depending on the level of study and the complexity of site-specific operating requirements. The basic approach is driven by clearly stated scope and objectives in the planning stage and comprehensive consultation with all parties.

Key process areas and process elements are identified and an inventory of current hazards and incidents are determined and compiled. Potential risk effects are initially evaluated against existing control measures and prioritised in terms of

E. (Ike) Isagon (✉)
Paterson & Cooke Canada Inc., 1351-C Kelly Lake Road, Unit #2,
Sudbury, ON, Canada, P3E 5P5
e-mail: Ike.Isagon@PatersonCooke.com

E. Yilmaz, M. Fall (eds.), *Paste Tailings Management*,
DOI 10.1007/978-3-319-39682-8_9

frequency of occurrence and gravity of consequence. Risk ranking of these risk effects is then applied to critical risks for further mitigation and preventive measures. Diligent risk assessment and effective control measures are key components to achieving a safe and productive application of paste in underground backfill and surface tailings storage.

The author acknowledges the contributions of the various authors of the previously published papers that are listed at the end of the chapter.

1.2 Background

Risk is the possibility that certain potential hazards or threats will occur that may set back the achievement of stated objectives. Risk assessment is applied in order to evaluate the potential frequency of occurrence and consequence of these events. Risk ranking helps to determine the most critical events and prioritise efforts for further mitigation. The exercise requires a strong commitment from the risk assessment team and the full support of management to deliver according to the stated objectives for optimum safety, time and cost-effectiveness.

1.3 Need for Risk Assessment

The need for a diligent risk assessment process has never been greater. Projects continue to struggle to obtain capital funding and owners are under pressure to do more with less. A risk assessment with stakeholder involvement will provide collective feedback to weigh against cost-cutting measures so that an owner can make an informed decision based on risk profile. If done properly, the risk assessment provides objective support for the financial decisions on a project.

2 Risk Assessment

2.1 Key Steps

The following steps are typically used in mining projects, with additional detail provided in the following subsections:

1. Planning stage to state clear objectives, to define scope and to allocate resources
2. Identification of the key process areas and process elements
3. Compilation of an inventory of current potential threats and hazards
4. Determination of risk effects from potential threats and hazards
5. Evaluation of risk effects against existing control measures based on frequency and consequence

6. Ranking of risks and development of post-control measures to mitigate critical risks
7. Conducting of periodic updates to the risk assessment process, as required.

2.2 *Planning*

A risk assessment process is initiated with a clear statement of its objective, context and scope. A team of appropriate and qualified resources are required to conduct an effective risk assessment process. In an underground backfill application, the main objective is to provide stable ground conditions in order to optimise safety, mining recovery and productivity. In surface tailings deposition, the main objective is to optimise mine life through a stable tailings deposition programme with strict environmental regulatory compliance. Such a programme aims to reduce the footprint of the tailings containment facility with minimal 'free' water storage and reduced environmental impact. There are potential hazards arising from key process areas and process elements in underground and surface deposition applications that could cause potential risks to safety, health and operations. The consequence could result in a severe failure to achieve the stated objectives.

2.2.1 Scoping Criteria

The scope intended for a risk assessment exercise could be driven by strategic, operational, functional or compliance (regulatory) issues, depending on what is considered to be most critical to the project or current operations.

The characteristics of a well-scoped risk assessment exercise are as follows:

- Clearly stated and specific objective
- Competent and available resources allocated
- Cost- and time-effective completion
- Factual and evidence-based risks identified
- Adaptable, consistent and practical for execution
- Generates value for key stakeholders
- Compliance to company policies and standards
- Compliance to health, safety and environmental regulations
- Periodically monitored and updated

2.2.2 Resource Requirements

The team assigned to perform a risk assessment exercise should be selected based on the following criteria:

- Ownership and commitment to the process
- Front-line familiarity to key process areas and process elements

- Subject-matter expertise
- Availability to spend the necessary time
- Multidisciplinary background to provide a holistic understanding

The team members are expected to be professional, constructive and actively involved in developing the process that includes risk assessment design, analysis and evaluation with clear recommendations for control measures.

2.3 *Identification of Key Process Areas and Elements*

Typical key process areas and elements involved in surface deposition and underground backfill operations using paste are the following:

- Material properties, sourcing and delivery
- Preparation system (processing plant)
- Surface pipeline and deposition system
- Underground distribution and placement system
- System management

Process elements are subprocess units that comprise each key process area. The potential risks identified from each process element include possible deviations from the key design and performance criteria. Out of specification material properties, flow rates and consequentially higher operating pressures are examples of elements that could deviate from the original design or performance criteria. The occurrence of these deviations may have varying consequences on operations when not properly monitored and mitigated timorously.

2.3.1 Material Properties, Sourcing and Delivery

The tailings or other raw materials and additives involved in surface transport or underground fill production are characterised to determine their physical and chemical properties. Understanding these properties is critical to the safe and reliable production of a quality product (tailings for deposition or backfill as required) in a specific application. With backfill in particular, the sourcing, availability and delivery of these materials could drive the sustainability and economics of a paste backfill operation. Typical sources and types of fill material and additives are the following:

- Run-of-mill tailings from ore beneficiation plants
- Reclaimed tailings from old ponds
- Aggregate sand from alluvial sources
- Crushed aggregates from waste rock or borrow pits
- Blended materials (tailings, alluvial sand, crushed aggregates)
- Binder (various cement types or blends)
- Additives (flocculants, rheology modifiers, etc.)

Process Elements

Examples of process elements often associated with the feed material are:

- Fill material characterisation (physical, chemical, etc.)
- Binder type and source
- Additives type and source
- Supply and delivery system of each ingredient
- Quality control and inspection of each ingredient
- Procurement contracts and monitoring systems
- Economic upsets in operating costs

Potential Risks

Examples of potential risks often associated with the feed material are:

- Excessive variation in material properties due to mineralogical or blending issues
- Underestimated availability or utilisation of key process circuits
- Limited resource of required materials
- Out-of-spec binder and additive types
- Erratic delivery of ingredients
- Poor quality control at source of materials
- Inadequate contract provision on quality and quantity of fill materials and additives
- Increased operating cost due to pricing and supply constraints

2.3.2 Preparation System

The paste fill or paste preparation system is developed from approved design criteria that set the guidelines on inputs to the process, material properties and standards for the design of the plant. Designers build from the design criteria to develop the process flow sheet, material mass balance, piping and instrumentation diagrams, general plant arrangement, equipment type and specifications, instrumentation, control, and civil, electrical and mechanical design.

The actual performance of a plant operation is measured in terms of its compliance with the approved design criteria. In practice, the plant performance could deviate from the approved design criteria and it is these deviations that impose risk to the plant performance and reliability which could result in varying impacts on the mining operations.

Process Elements

Examples of process elements often associated with the high solids concentration paste systems are:

- Process flow sheet
- Material mass balance and recovery
- Equipment type and specification

- Mechanical and piping
- Electrical
- Civil, structural and architectural
- Backup services for power and water supply
- Operating and maintenance programme
- Instrumentation and control system
- QA/QC programme
- Communication system

Potential Risks

Examples of potential risks often associated with high solids concentration paste systems are:

- Change in rheology due to excessive variation in ore mineralogy or blending
- Misunderstood underground availability and effective backfilling hours
- Undersized plant capacity
- Mismatched plant capacity and underground distribution system capacity
- Inconsistent thickener underflow due to erratic slurry feed
- Higher moisture content in filter cake than specification
- Poor mixing of binder and tailings, particularly filter cake
- Inability of plant to cope with surge demands in backfill
- Inability of plant to produce desired paste quality as per specification
- Poor operating system control due to inadequate instrumentation
- Non-existent or inadequate backup services on power and water supply
- Lower plant availability and utilisation due to high downtime and unscheduled maintenance
- Inadequate operating resource and training programme
- Poor communication system

2.3.3 Surface Pipeline and Deposition System

Depending on a number of project parameters, paste tailings may be a more effective form of surface tailings management due to its lower moisture content (higher water recovery in the mill) and lower volume per unit weight of material placed. The surface paste tailings are traditionally in uncemented form, although some operations add low dosages of binder, depending on the structural and geochemical stability requirements. The transportation and deposition system involves a pumping facility, pipelines and transfer stations and in some cases may involve trucking or re-handling of the paste. Some operations may consider pressure filtering and conveying or trucking their tailings as filter cake to the tailings area. The criteria for efficient surface deposition will depend on site-specific conditions which differ significantly from the underground backfill requirements.

The recovery of process water is optimised at the plant with minimal handling or evaporation of free water at the tailings area. Every operation is unique but traditionally pumping paste horizontally to a deposition area requires high-pressure positive displacement pumps due to the high viscosity of the paste.

Process Elements

Examples of process elements often associated with the surface deposition are:

- Tailings dewatering and water treatment facility
- Pump station facilities
- Surface pipelines
- Instrumentation and control
- Tailings area storage capacity and deposition strategy
- Progressive closure plan
- Mass consolidation and water desiccation rate
- Permitting and other waste deposition regulations for environmental protection
- Tailings area infrastructure and confinement design for stability
- Backup services on power and water supply
- Maintenance and inspection programme
- Monitoring and communication system

Potential Risks

Examples of potential risks often associated with the surface deposition are:

- Higher downtime due to lack of critical spares
- Pipe blockage due to poor design, dead legs or erratic flow
- Short pipe life due to high pipe wear erosion pattern
- Pipe leakage or failure
- Poor communication
- Inadequate instrumentation for flow monitoring and control
- Inadequate storage capacity of tailings area to sustain life-of-mine production
- Slower paste consolidation due to poor paste quality
- Tailings dam leakage or failure due to inadequate dam structure
- Inadequate backup services on power and water supply
- Scaling in pipes due to inadequate flushing

2.3.4 Underground Distribution System

The delivery of paste fill to underground stopes requires a balancing act which may rely on either gravity flow or pumping depending on many factors. In some cases a combination of both methods is used depending on the stope locations and distances from the paste fill preparation plant. The backfill distribution system is comprised of surface boreholes, fill stations, primary-level pipelines, inter-level boreholes and secondary feeder pipes to production stopes.

Surface boreholes are typically caused due to poor ground condition and possible groundwater infiltration from near-surface water sources. Boreholes drilled in poor ground may require grouting in order to stabilise the hole and control groundwater infiltration. Ceramic-lined casing is gaining in popularity as total cost of ownership for the life of the mine gets factored in more frequently.

Inter-level pipe loops are often used in order to balance excess head in the system which may cause slack flow and premature pipeline wear. Frequent switching over of the lines is a function of the stope size and the length of the fill cycle; cut-and-fill operations may see daily redirection of the piping system. The backfill lines are normally flushed before and after each pour. The placement of paste in underground stopes is based on comprehensive pour procedures, with due emphasis on fill recipes and curing periods, barricade installations, and flushing and drainage systems. Proper mix design, planning and scheduling of underground backfill pours are critical elements to achieving the safe production requirements.

Process Elements

Examples of process elements often associated with the underground distribution are:

- Borehole design and site selection
- Primary distribution pipes and accessories
- Secondary distribution pipes and accessories
- Transient analysis and design
- Fill stations
- Pipe support system
- Instrumentation control
- Fill plan and schedule (as per mining sequence)
- Pour procedures
- Barricade type and design
- Drainage system
- Flushing system
- Maintenance and inspection programme
- Monitoring and communication system

Potential Risks

Examples of potential risks often associated with the underground distribution are:

- Unstable boreholes sited in poor ground condition
- Plugged boreholes and pipelines with no provision for spares or twinned lines
- Scaling in pipes due to inadequate flushing
- Hydraulic imbalance (slack flow and free fall)
- Short-term oriented distribution system vs. life-of-mine planning
- Higher downtime due to lack of critical spares
- Excessive pipe erosion and corrosion patterns

- High inventory of unfilled stopes due to poor fill planning and scheduling
- Lower mining recovery due to dilution from fill
- Loss of production due to fill failure
- Inadequate backup services on power and water supply
- Poor communication

2.3.5 System Management

Tailings management, whether for backfilling or surface deposition, is as critical within the mining cycle as any other process. Like drilling, blasting or mucking, the system required to manage these operations includes planning, engineering, scheduling, maintenance and services. The coordination role requires active liaison of various tasks spanning several functional areas such as milling, sand or tailings delivery, paste preparation, surface pipelines and deposition systems, underground distribution and stope placement, as well as maintenance and technical support. In a typical organisation, the paste plant and surface tailings deposition system work hand in hand; tailings are either directed underground when backfill is needed or diverted to surface when no backfill is required.

At most operations, the backfill system is part of milling and the tailings management department, yet uniquely the mining department retains responsibility for underground delivery and placement. The role of the backfill or tailings engineer as a coordinator is critical in maintaining the flow of information between all stakeholders.

Process Elements

Examples of process elements often associated with the management of the system are:

- Operating organisation (staffing, leadership and ownership)
- Technical and engineering support and services
- Operators' skill sets, responsibilities and training programme
- Budget support for replacement capital and operating cost
- Procurement policy and contract management
- Health and safety policies and programmes

Potential Risks

Examples of potential risks often associated with the management of the system are:

- Lack of leadership and ownership of the unit operations
- Lack of dedicated and qualified resources
- Inadequate operator training programmes
- Lack of critical spares inventory
- Inadequate backfill QA/QC programme
- Unsustainable supply of strategic fill materials and additives
- Inadequate budget support for replacement capital and continuous improvement
- Poor health and safety plan

3 Risk Analysis and Ranking

Once the framework for the risk assessment is in place, including identification of the key process areas, the potential hazards and the control measures, the team must collectively assess each potential risk and rank it accordingly, based on likelihood of occurrence and consequence.

On a 5 × 5 table (ref. Table 9.1), the likelihood of occurrence could be a measure of its probability of occurrence ranging from rare (<10%) to almost certain (>90%). The consequence rating is value based ranging from minor to critical. The rating system can be site specific depending on the scale of mining operations and value of impact caused by the risk. Table 9.1 provides an example of a risk ranking matrix. Table 9.2 provides commonly used consequence types and example rankings.

4 Mitigation Strategies

In each key process area, there are various forms of control measures that can be built into the design and/or installed subsequently as part of operations. A periodic update of the risk management programme is necessary in order to determine current critical risks that require post-control mitigation measures. The recommended measures are then prioritised by management for execution in order to achieve a cost- and time effective mitigation of the risks.

Table 9.1 Risk ranking matrix example (5 × 5 Table)

OCCURRENCE (LIKELIHOOD)	Probability of Occurring		Risk Ranking				
	Almost Certain	>90%	5	10	15	20	25
	Likely	55%-90%	4	8	12	16	20
	Moderate	20%-55%	3	6	9	12	15
	Unlikely	10-20%	2	4	6	8	10
	Rare	<10%	1	2	3	4	5
	Impact >>>>		Minor	Moderate	Significant	Major	Critical
			CONSEQUENCE				

High

Medium

Low

Table 9.2 Risk assessment matrix of consequence example

Consequence (Where an event has more than one 'consequence type', choose the 'consequence type' with the highest rating)					
Risk category	1—Insignificant	2—Minor	3—Moderate	4—High	5—Major
Project schedule	<1 month increase in project duration	1 ≤ 3 months increase in project duration	3 ≤ 6 months increase in project duration	6 ≤ 12 months increase in project duration	>12 months increase in project duration
Project cost	May result in overall project budget overrun less than 3%	May result in overall project budget overrun of 3–10%	May result in overall project budget overrun of 10–25%	May result in overall project budget overrun of 25–50%	May result in overall project budget overrun over 50%
Quality and technical integrity	No significant impact on quality of deliverables or effect on production	Quality issues that can be addressed prior to handover or could affect production by more than 1% and less than 3%	Quality issues that can be addressed during ramp-up or could affect production by more than 3% and less than 10%	Quality issues that require significant intervention to maintain performance or could affect production by more than 10% and less than 30%	Quality issues that require significant intervention to achieve performance or could affect production by 30% or more
Natural environment	Events with no negative impact on the environment	Events with low impact on the environment. Impact restricted to company premises	Events causing short-term damage to the environment (may exceed company premises) but not dysfunction of ecosystems	Events causing damage to the environment may exceed company premises and dysfunction of ecosystems. Recultivation requiring 2 years or less	Events causing damage to the environment may exceed company premises and dysfunction of ecosystems. Recultivation requiring more than 2 years

(continued)

Table 9.2 (continued)

Consequence (Where an event has more than one 'consequence type', choose the 'consequence type' with the highest rating)					
Risk category	1—Insignificant	2—Minor	3—Moderate	4—High	5—Major
Health and safety	Events which do not require first-aid treatment	Events resulting in injuries which require first-aid treatment or medical help on site. Short-term disability	Events which require casualty treatment. Long-term injuries	Events resulting in permanent loss of function or disability, or a single fatality	Events resulting in multiple fatalities or numerous cases of permanent loss of function or disability
Legal and regulatory	Events not subject to regulatory scrutiny. Events not triggering dispute	Events which may trigger administrative action or investigation against the company. Disputes and claims which may be resolved without litigation	Decisions by regulators which require non-standard actions within the company to ensure compliance. Disputes or claims which may require resolution by litigation or arbitration	Regulatory sanctions having significant impact on company activity or cost of ensuring compliance. Litigation or arbitration in local jurisdiction	Revocation of licenses or permissions, which cause long-term disruption of company operations or very high cost of adapting to regulatory requirements. Criminal sanctions against company directors and staff. Litigation or arbitration in foreign jurisdictions
Social	Minor disturbance of culture/social structures	Some impacts on local population, mostly repairable. Single stakeholder complaint in reporting period	Ongoing social issues. Isolated complaints from community members/ stakeholders	Significant social impacts. Organised community protests threatening continuity of operations	Major widespread social impacts. Community reaction affecting business continuity. 'License to operate' under jeopardy

Consequence (Where an event has more than one 'consequence type', choose the 'consequence type' with the highest rating)					
Risk category	1—Insignificant	2—Minor	3—Moderate	4—High	5—Major
Reputation	Minor impact; awareness/concern from specific individuals	Limited impact; concern/ complaints from certain groups/organisations (e.g. NGOs) period	Local impact; public concern/adverse publicity localised within neighbouring communities	Suspected reputational damage; local/ regional public concern and reactions	Noticeable reputational damage; national/international public attention and repercussions
Sustainability	<3% increase in OPEX	3–10% increase in OPEX	10–25% increase in OPEX	25–50% increase in OPEX	>50% increase in OPEX
Likelihood					
Likelihood ranking	1—Rare	2—Unlikely, 3–10%	3—Possible, 10–30%	4—Likely, 30–90%	5—Almost certain, >90%
Likelihood	The event is not foreseen to occur within the project life cycle	The event is expected to occur in exceptional circumstances within the project life cycle	The event is expected to occur at least once within the project life cycle	The event is expected to occur sometime within the project life cycle	The event is expected to occur repeatedly within the project life cycle

4.1 Types of Mitigation

Examples of typical mitigation measures are:

- Process or technology innovation
- Changes to material inputs
- Revision of procedures
- Operators and maintenance crews' training upgrade
- Health and safety improvement
- Environmental and regulatory compliance
- Productivity and efficiency improvement (cost)
- Equipment or plant upgrade
- Review of critical spares inventory policy
- Organisational re-structuring

5 Risk Assessment Exercises

The following tables provide example registries of some possible risk assessment exercises conducted on surface or underground paste tailings projects. The potential effect or impact of each risk is completely site specific and should be taken for what they are—a sampling of typical elements, risks and effects used when analysing a project or operation's risk profile.

In the above registries, Tables 9.3 and 9.4 are used to rate the frequency and consequence. The combined score for each risk determines its criticality, classified into low, medium and high as shown in Table 9.1. Table 9.2 shows a more detailed matrix of the type and impact of the consequence. In these exercises, the recommended mitigation measures would be compiled into a risk mitigation plan that is submitted to the project owner for consideration at the next level of study. In practice, the role of the risk assessment team ends upon completion of the risk assessment exercise and submission of the report. The respective owner of each key process area is responsible for the execution of the mitigation plan and the subsequent follow-up.

6 Opportunities and Challenges to Effective Risk Assessment

A risk assessment is intended to achieve its stated objective on a specified timeline and budget. A safe and inclusive atmosphere with input encouraged from all the team members is the appropriate environment to ensure an effective risk assessment. Well-prepared participants who have read the background material and are keen to solve the problem at hand will ensure that the most value is gained from the time spent together. Conversely, a lack of commitment and poor participation will result in a failed risk assessment.

Table 9.3 Example of risk registry for paste underground delivery system

#	Baseline element	Potential risk	Potential effect	Current design control measure	Freq.	Cons.	Risk level		Recommended mitigation	Freq.	Cons.	Risk level		Comments
1	Surface borehole	Drilling borehole in poor ground conditions, deviation from layout	May require re-drilling or change of collar location, upset schedule and budget	Borehole site close to ventilation raise with same geology	3	3	9	Medium	Detailed geological information required during detailed design. Drill test holes as required	1	3	3	Low	Risk level reduced from medium to low
2	Hydraulic balance	Change in slurry properties due to limited production or temporary shutdown from one mine source	Hydraulic imbalance resulting to slack flow and free fall	Potential variation in rheology taken into consideration during design	3	3	9	Medium	Conduct sensitivity of hydraulic model based on management emergency production strategy	1	3	3	Low	Risk level reduced from medium to low
3	Fill stations	Over the life of mine, inter-level fill stations may end up in abandoned mine levels	Short effective life of fill stations would incur unnecessary mine maintenance and higher replacement cost	Bypass borehole included in design	3	2	6	Medium	Align design with life-of-mine production outlook	1	2	2	Low	Risk level reduced from medium to low
4	Maintenance and inspection	High erosion and corrosion patterns in pipes and boreholes	Premature pipe replacement	Pipe specified for abrasion and corrosion resistance; included contingencies for ground and pipe support	2	3	6	Medium	Regular NDT testing and planned inspection required during operation	1	3	3	Low	Risk level reduced from medium to low

(continued)

Table 9.3 (continued)

#	Baseline element	Potential risk	Potential effect	Current design control measure	Freq.	Cons.	Risk level		Recommended mitigation	Freq.	Cons.	Risk level		Comments
5	Critical spares inventory underground	Critical spares not in budget	High downtime due to lack of spares; requires storage facility for critical spares	Use of uniform-size pipe optimises spares availability	2	2	4	Low	Sustaining capital to include critical spares and storage requirement	1	2	2	Low	Risk level remains low
6	Communication system	Lack of communication protocol between plant operator and underground crews	Poor communication could result in scheduling, health and safety issues	Instrumentation provided for continuous monitoring of system	3	2	6	Medium	Outline roles of key underground crew; develop communication protocol	1	2	2	Low	Risk level reduced from medium to low
7	Surface boreholes and pipelines	Plugging incidents may occur during pour due to various reasons	Shutdown in plant and underground filling operations due to plugged lines	Provided underground instrumentation for pressure and flow monitoring	3	3	9	Medium	Provide plugging and unplugging procedures in communication protocol. Provide necessary equipment	1	3	3	Low	Risk level reduced from medium to low
8	Flushing system	Inadequate flushing of boreholes and pipelines before and after each pour	Scaling and flow obstruction overtime. Excessive flushing could flood mine workings	Provided adequate storage of process water for flushing; automated control in pumping flush water	2	3	6	Medium	Regular update of flushing procedures according to actual experience from various stope destinations	1	3	3	Low	Risk level reduced from medium to low

Table 9.4 Example risk registry for surface paste deposition system

#	Baseline element	Potential risk	Potential effect	Current design control measure	Freq.	Cons.	Risk level		Recommended mitigation	Freq.	Cons.	Risk level		Comments
1	Permitting an environmental protection regulation	Delay in environmental permitting for the project	Upset in project schedule and budget	Tailings pond design and permitting package are in critical path	3	3	9	Medium	Allow contingencies in schedule and budget in the detailed design; closer liaison with permitting agency	1	3	3	Low	Risk level reduced from medium to low
2	Hydraulic balance	Change in slurry properties due to limited production or temporary shutdown from one mine source	Change in slurry flow behaviour causing hydraulic imbalance	Potential variation in rheology taken into consideration during design	3	3	9	Medium	Conduct sensitivity of hydraulic model based on management emergency production strategy	1	3	3	Low	Risk level reduced from medium to low
3	Pumps and pump station	No spare pump in case of pump breakdown and maintenance	No pumping of paste will cause shutdown of paste plant and possibly mill if tailings cannot be bypassed	Spare pump provided	2	2	4	Low	Allow bypass of run-of-mill tailings to a spare dewatering pond	1	2	2	Low	Risk level remains low
4	Maintenance and inspection	High erosion and corrosion patterns in pipes	Premature pipe replacement	Pipe specified for abrasion and corrosion resistance	2	2	4	Low	Regular NDT testing and planned inspection required during operation	1	2	2	Low	Risk level remains low

(continued)

Table 9.4 (continued)

#	Baseline element	Potential risk	Potential effect	Current design control measure	Freq.	Cons.	Risk level		Recommended mitigation	Freq.	Cons.	Risk level		Comments
5	Critical spares inventory	Critical spares not in budget	High downtime due to lack of spares; requires storage facility for critical spares	Use of uniform-size pipe optimises spares availability	2	2	4	Low	Sustaining capital to include critical spares and storage requirement	1	2	2	Low	Risk level remains low
6	Communication system	Lack of communication protocol between plant operator and surface crew	Poor communication could result in scheduling, health and safety issues	Instrumentation provided for continuous monitoring of system	3	2	6	Medium	Outline roles of key surface crew; develop communication protocol	1	2	2	Low	Risk level reduced from medium to low
7	Water treatment facility	No surge capacity to contain surface run-off during precipitation	Possible spill of waste water with hazardous elements	Provided surge capacity in water storage and treatment facility	2	3	6	Medium	Provide drainage system around the facility	1	3	3	Low	Risk level reduced from medium to low
8	Flushing system	Inadequate flushing of pipelines	Scaling and flow obstruction overtime	Provided adequate storage of process water for flushing; automated control in pumping flush water	2	3	6	Medium	Regular update of flushing procedures according to actual experience	1	3	3	Low	Risk level reduced from medium to low

A risk assessment exercise has the potential to create opportunities with unintended positive outcomes, and such risks may not require any mitigation measures. However, potential risks with negative consequence require an assessment that could lead to mitigation depending on priority.

The risk assessment exercise is only one tool and may not necessarily guarantee an effective mitigation strategy depending on its conclusion and subsequent follow-through for mitigation.

Some of the challenges to an effective risk assessment are the following:

- Participants may feel that they are just checking a box—no real action expected
- May provide limited value to stakeholders
- Inconclusive analysis and misinterpretation
- Poor execution plan and limited management support
- Lack of strong commitment, ownership and sponsorship from management
- Improperly defined objectives and deliverables
- Routine exercise with same repeated failures
- Health and safety risks not properly assessed

A proper risk assessment is considered a crucial aspect of the viability of a project and is relied on to guide critical project decisions that are made around capital cuts and short execution schedules. Getting the right people involved and using an effective facilitator with a knowledgeable background of paste-specific operations can help improve the effectiveness of the assessment to derive real value from the investment in time and resources.

References[1]

Health and Safety Executive (2014) Controlling the risks in the workplace, The HSE Publications. http://www.hse.gov.uk/risk/controlling-risks.htm. Accessed 15 Dec 2014

KlohnKrippen Berger (2012) Award of excellence: tailings management at greens creek mine, Canadian consulting engineer. http://www.canadianconsultingengineer.com/news/award-of-excellence-tailings-management-at-greens-creek-mine/1001820155/?&er=NA. Accessed 15 Dec 2014

Lyle G, Foster P (2014) (MIRARCO 2014) Safety and Risk Management Using Bow Tie Analysis. In: CIM AGM, Vancouver, May 2014

PricewaterhouseCoopers (2008) A practical guide to risk assessment. http://www.pwc.com/en_us/us/issues/enterprise-risk-management/assets/risk_assessment_guide.pdf. Accessed 15 Dec 2014

[1] The following documents were relied on or referred to in the development of this chapter.

Chapter 10
Cayeli Paste Backfill System and Operations

Erol Yilmaz and Mehmet Yumlu

1 Introduction

The mining and milling operations produce large volumes of mine waste rocks and tailings annually. It is apparent that open pit mining causes more environmental contamination than underground mining due to the use of lower grade deposits, the creation of large open voids, and the release of hazardous matters into the environment. The safe treatment and disposal of tailings produced pose enormous challenges that require a multidisciplinary study approach, because they can lead to environmental pollutions such as acidic water generation, tailings dam failures, and groundwater contamination. Hence, increasingly strict environmental legislation and cost-competitiveness furthermore dictate use of technically suitable, economically viable, environmentally acceptable, and socially responsible techniques.

Certainly, novel approaches have been developed in recent years. A common element of these approaches is that tailings are thickened or de-watered prior to final disposal in dry form. One new approach is to use paste backfill which offers many advantages from the geomechanical, operational, financial, and environmental perspectives (Hassani and Archibald 1998; Yilmaz 2015). These impacts range from improving working conditions safety to minimizing surface expression of local excavations instability. Due to shorter time required to fill an excavation, paste-filled stopes can be cycled quicker, leading to improved cash flows. The paste backfill is a fundamental part of the Cayeli underground mining operations. Each year, large volumes of underground openings or stopes are created to extract ore first and then filled with paste. This chapter provides an overview of paste backfill system and operations implemented at Cayeli mine.

E. Yilmaz (✉)
Cayeli Bakir Isletmeleri A.S, PO Box 42, Madenli Beldesi, Cayeli, Rize 53200, Turkey
e-mail: yilmazer@fqml.com

M. Yumlu
AMC Consultants Pty Ltd, Level 19, 114 William Street, Melbourne, VIC 3000, Australia

E. Yilmaz, M. Fall (eds.), *Paste Tailings Management*,
DOI 10.1007/978-3-319-39682-8_10

2 Cayeli Cu-Zn Mine

The Cayeli underground copper and zinc mine, operated by Cayeli Bakir Isletmeleri A.S., is a wholly owned subsidiary of First Quantum Minerals Ltd. Cayeli is located at 8 km south of the Black Seacoast of northeastern Turkey. The mine produced its first concentrate in 1994 and is expected to operate until at least 2019. Cayeli extracts about 1.2 million tons of ore per year to produce about 220,000 tons of copper and zinc concentrates. A full description of the Cayeli mine operation is given in Yumlu and Guresci (2007).

2.1 Mining and Processing Methods

Cayeli's mine design is based on underground bulk mining methods with the use of delayed backfill to extract ore sequentially. The access to the mine is provided by a single production shaft located on footwall side of the orebody and a service ramp. The main levels divide the orebody into mining blocks. In the upper parts of the mine, the sublevel heights are averagely 20 m which allows a 15 m high stope and 7 m wide bench for production drilling. In the lower parts of the mine, the sublevel heights are averagely 15 m which allows a 10 m high stope and 10 m wide bench for production drilling. The primary mining method is retreat transverse and longitudinal long-hole stoping with paste backfill and loose or consolidated waste rock fill. Stopes are mined in primary, secondary, and tertiary sequencing.

During mining, the ore is separated into six different types and hoisted to surface, where it is transferred into one of the eight storage bins. Before processing, the ore from surface storage bins is blended to achieve optimal metallurgical results. Ore processing includes three stages of crushing primary and secondary ball mill grinding, conventional flotation using standard or column cells, and water removal by thickening and pressure filtering to produce copper and zinc concentrates. Cayeli produces a copper concentrate of 150,000 tons per year with 23% Cu grade and a zinc concentrate of 90,000 tons per year with 49% Zn grade. Yearly, the mean feed grades were 3.2% Cu and 6.3% Zn. Concentrate grade averaged 20% for Cu at a 76% recovery and Zn concentrate grade averaged 49% at a 71% recovery. In the last two decades, the average recovery of copper has been 81% and of zinc 72%.

2.2 Tailings and Wastewater Management

After obtaining the Cu-Zn concentrates, a large amount of concentrator tailings with a high acid-generating potential is produced. At the mine site, a submarine tailings disposal system was used whereby tailings were transported by pipeline along a river to be discharged into the anoxic environment of the Black Sea. Since 2000, this system has partially been replaced by paste backfill operation to prevent environmental impact for short- and long-term perspectives. Tailings are used as a paste

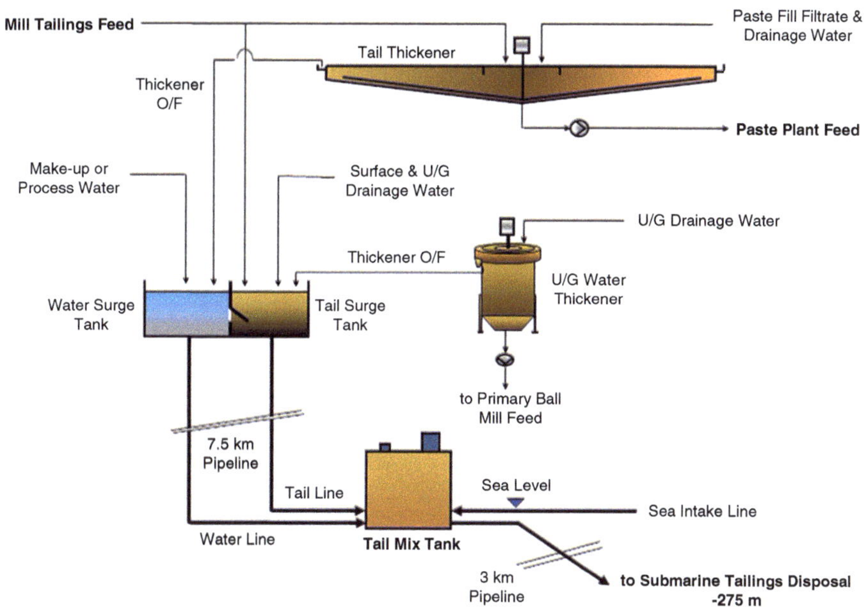

Fig. 10.1 Schematic view of tailings and waste water management at the Cayeli mine

backfill which is pumped underground or they are pumped via a pipeline to a submarine tailings disposal which is located 3 km offshore at a depth of 385 m. There are two 7.5 km long overland pipelines: one for tailings slurry and the other for excess water from thickener overflows. Mine water and sewage plant effluent are also added water side. Both tail line and water line are connected at the deaeration tank located in Black Sea coast. Tailings gravitate down to 275 m sea level via an out-fall pipeline by density differences of tailings and seawater. Figure 10.1 shows a flow sheet of tailings and waste water management implemented at the mine.

3 Cemented Paste Backfill Design

One of the most important parameters in designing a paste backfill operation for underground mines is compressive strength. Other important design issues include material properties and flow and dewatering characteristics. This subsection will provide the main required details.

3.1 Backfill Mix Design

The design of the CPB ingredients (i.e., tailings, cement, and mixing water) plays a key role in its resulting strength and stability performance. To prepare paste backfill mixes, Cayeli uses total tailings without extracting any fines content. Depending on the ore feed blend, the tailings are categorized as spec, clastic, bornite yellow ore

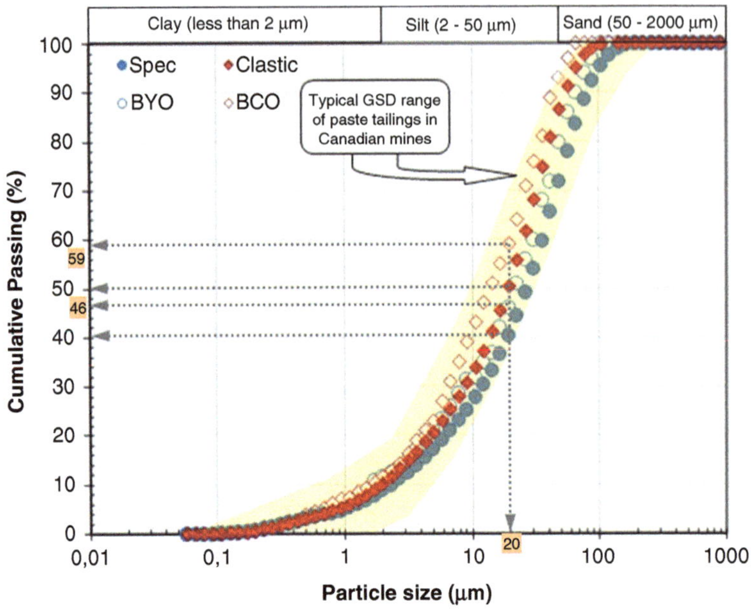

Fig. 10.2 Particle size distribution of Cayeli tailings: Spec, Clastic, BYO, and BCO

BYO, and bornite clastic ore BCO. Tailings were sized via a Malvern Mastersizer particle size analyzer and the results obtained are shown in Fig. 10.2. The grain size distribution curves of the tailings are closely similar. All the tailings were found to have 40–59 wt% finer than 20 μm, which indicates that the tailings can be classified as medium-size tailings in accordance with the classification by Landriault (2001). One can also see that the fraction under effective diameter (10 μm) is fine, showing considerable fractions of fine particles. It is well known that the significant presence of the fines within the tailings normally requires higher cement content in order to reach equal or higher strengths when comparing to tailings without containing too much fines.

The specific gravity of Cayeli tailings varies between 3.2 and 4.1 depending on the type, mineralogy, and properties of the blended ore. Spec tailings had the highest specific gravity of 4.06. The major mineral phase identified within the tailings is pyrite (73.3–78.4 wt%). A minor detection of sphalerite (3.4–5.2 wt%), chlorite (2.4–2.7 wt%), and kaolinite (1.6–2.4 wt%) were found in the tailings together with trace amounts of biotite, dolomite, quartz, and barite (all less than 3 wt%). Table 10.1 lists the mineralogical composition of the studied tailings.

Moreover, Fig. 10.3 shows photos of fine sections conducted to scrutinize the tailings' mineralogy. Results indicate that pyrite is the most abundant mineral in the tailings, following by the other minerals, such as chalcopyrite, bornite, and sphalerite. The amount and type of such sulfides in the tailings may affect both the short- and long-term strength development of paste backfill samples and the selection of binder type, considering the water retention potential of the phyllosilicates and the

Table 10.1 Compositional and quantitative analysis (percentage) of Cayeli tailings

Mineral	Spec	Clastic	BYO	BCO
Pyrite	73.3	78.4	74.8	76.2
Sphalerite	5.2	3.7	4.1	3.4
Chlorite	2.4	2.6	2.5	2.7
Kaolinite	2.3	2.4	1.6	2.1
Biotite	1.9	1.7	1.9	1.6
Dickite	2.7	1.3	1.4	2.0
Chalcopyrite	1.3	1.4	1.5	1.4
Dolomite	2.6	2.2	3.1	3.4
Quartz	1.5	1.9	2.8	2.3
Smithsonite	0.6	0.4	0.6	0.5
Barite	0.4	0.2	0.3	0.6
Nantokite	0.3	0.2	0.2	0.4
Total clays	5.5	3.6	5.2	3.4

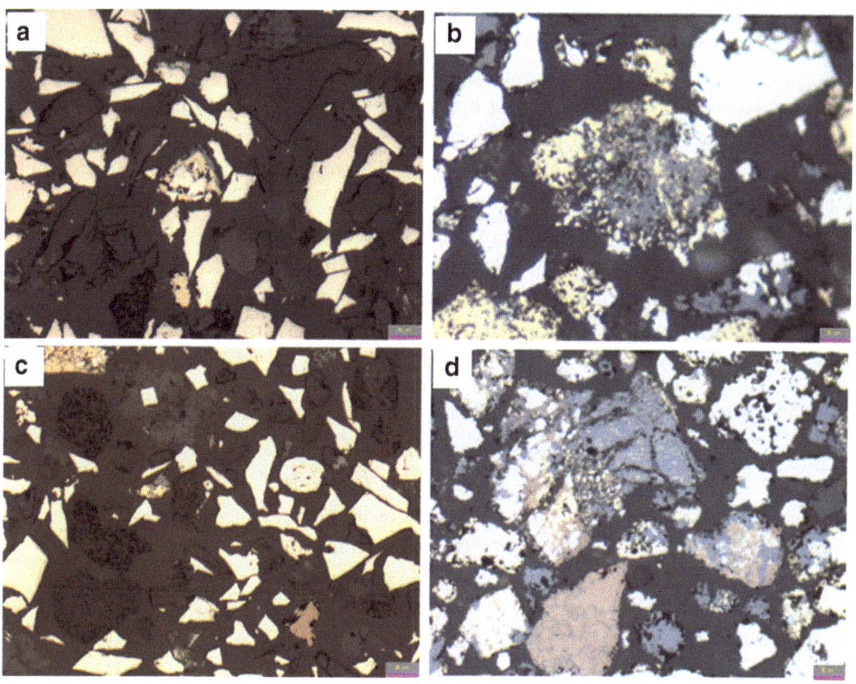

Fig. 10.3 Fine sections of Cayeli tailings: (**a**) Spec, (**b**) clastic, (**c**) BYO, and (**d**) BCO

sulfate generation potential of the iron sulfides. Minerals such as sericite, micas, and clay can reduce backfill strength and stability due to their water-absorbent mineral layers. Sulfide minerals raise the specific gravity, while the abrasiveness of silica minerals causes serious pipeline wear (Benzaazoua et al. 2002; Fall and Benzaazoua 2005; Potvin et al. 2005).

At Cayeli mine, the binding agent used for paste backfill preparation is Portland composite cement (CEM II/A-M (P-LL) 42.5R). To produce this cement type, pozzolanas and limestone are added to clinker as additive. Pozzolana is the most important natural pozzolanic material that is used as cement additive. Total additive in CEM II/A-M (P-LL) 42.5R cements is between 12 and 20 wt%. The specified cement contents vary from 5 to 8.5 wt% depending on the stope type, stage of filling, and whether (or not) the stope will be undercut. The initial plug (at least 3 m above stope brow) is filled with 8.5 wt% cement if it will be undercut or 6.5 wt% for no undercutting. The rest of the stopes during filling of stage 2 are either filled with 5 wt% or 6.5 wt% cement contents depending on the stope type. The average cement content is 6.5%.

Cayeli had used CEM V/A (P-S) 32.5R prior to March 2011, and CEM II/A-M (P-LL) 42.5R from March 2011 through to September 2011 both supplied from the Unye Cement. Since September 2013, the Cayeli mine has been using CEM II/A-M (P-LL) 42.5 supplied from the Askale Cement. CEM V/A (P-S) 32.5R had higher clinker ratio and contained blast furnace slag while CEM II/A-M (P-LL) 42.5 had lower clinker ratio and contained natural pozzolanic material (trass) instead of slag. The main reason why the Cayeli uses different types of cement is the change in the tailings created as the mine is gone deeper and deeper. Cayeli undertakes a comprehensive test work to govern the best binding agent interacting with variable tailings to be used for paste backfill mixes. Table 10.2 lists the main properties of the presently used cement (CEM II/A-M (P-LL) 42.5R), supplied from the Askale cement factory.

As mixing water, a combination of process water and/or tap water is used in order to reach the targeted slump (19.5 cm or 7.7 in.) which can facilitate paste transport to underground stopes through either gravity or pumping. The process water is relatively aggressive, with sulfate (SO_4^{2-}) content of 2512 ppm. The relatively high

Table 10.2 The chemical, physical, and mechanical requirements of CEM II/A-M (P-LL) 42.5R

Chemical requirements	Value (%)	Physical requirements	Value
Total SiO_2	20.76	Residue on 45 μm sieve	1.2%
Insoluble residue	6.05	Residue on 90 μm sieve	–
Al_2O_3	4.96	Blaine's specific surface	4748 cm^2/g
Fe_2O_3	4.13	Specific gravity	2.99
CaO	56.73	Setting time – initial (min 60)	155 min
MgO	1.72	Setting time – final	220 min
SO_3 (max 4%)	2.96	Water requirement	34.8%
Loss on ignition	5.59	Le Chatelier's soundness	1 mm
Na_2O	0.38	Mechanical requirements	Strength (MPa)
K_2O	0.77		
Cl^- (max 0.1%)	0.0218	1-day cured strength	15.3
Total	98.01	2-day cured strength (min 20)	27.7
Free lime	1.51	7-day cured strength	47.8
Total additives (12–20%)	14.98	28-day cured strength (min 42.5)	48.2

Table 10.3 Chemical and geochemical analysis of process and tap waters

Geo-chemical parameter	Process water	Tap water
Calcium Ca (wt%)	334	52.4
Sulfate content SO_4^{2-} (ppm)	2512	142
Electrical conductivity EC (mS cm^{-1})	8.53	0.25
Redox potential Eh (V)	0.16	0.39
pH	7.83	6.44

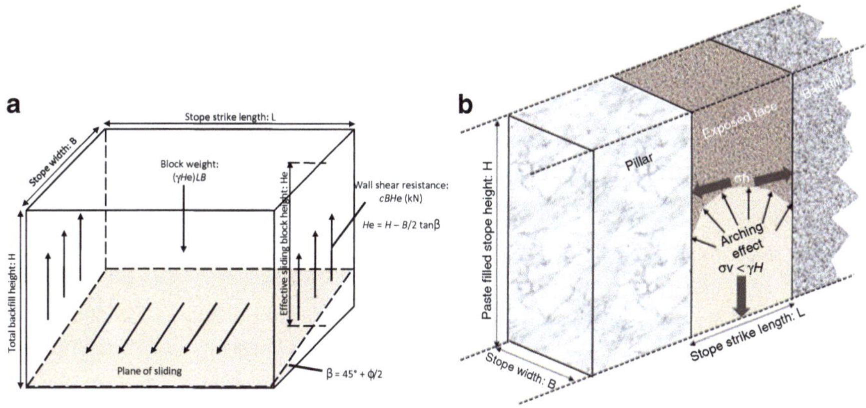

Fig. 10.4 Schematic of stope exposures: (**a**) exposed frictional fill face; (**b**) narrowly exposed fill face

concentration of calcium Ca (334 ppm) is due to lime added during processing. The pH, Eh (redox potential), and EC (electrical conductivity) of the mixing waters were also analyzed and the results are listed in Table 10.3.

3.2 *Target Strength Design*

The target design backfill strengths depend on several factors, such as the type, size, and number of exposures and the depth of the backfill within the stope. The strength requirements to keep stability of horizontal exposures are habitually higher than vertical exposures. Moreover, the stopes need higher strength at the base compared to lower strength in the upper parts of the stopes. To determine target design strengths, Cayeli presently uses cases of an exposed backfill where the two opposite sides of the backfill are against stope walls or exposed frictionless backfill face (Fig. 10.4a). It also uses case of narrowly exposed backfill face which accounts for arching effects on confined backfill by adjacent stope walls (Fig. 10.4b).

The sublevel vertical distance is dictated by the stope height. In the upper parts of the mine above 800 level, the sublevel is 20 m apart, allowing the development of a 15 m high by 7 m wide stope bench for production drilling. In the lower part of the

Table 10.4 Target design strengths of the stopes above and/or below level 800

Considered stope models	Min. strength (kPa)	Max. strength (kPa)
Above level 800		
Narrowly exposed fill face (Askew et al. 1978)	226	578
Exposed friction fill face (Mitchell et al. 1982)	313	333
Exposed frictionless fill face (Mitchell et al. 1982)	298	694
Average results	279	535
Below level 800		
Narrowly exposed fill face (Askew et al. 1978)	213	425
Exposed friction fill face (Mitchell et al. 1982)	235	266
Exposed frictionless fill face (Mitchell et al. 1982)	278	595
Average results	242	429

Table 10.5 The designed cement contents of paste-filled stopes and their resultant strengths

Stopes name	Status	Part I (B_w/UCS)	Part II (B_w/UCS)
Primary stopes (open or blind)	Work under stope	8.5% (1400 kPa)	5.5% (700 kPa)
	No work	6.5% (950 kPa)	5.5% (700 kPa)
Secondary/tertiary stopes (open or blind)	Work under stope	8.5% (1400 kPa)	4.5% (500 kPa)
	No work	6.5% (950 kPa)	4.5% (500 kPa)

Waiting time between Part I (first 8 m of the stope) and Part II (the remaining 7 m of the stope) is 2 days
After 7 days, access to mirror from side wall of paste backfill and work under paste-filled stope after 14 days

mine below 800 level, the sublevels have been developed 15 m apart, allowing the development of a 10 m high by 10 m wide bench for production drilling. Cayeli uses a fill pressure coefficient K of 0.45; a fill bulk unit weight γ of 24.5 kN/m^3; a fill internal friction angle, ϕ, of 34°; a fill cohesive strength, c, of 45 kPa; and a factor of safety, FS, of 1.7. Table 10.4 lists results of target design strengths.

Table 10.5 summarizes the existing cement contents specified for the underground stopes and the resultant design strengths. The target design strengths for binder content of 4.5, 5.5, 6.5, and 8.5 wt% were 500, 700, 950, and 1400 kPa, respectively, depending on the type of tailings, the stope type, the paste backfill properties, and backfill pouring strategies. In general, the current average cement consumption is 6.5 wt%. Considering that the average measured wet paste density of 2.3 tons/m^3, and solid density of 77%, this corresponds to 120 kg of cement per cubic meter. However, the total average cement contents can be potentially reduced to approximately 5–6% by optimizing cement types and rates used for paste backfill mixes and adopting the higher capacity barricade designs (e.g., using buttressed or arched designs).

Figure 10.5 provides stability charts determined for vertical and undercut exposures for a range of backfill exposure sizes, based on the assumed backfill properties and factor of safety of 1.3 and 1.5 for vertical and undercut exposures, respectively. For vertical exposures at a factor of safety of 1.3, a target design strength of 300–400 kPa is required for stope spans of up to 25 m long, up to 10 m wide and exposed

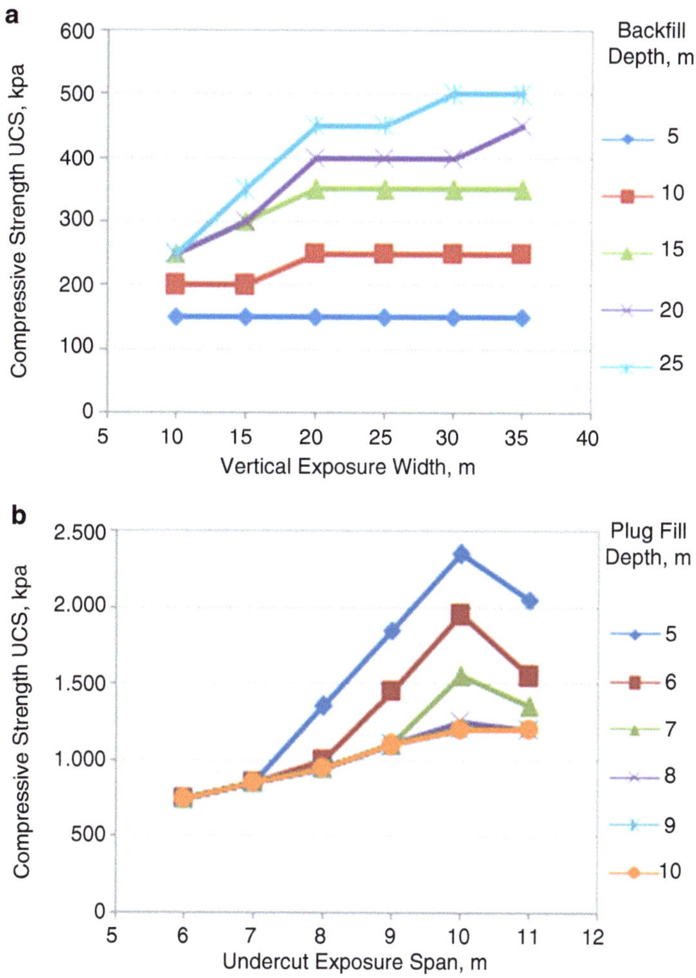

Fig. 10.5 Paste backfill vertical (**a**) and undercut (**b**) exposure stability charts

backfill height of up to 25 m. For a minimum backfill plug depth of 10 m, and undercut spans of 7.0 m up to 10 m at a factor of safety of 1.5, a target design strength of 900–1200 kPa is required.

3.3 Quality Control Testing

A well-established quality control/quality assurance is critical to ensure that the desired paste backfill strengths are achieved, at acceptable cement contents. In this view, Figs. 10.6, 10.7, 10.8, and 10.9 provide results of paste backfill optimization

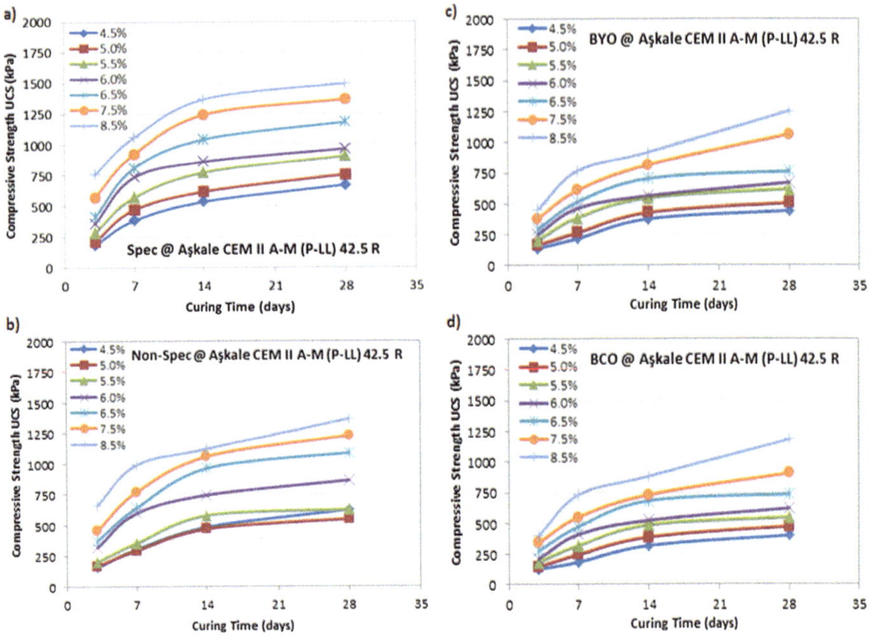

Fig. 10.6 Optimization of Askale CEM II cement rates used in CPB with different tailings types: (**a**) Spec tailings, (**b**) non-spec tailings, (**c**) BYO tailings, and (**d**) BCO tailings

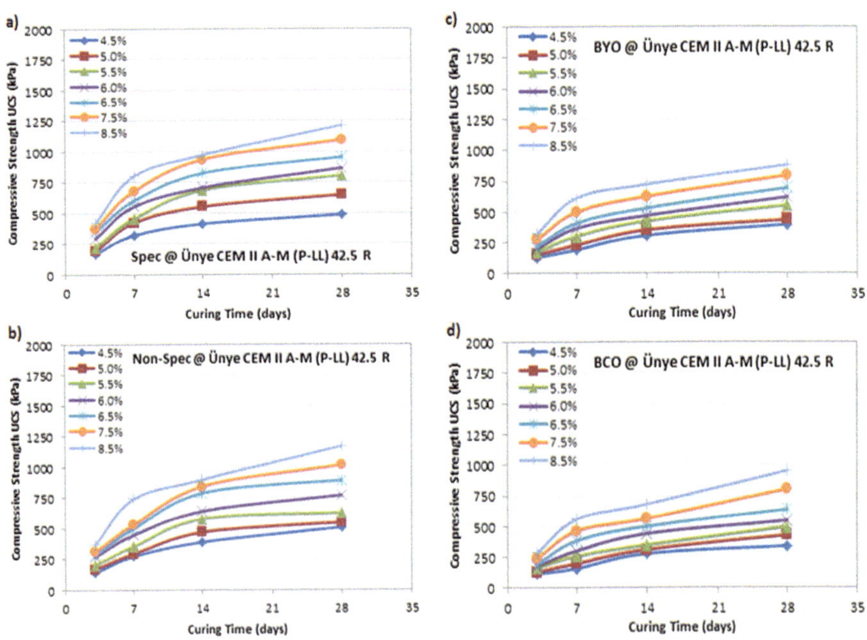

Fig. 10.7 Optimization of Unye CEM II cement rates used in CPB with different tailings types: (**a**) Spec tailings, (**b**) non-spec tailings, (**c**) BYO tailings, and (**d**) BCO tailings

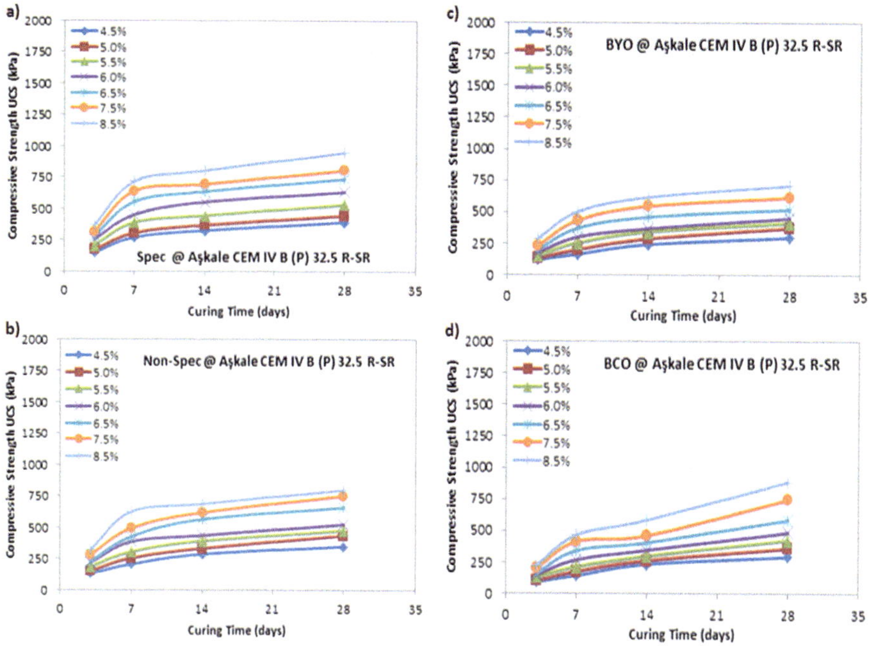

Fig. 10.8 Optimization of Askale CEM IV cement rates used in CPB with different tailings types: (**a**) Spec tailings, (**b**) non-spec tailings, (**c**) BYO tailings, and (**d**) BCO tailings

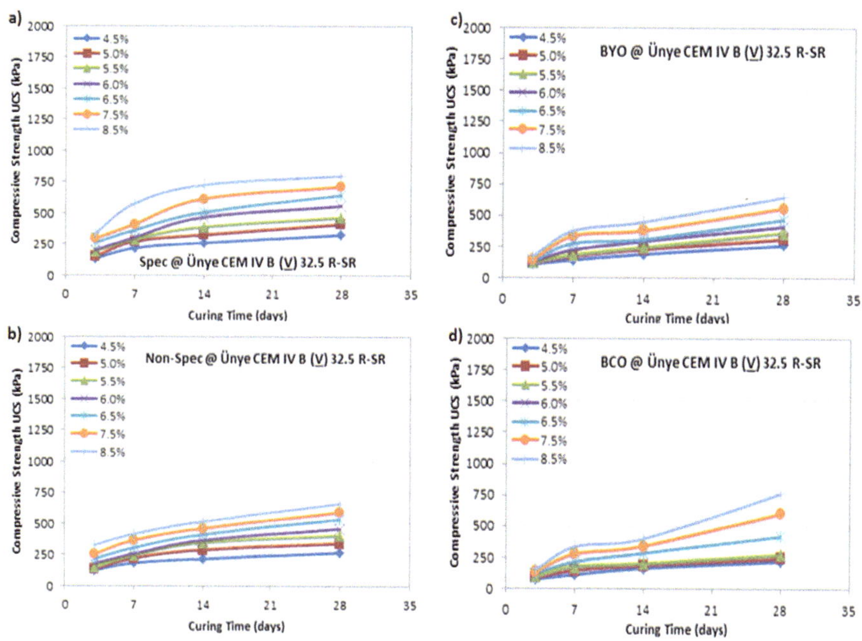

Fig. 10.9 Optimization of Unye CEM IV cement rates used in CPB with different tailings types: (**a**) Spec tailings, (**b**) non-spec tailings, (**c**) BYO tailings, and (**d**) BCO tailings

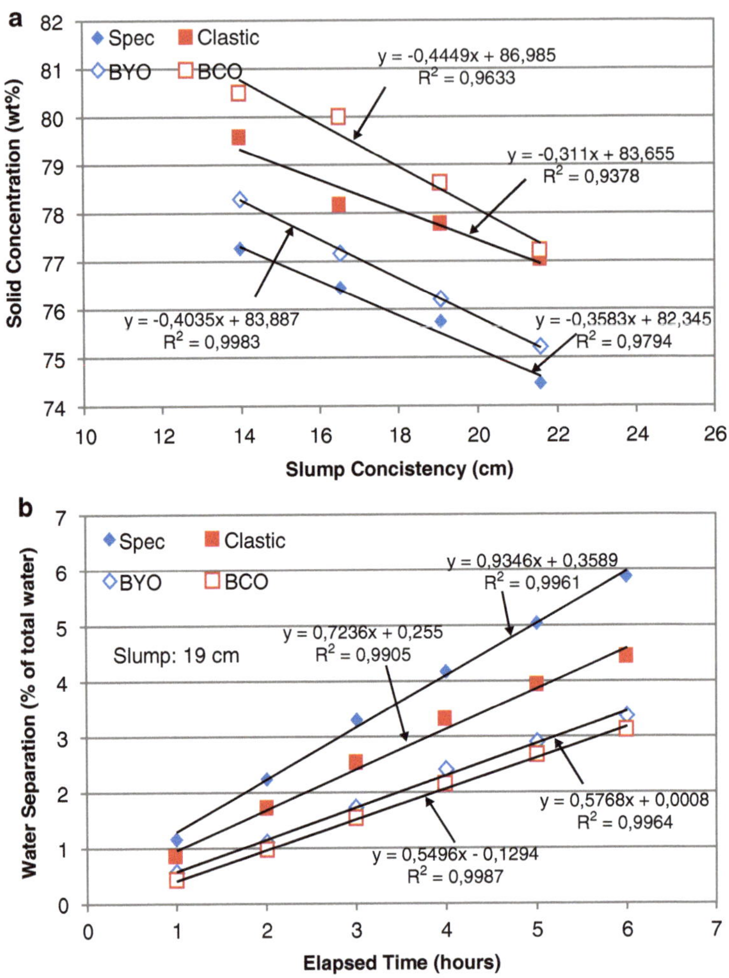

Fig. 10.10 Plots of (**a**) solid content versus paste slump and (**b**) water separation versus time for different non-cemented tailings

testing in terms of tailings type, cement type, and cement contents vs. compressive strength and curing time. It is fairly clear that for a given binder type and content, spec tailings always give the highest strengths among other tailings: clastic, BYO, and BCO. The highest strength gains are obtained from spec tailings with Askale CEM II while the lowest strengths are obtained from BCO tailings with Unye CEM IV. After the optimization tests, it was decided to use the Askale CEM II/A-M (P-LL) 42.5R for reducing significantly the paste backfill cement-related costs.

The rheological index testing consists of a series of modified slump cone tests designed for assessing the colloidal properties of non-cemented tailings which were mixed at slump values of 14–21.5 cm. Slumps corresponding to the solids content are plotted on a graph to produce relational curves in which the slump decreases as solid content increases (Fig. 10.10a).

Moreover, water separation tests were carried out on non-cemented spec, clastic, BYO, and BCO tailings. Figure 10.10b also shows variation in water separation with time for different tailings. Spec tailings provided higher water separation than other tailings types. For an elapsed time of 6 h, the rate was 6% of total water in spec tailings. From the performed tests, one can comment that non-cemented bornite clastic ore tailings have bled apparently lowest water than other tailings due to the nature of its water-retention minerals. Note that water separation tests were conducted on non-cemented tailings having 19 cm slump. The water separation of tailings can also be determined using the equations given in Fig. 10.10b.

4 Paste Backfill Plant

The paste plant was commissioned in 1999 and is designed to deliver backfill at a maximum production rate of 55 m^3/h. At full plant capacity, only 90 dtph (dry solids flow rate) of tailings out of 127 dtph of available tailings is used. However, the plant is currently only operating at an average 37 m^3/h utilizing only 57 dtph of tailings. At the current rate, less than half of the available tailings are used. Table 10.6 demonstrates CPB system rates, including fill mix design, throughput rates, and component quantities. The tailings generated from concentrator are pumped to the paste thickener at 20 wt% solids. The density of tailings in the 600 m^3 capacity is 16 m dia. High rate thickener is increased from 20–30 to 65–75 wt% solids. The underflow from the thickener is pumped over a distance of 300 m to a 300 m^3 capacity agitating tank.

Depending on the backfill rate, the agitating storage tank underflow is pumped to one or both of the two vacuum disc filters. The filters reduce the moisture content of the tailings producing a cake to 79–83 wt% solids. The filter cake discharged from the disc filters and drops to a reversible belt conveyor, which feeds to the conditioning mixer. The filter cake with tailings slurry from the agitating slurry tank is mixed in a conditioner mixer before being discharged into the paste mixer. At the mixer additional makeup water and cement are added and mixed to desired slump (19–22 cm). The cement content is governed by the specific requirements of each stope and varies from 4.5 to 8.5%. Average annual cement content is 130 kg/m^3 equivalent to 6.5% cement based on the current fill recipe. The formed CPB is delivered underground by an 80-bar capacity Putzmeister positive-displacement pump through a network of surface and underground pipeline (Fig. 10.11). A hydraulic switch-over valve diverts paste between the two operating reticulation routes: one serving above 800 level and one below 800 level.

4.1 Tailings, Cement, and Water Supply

The plant has a capacity to process 90 dtph of total tailings. Depending on the ore feed blend the mill generates four types of tailings: spec, clastic, BYO, and BCO. It is apparent that the filtration as well as flow and strength properties vary according to tailings used for CPB. The mill generates 127 dtph of tailings at the current

Table 10.6 Paste backfill system and operating rates used at Cayeli mine

Mining and fill demand physicals						Backfill types				
Ore production	1,350,000 tpa					Rock fill		100,000 m³		20%
Ore density	3.31 tons/m³					Cemented rock fill		20,000 m³		5%
Void mined	410,000 m³pa					Paste fill		290,000 m³		75%
Void fill ratio	100%					Total		410,000 m³		100%
Backfill required	410,000 m³pa									
Backfill mix design (calculations for 1000 dry tons)							Fill quantities			
Material	SG	Dry bulk density	Mix	Wt (kg)	Vol (l)	kg/m³	Tons per hour	Tons per day	Tons per month	Vol (BCM) m³
Cement	3.15	1.10	5%	50	16	85	3.2	64	1748	1589
Aggr (−10 mm)	2.85	2.20	–	–	–	–	–	–	–	–
Aggr (−40 mm)	2.85	2.00	–	–	–	–	–	–	–	–
Tailings	3.50	2.41	95%	950	271	1621	61	1208	33,212	13,759
			100%			1706				
Dry solids				1000	287		64	1272	34,960	15,348
Water in backfill				299	299		19	380	10,443	10,443
Total wet fill				1299	586		83	1652	45,402	
Material flow physicals							Mix components and fill properties			
Number of shifts per day		3	8 h shifts				Filter cake moisture		15%	w/w
Operating shifts per day		3					Wet tails used		1907 kg/m³	
Mill feed rate		169 dtph		91%			Cement added		85 kg/m³	
Con mass pull rate		25%	w/w				Makeup water		218 kg/m³	
Dry tails available		127 dtph		45%			Plasticizer added		5211 kg/m³	

Mining and fill demand physicals							Backfill types				
Dry tails used for paste			57 dtph					Slurry added		–	
Cement feed rate			7 dtph					Dry bulk density		1.71 tons/m^3	
Backfill plant capacity			64 dtph					Wet bulk density		2.22 tons/m^3	
Backfill pulp density			77%	w/w				Pulp density		49%	v/v
Total water flow			19 tph					Porosity		51%	–
Slump			195 cm					Void ratio		1.04	–
Backfill system design rates											
Instantaneous design rate					These are the plant and component design throughput rates						
100% utilization											
dtph	wtph	m^3/h	dtpd	wtpd	m^3/day						
64	83	37	1532	1990	898						
Daily average rate					These are the targets that the operators should achieve each day						
83% utilization											
dtph	wtph	m^3/h	dtpd	wtpd	m^3/day						
53	69	31	1272	1652	745						
Monthly scheduling rates					These are the rates used for scheduling of paste placement						
75% utilization											
dtph	wtph	m^3/h	dtpd	wtpd	m^3/day	dtpm	wtpm	m^3/m			
48	62	28	1149	1493	674	34,960	45,402	20,486			
Annual system performance					These are the expected annual performance achievements						
72% utilization											
dtph	wtph	m^3/h	dtpd	wtpd	m^3/day	dtpm	wtpm	m^3/m	dtpa	wtpa	m^{3}pa
46	60	27	1103	1433	647	33,558	43,582	19,665	402,697	522,983	235,982

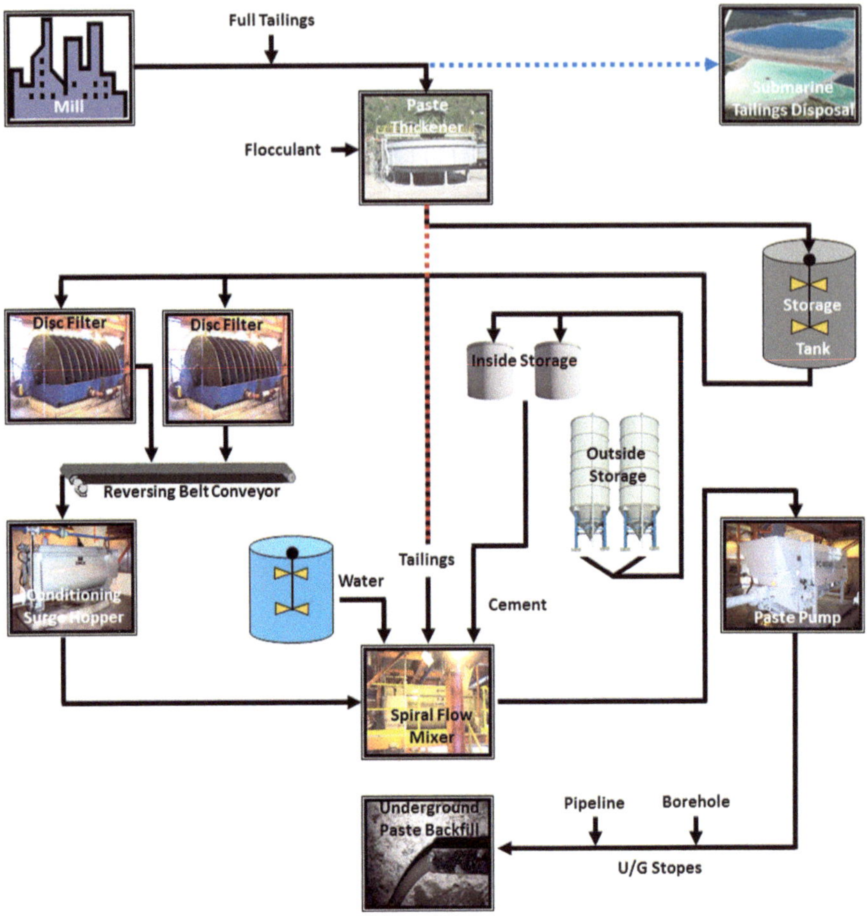

Fig. 10.11 The flow sheet of the Cayeli paste backfill plant

production rate of 1.35 Mtpa. Approximately 53 wt% of the processing tailings generated are used for underground paste backfilling while the remaining tailings (47 wt%) cannot be processed by the paste plant, and thus are deposited by submarine tailings disposal system. Figure 10.12 shows the usage amounts of the created tailings as both cemented paste backfill (CPB) and submarine tailings disposal (STD) techniques.

The cement dosing system includes two inside cement silos (each having 6-ton capacity), screw feeders, and K-Tron impact cement feed flow scale. The cement is delivered by a screw feeder at a rate varying from 1 to 10%. The total cement silo storage capacity of two outside cement silos is 100 tons (each 50 tons). On average, four truck deliveries (25 tons per truck) are required when pouring a standard rate of 658 m^3 of cemented paste backfill at current average cement rate of 130 kg/m^3. The need of fresh and process waters as mix water is met by the wells constructed near the river.

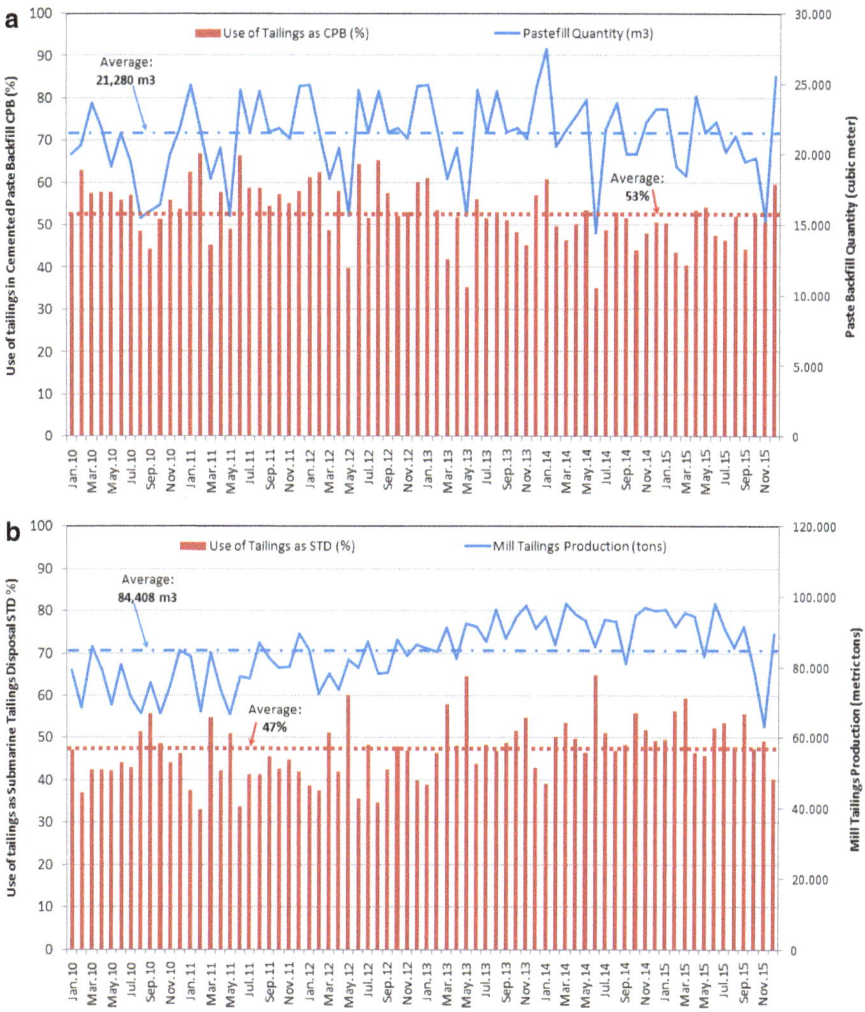

Fig. 10.12 Use of tailings as (**a**) underground paste backfill (CPB) and (**b**) submarine tailings disposal (STD)

4.2 *Mixing and Pumping Processes*

Cayeli uses a Simem conditioner and mixer to condition and mix the paste backfill ingredients to produce a homogeneous paste at the end of mixing. The mix products enter the mixer at one end and mixed thoroughly before discharged into the pump feed hopper at the other end. Plant operation in auto mode is based on the mixer power draw and therefore routine cleaning and planned maintenance of mixer are essential for optimal performance. Current practice is to clean the mixer shaft and paddles once every 24-h operation. Operators use the mixer power draw (100–120 kW for 19–22 cm cone slump) to infer paste slump in the mixer. When operated in auto mode the mix ingredients are also batched according to the power draw on the mixer. Higher power draw means that the paste in the mixer has higher solids density (denser) or has high yield stress. Similarly very low power draw means that the paste in the mixer has low solids density and has very low yield stress. The mixer power draw is an indirect indicator of the consistency of the paste in the mixer.

The Putzmeister HSP 2180 pump is rated at 80 bar at 60 m^3/h. The current operating range has been throttled to maximum 50 bar. The steel reticulation lines are rated 80 bar and the Victaulic flanges are rated at 70 bar. Note that the pump can operate up to 60 bar pressures to allow higher density low-slump paste to be delivered. The operational issues and downtime have been attributed to the systematic cleaning of the mixer and blockages in the cement dosing system. The pump maintenance includes changing of the poppet seat valves every 3000–3500 h and the re-haul of the power pack every 4–5 years.

4.3 Plant Control and Automation

The paste plant is operated by permanent operators allocated to the plant. Plant management belongs to the mill department. The plant is supervised by paste supervisor, an engineer from the mill. Plant operation is automated and controlled using Siemens automation system with direct interaction by the operator.

The plant can be run in auto or manual modes, including remote operation from the mill. The auto mode relies on the mixer power draw for controlling the slump. Underground barricade and reticulation discharge-point observation cameras and barricade earth pressure cells are connected to the automation through fiber-optic data cable. Currently there are no underground reticulation pressure gauges neither installed nor remotely operated hydraulic dump valves that are connected to the automation system at the plant control room.

4.4 Plant Shutdown Distribution

Figure 10.13 shows a sample paste plant shutdown distribution on a monthly basis. There are eight items considered for paste backfill plant shutdown. The monthly average paste fill shutdown distribution items are underground (52%), flush (11%), operational (5%), instrument (6%), no tailings (6%), mixer cleaning (12%), IT (2%), no tailings (7%), and mechanical (6%). It is clear that the greatest causes of plant shutdowns are based on underground works, such as stope changes, pipeline changes, and damaged pipe repairs. Operational works are considered mainly as filter clothes change and pump cleaning. Even though the planned paste backfill mechanical

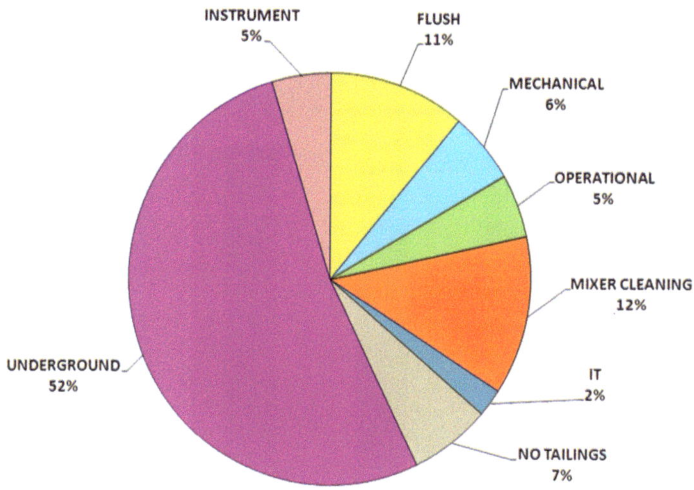

Fig. 10.13 A sample plant shutdown distribution on a monthly basis

maintenance take place on a monthly basis, some unexpected works are still available in the plant. These are Putzmeister pump failure, mixer cutter failure, belt conveyor failures, etc. To clean pipelines, flushing is also made during each stope change.

4.5 Plant Availability and Utilization Rates

Figure 10.14a, b shows the paste plant production rate and cement rate over the last 5 years. It is clear that the yearly average paste backfill production rate and cement rate are 38 m^3/h and 6.8 wt%, respectively. Figure 10.15a, b shows the plant availability and utilization over the last 6 years. The plant availability varies between 88 and 96% and plant utilization varies between 69 and 81%. The plant performance in terms of both availability and utilization rates is comparable to some of the best performing plants around the world. This is a very good indication of the paste plant being generally kept in good operating condition as well as high underground availability for stope fill assisted by the dual-backfill reticulation route and the smaller stope sizes requiring a high number of stopes being available at any given time.

5 Paste Reticulation and Barricade Systems

5.1 Paste Reticulation System

Cayeli uses a network of surface and underground pipelines and series of boreholes to deliver CPB mixtures to underground stopes. There are two independent operating paste reticulation routes: decline route and surface borehole route. The paste

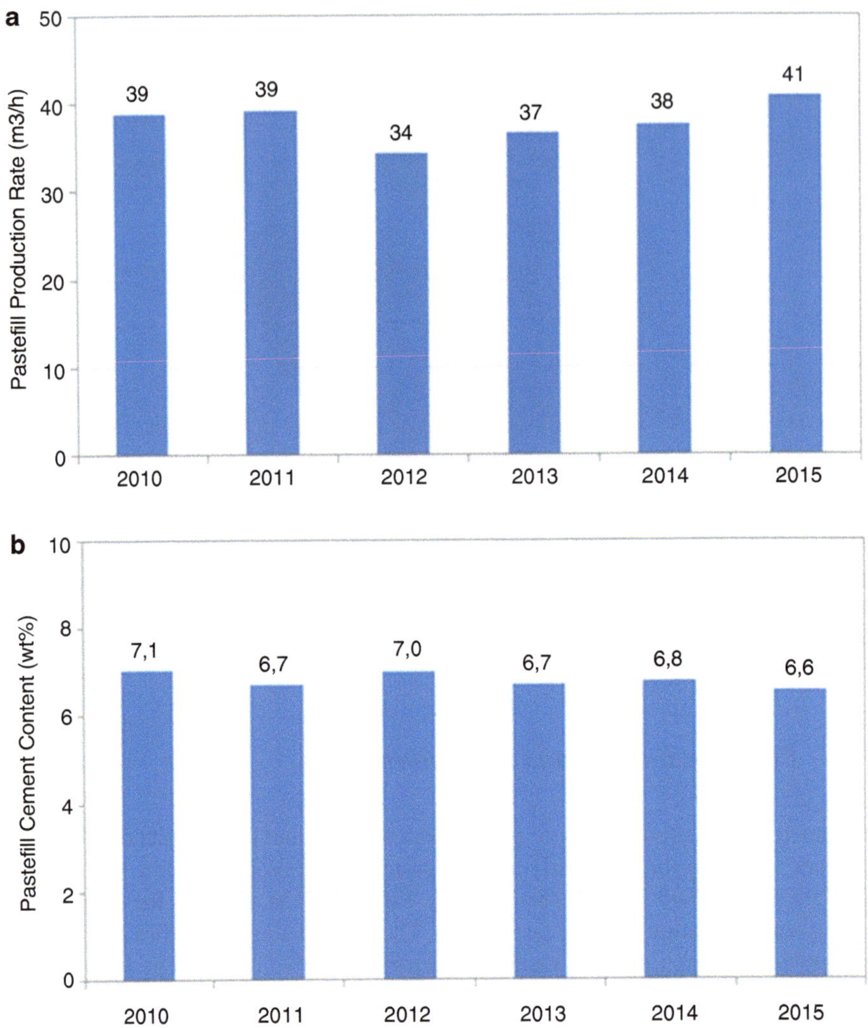

Fig. 10.14 The paste backfill plant performance: (**a**) production rate; and (**b**) cement rate

backfill reticulation system at the mine requires pump delivery and comprises two separate operating routes: one dedicated to production areas below 800 level and another for stopes above 800 level. The surface borehole route delivers paste backfill through a 300 m deep surface borehole and two separate 100 m deep inter-boreholes below 900 level to 800 level and 700 levels. The decline route delivers backfill laterally for 40 m on surface and then paste runs along the main decline access for a distance of 100 m at 1:7 gradient before being directed to a system of short internal boreholes and lines below level 1060. Figure 10.16 shows a schematic view of the Cayeli backfill reticulation system. The blue line shows the decline route and the brown line represents the surface borehole while the pink route to 880

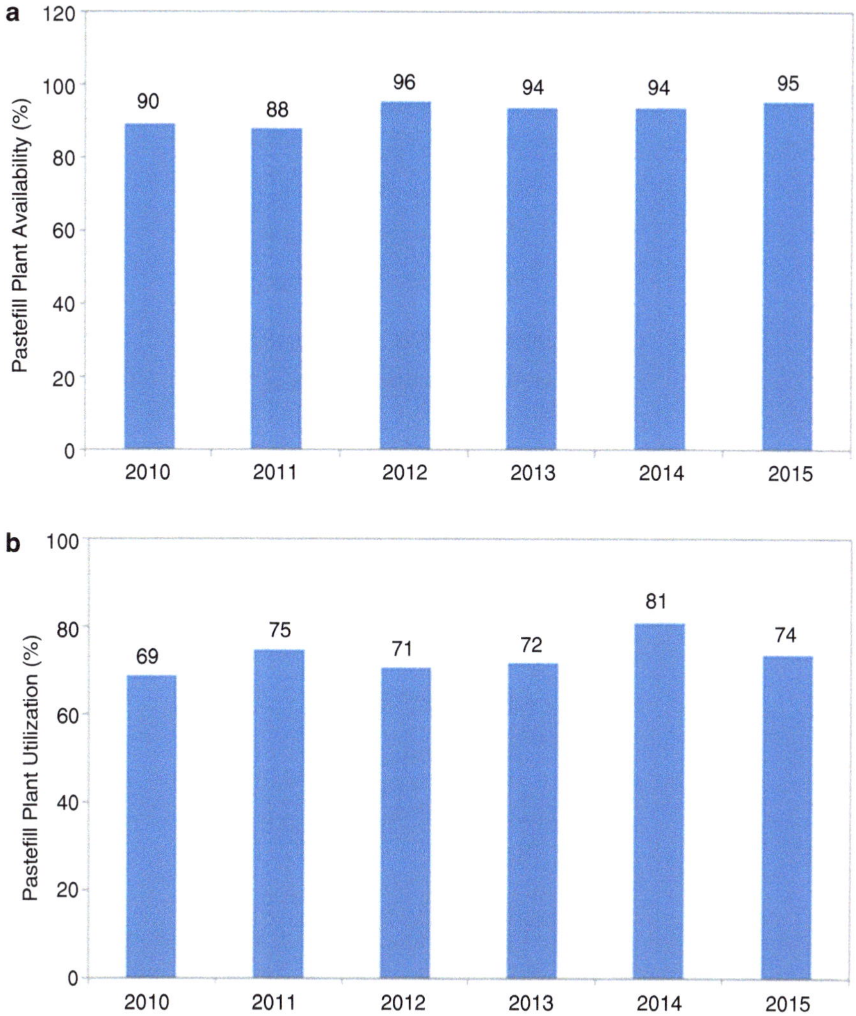

Fig. 10.15 The paste backfill plant performance: (**a**) availability; and (**b**) utilization

level is plugged and has been abandoned. Experiences indicate that the reticulation system has been operating at a slump of 19–22 cm with the pump delivery pressures varying from 20 to 40 bar significantly below the 80 bar pump capacity.

5.2 *Paste Flow Modeling*

Cayeli has developed a flow model tool that allows detailed paste backfill flow modeling for reticulation planning and design. Cayeli has also modeled filling of stopes in different parts of the mine through both reticulation routes using assumed

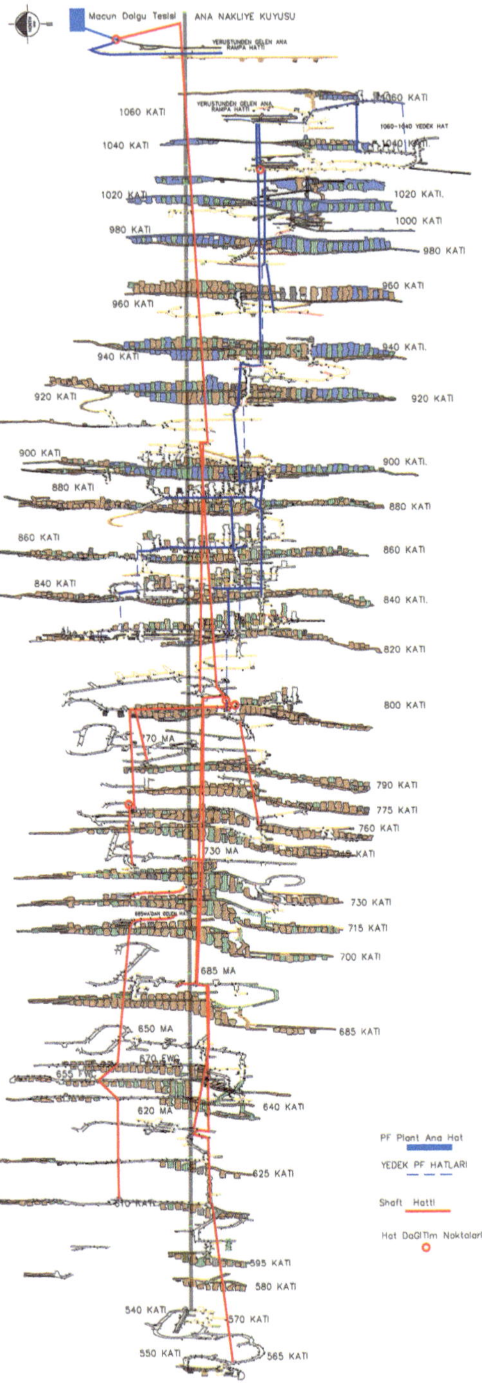

Fig. 10.16 Schematic view of paste backfill reticulation system

viscosity and yield stress properties. Generally, the conic slump varied from 19 to 22.5 cm while the cylinder slump varied from 7 to 8 cm. A yield stress of 200–300 Pa is estimated based on the measured wet paste bulk density of 2.24 t per cubic meter. 200 Pa yield stress is used for the flow model analysis. The flow model was run for a minimum and maximum production rate of 20 and 55 m^3/h representing the two extremes of the operating range of the paste plant.

The flow modeling results for surface borehole route show that pump delivery pressure of less than 20 bar is estimated at plant capacity of 55 m^3/h (Fig. 10.17a). The lateral pipeline pressures are 15 bar indicating that the 20 bar poly pipes are sufficient for maximum backfill production rate at 200 Pa yield stress. However the poly pipes will not be enough to operate at higher yield stress above 250 Pa. Results show slack flow due to free fall in all boreholes. Pump delivery pressure of below 15 bar is estimated at 20 m^3/h delivery rate (Fig. 10.18a). At this low production rate, the surface borehole and the 900–800 level borehole experience free fall. The lateral pipeline pressures are estimated to be 10 bar below the 20 bar poly pipe's capacity.

The flow modeling results for decline route show that pump delivery pressure of 38 bar is estimated at full plant capacity of 55 m^3/h and yield stress of 200 Pa (Fig. 10.17b). At this fill rate, there is full pipeline flow with no free fall in the system. The lateral pipeline pressures are close to 20–25 bar indicating that the 20 bar poly pipes may not be adequate for the maximum paste fill production rate. Pump delivery pressure of 28 bar is estimated at 20 m^3/h delivery rate and at yield stress of 200 Pa (Fig. 10.18d). At this low production rate, there is full pipeline flow with no free fall in the system. The lateral pipeline pressures are estimated to be close to the 20 bar poly pipe's capacity, indicating likely pipeline failures at this rate. At lower 175 Pa yield stress however, there is free fall occurring at the 1060 borehole.

5.3 *Barricade Design and Pressure Monitoring*

Cayeli uses simple planar shotcrete barricades as fill-retaining structures. The current method for the construction of shotcrete barricades includes timber frame with plywood, embedded reinforcement rebars tied to the cemented rebar shear pins drilled along the perimeter, and sprayed shotcrete to a nominal depth of 300 mm. Cayeli barricades are planar in shape and vary in width of 5–10 m and height of 5–6 m. The barricade has a working capacity of less than 50 kPa and ultimate failure capacity of 100–120 kPa. The capacity of planar barricades is low. To ensure safety, the limited barricade capacity has dictated a strict fill placement regime. For Stage 1 filling, the filling rate is capped at max rise of 0.35 m/h and the cement rate is kept at 6.5 wt% regardless of stope type. The filling rate during Stage 2 is also controlled at maximum rise of 0.43 m/h. Taking into account the relatively small average stope sizes in the mine, records indicate that on average stopes are filled at average filling rate of 37 m^3/h significantly below the plant capacity of 55 m^3/h.

The barricade pressures depend on a variety of factors including the size of the stope, size and location of the barricades, type of tailings, slump, cement type and

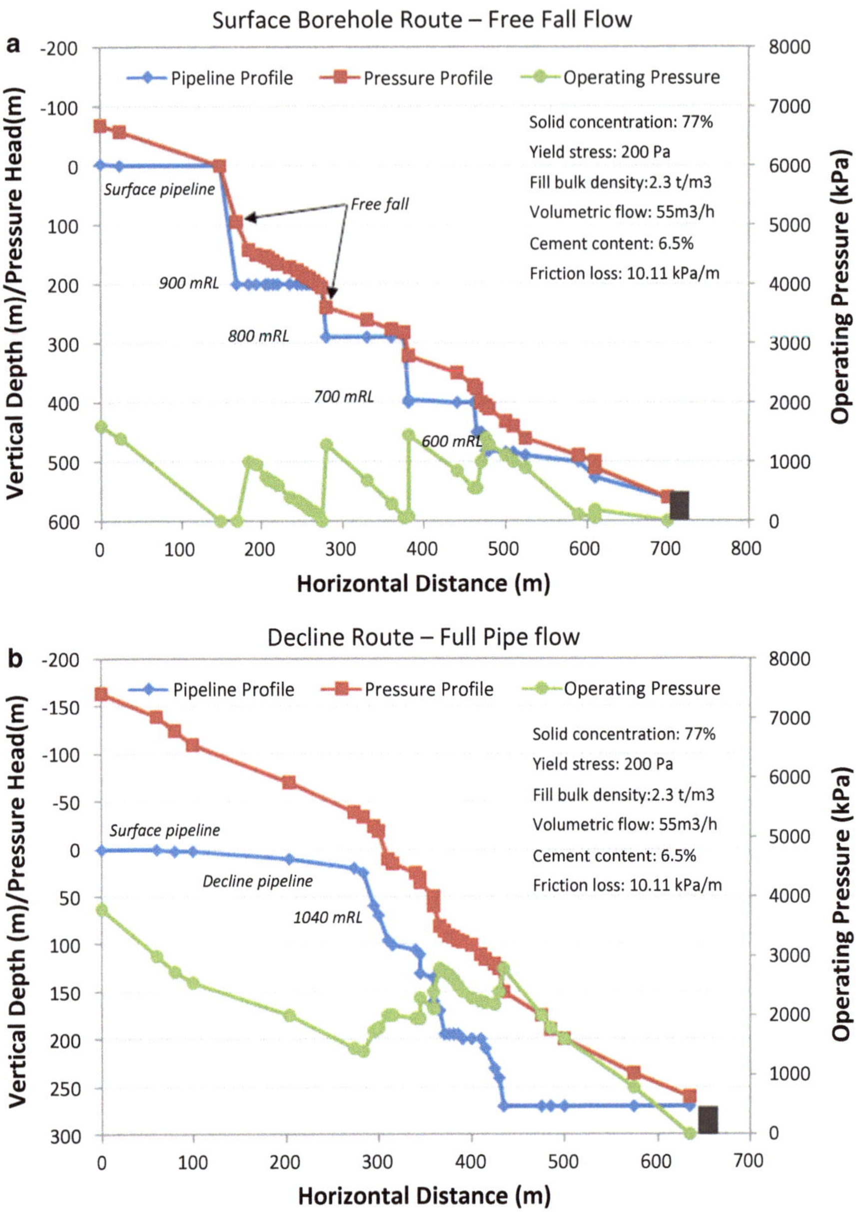

Fig. 10.17 Paste flow models: (**a**) surface borehole route at 55 m^3/h; and (**b**) decline route at 55 m^3/h

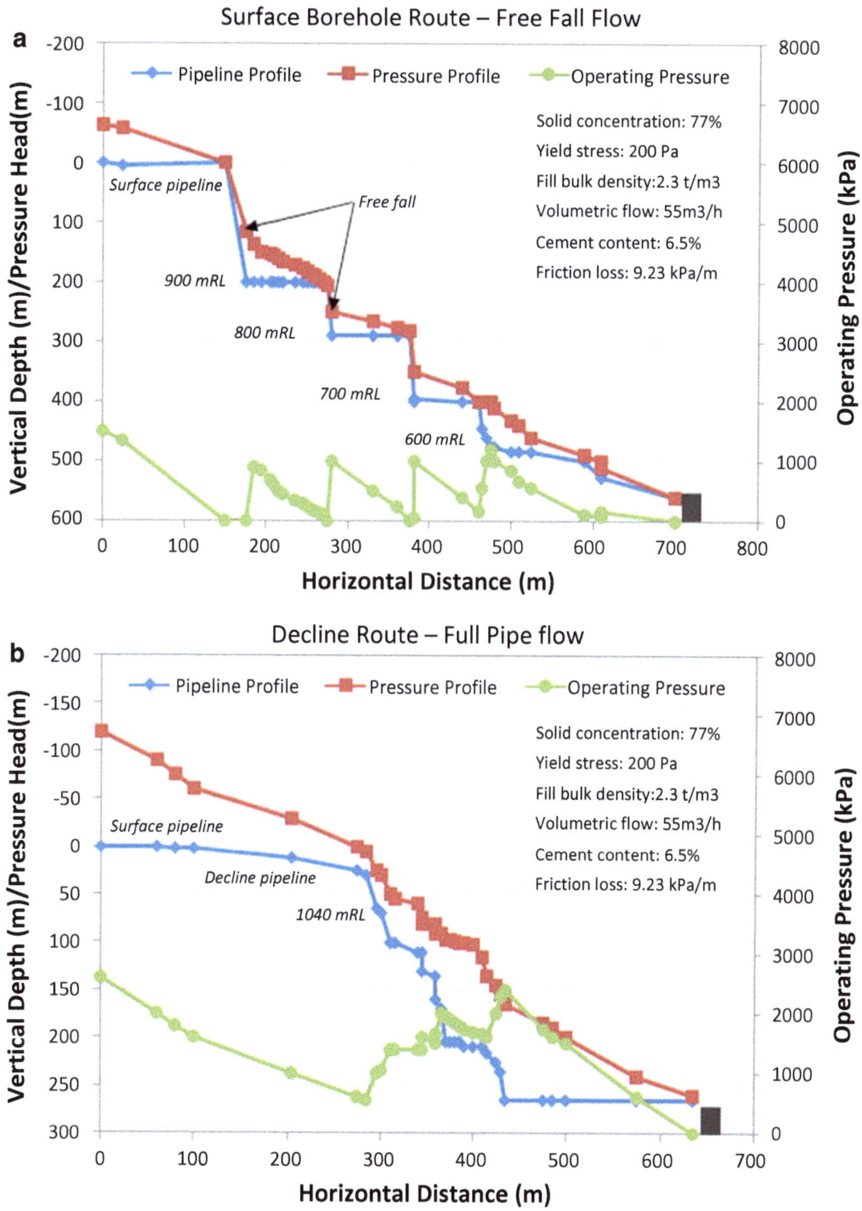

Fig. 10.18 Paste flow models: (**a**) surface borehole route at 20 m^3/h; and (**b**) decline route at 20 m^3/h

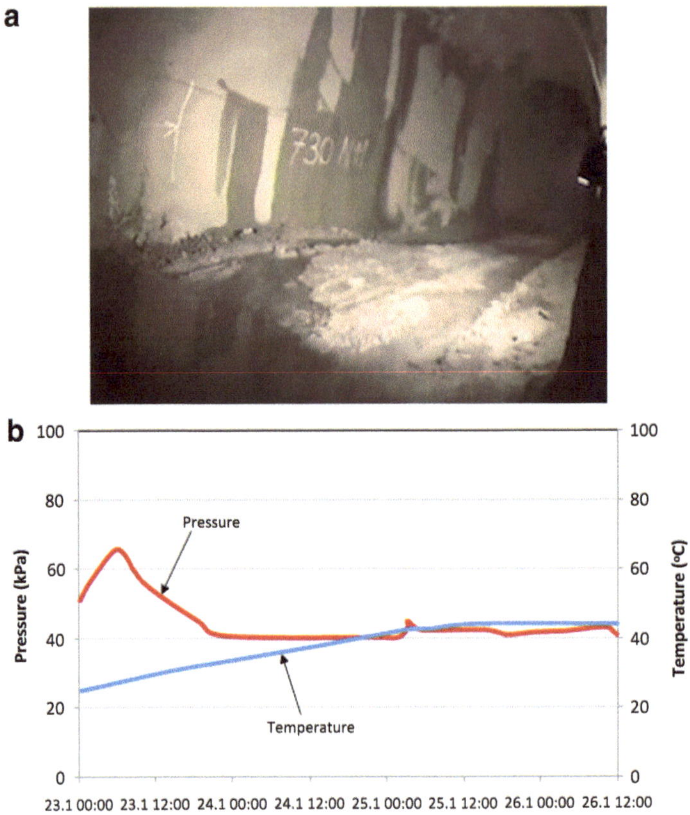

Fig. 10.19 Online barricade monitoring: (**a**) CCTV view; (**b**) pressure monitoring

cement content, filling rate, and fill placement regime. The barricade pressure monitoring practice is to maintain pressures below the barricade capacity. Cayeli has nominally set the barricade capacity at maximum 75 kPa, through combination of measures including controlling fill rise rate, allowing sufficient plug cure time (minimum 3 days) and real-time CCTV monitoring (Fig. 10.19).

6 Fill Management System

Paste backfill operations are site specific and complex which involve interdependent activities that require interaction between mine planners, geotechnical engineers, underground service crew and supervisors, plant operators and supervisors, and maintenance personnel, including fill material suppliers, contractors, and consultants.

Developing a fill management plan that overarches all disciplines and covers all of the major site-specific operational and technical features is vital in establishing a common knowledge and maintaining best practice operation. It is imperative that all key stakeholders have a good common understanding of the key aspects of the entire paste backfill operations from the production stage on surface through to the final exposure of backfill mass in open stopes underground. This plan provides the background information and the details of operating and control strategy for the paste backfill plant and underground distribution and placement. The plan will be used by operators, supervisors, engineers, and maintenance personnel.

6.1 *Paste Backfill Costs*

For a given recipe, the CPB costs can be 12–16% of total mining costs at Cayeli mine. The current cement cost alone is about 10% of total mining cost (assuming current average cement addition rate—equivalent to 130 kg/m^3 and mining cost of $28 per ton of ore, and bulk cement cost of $70 per ton cement). Depending on the recipe, cement costs can constitute up to 55–65% of the total backfilling costs. It is thus essential to precisely calculate the target design strengths and optimize backfill recipes to encounter the target design strengths. Table 10.7 shows preliminary paste backfill operating costs.

6.2 *Underground Void Management*

The backfill reports show a database of all stopes that have been mined and backfilled with various backfill types. The database tracks weekly open stope volumes as well as the type and amount of backfill placed. Figure 10.20 shows the paste backfill production over the last 6 years. The records indicate that the yearly paste fill production varies from 1200 to 25,000 m^3 with an average yearly production rate of 20,000 m^3. The recent total annual backfill has averaged 342,000 m^3 sufficient for production rate of 1.35 Mtpa. Paste fill constitutes 242,000 m^3 with remaining 100,000 m^3 for waste fill. A cumulative net void of approximately 19,000 m^3 is very well managed in the mine by implementing paste backfill technology. Records also indicate that the average stope size has decreased from 8000 tons in 2008 to approximately 5000 tons in 2011. An average stope is completed in 30 days including drilling, blasting, mucking, and backfilling. The average stope volume of 2500 m^3 is equivalent to 10 m wide and 15 m long mined over a 15 m sublevel (floor to floor) and to meet the production target 150 stopes were mined annually.

Figure 10.21 shows the amounts of the cement used for paste backfill production over the last 6 years. The records indicate that the yearly cement consumption varies

Table 10.7 Paste backfill operating cost estimation

Paste operating costs	Optimized[a]	Corrected[b]	Current[c]	Current recipe		
Cement content (wt%)	5.0	6.5	7.7	Ore/tails density	3.8	tons/m^3
Cement rate (kg/m^3)	91	118	140	Wet bulk density	2.3	tons/m^3
Cement cost ($/$m^3$)	6.36	8.27	9.80	Dry bulk density	1.8	tons/m^3
Paste production ($/ m^3)	3.00	3.00	3.00	Paste pulp density	78	wt%
–Energy						
–Labor				Original recipe		
–Materials						
–Maintenance				Ore/tails density	4.4	tons/m^3
Reticulation system ($/$m^3$)	1.00	1.00	1.00	Wet bulk density	2.7	tons/m^3
–Boreholes/pipelines				Dry bulk density	2.2	tons/m^3
–Labor				Paste pulp density	80	wt%
–Materials						
Paste backfill barricades ($/$m^3$)	1.00	1.00	1.00	Cost parameters		
–Materials						
–Equipments				Cement cost	70	$/tons
–Labor				Mining cost	26	$/tons
–Shotcrete						
Total cost ($/$m^3$)	11.36	13.27	14.80			
Cement costs (%)	56	62	66			
Paste cost per ton of ore ($)	3.25	3.79	4.23			
Paste costs as a % of mining cost	12	15	16			

[a]Cement dosing based on the wet density of 2.7 tons/m^3 and solids density of 80%
[b]Cement dosing based on current recipe back-calculated using actual average cement consumption in 2013
[c]Cement dosing optimized based on AMC target design strengths and improved barricade capacity

from 30,964 tons to 36,535 tons with an average yearly production rate of 33,000 ton. The recent total annual cement consumption has averaged 33,328 tons which corresponds to a total paste backfill amount of 250,480 m^3. The average cement content has also been kept as 6.6 wt%.

6.3 *Risks and Opportunities*

The operating with a single borehole is a major operational risk. Most backfill operations often operate with two main boreholes with one as spare. Reticulation redesigns and upgrade and operating strategy should be carried out as a priority.

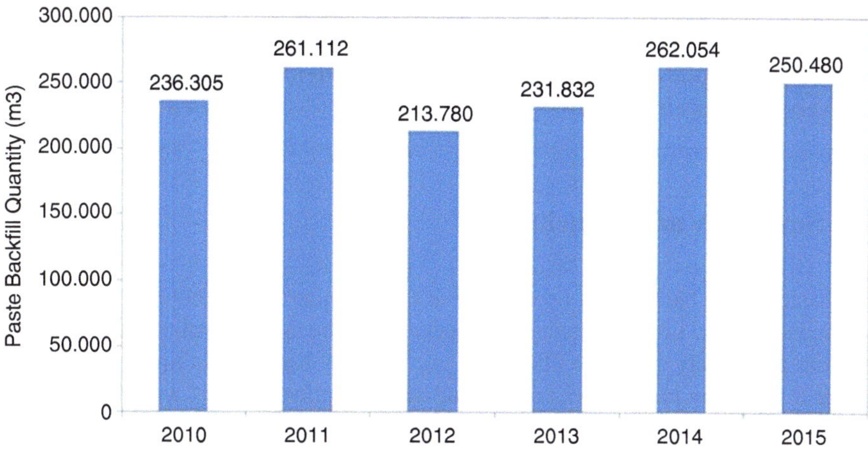

Fig. 10.20 Annual actual paste backfill production for last 6 years

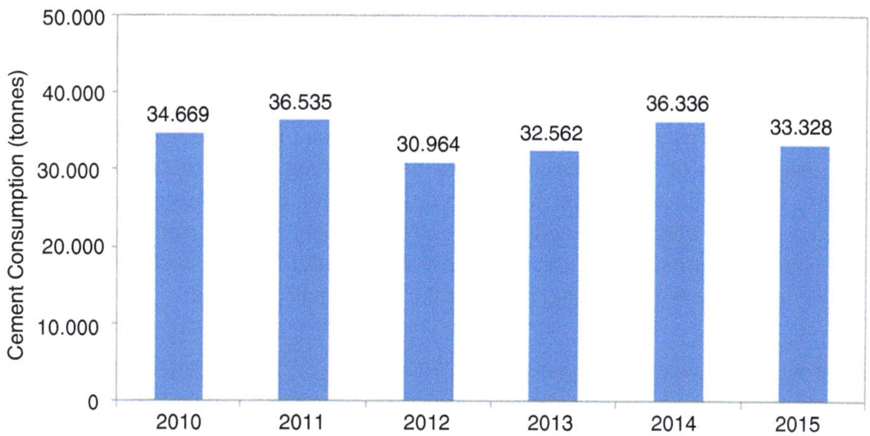

Fig. 10.21 Annual actual cement consumption volumes for last 6 years

Recent decreasing trend in the strength results as contained in the QA/QC is a concern and should be investigated. Paste backfill failures and significant dilution when undercut are economic and safety concerns and should be further evaluated. Lack of a full-time fill management plan is an operational risk. A detailed system audit should be carried out every 2 years in order to make sure that the system is safe, efficient, and up to date in terms of the application of industry best practice. A root-cause analysis and technical and operational review of all major incidents should be undertaken and logged in a central database. Paste plant operation shift records indicate that leakage of pipes at joints and wear at pipe bends and pipe bursts are not uncommon. Cayeli redesigns and upgrades the backbone reticulation system and develops an engineered flow model for fill delivery and placement.

7 Summary and Conclusions

Cemented paste backfill has been in use at Cayeli mine since 1999 and is a critical component of the underground mining operations. The mining method and sequence require all primary stopes be filled with good-quality backfill in a timely manner in order to maintain local stope stability and regional mine stability, and to maximize ore extraction. The use of the backfill at Cayeli mine serves four main functions: (1) to stabilize ground in order to permit the extraction of adjacent ore, (2) to fill voids to provide regional support, (3) as a working floor for the next lift of stopes, and (4) for disposal of development waste. The Cayeli mine uses a number of cemented fill methods, including cemented aggregate fill, cemented waste fill, and cemented paste backfill as the dominant filling method. Cayeli has successfully integrated backfilling into the mining cycle. Since start-up, more than 3.5 million m^3 of cemented paste backfill has been successfully placed into the mined-out underground stopes. Cemented stope pillars have proven to be stable and allow 100% extraction of secondary stopes. Paste filling operations are complex and require a vigilant approach and cooperation of many stakeholders to balance many competing aspects of the system for a safe and efficient filling operation. Cayeli has well-established procedures, a good-quality assurance/quality control program, efficient plant maintenance, and ongoing continuous improvement program in order to better assess methods of improving the performance and reducing the cost of paste backfill.

References

Askew JE, McCarthy PL, Fitzerald DJ (1978) Backfill research for pillar extraction at ZC/NBHC. In: The 12th Canadian Rock Mechanics Symposium, Montreal, Canada, p 100–110

Benzaazoua M, Belem T, Bussiere B (2002) Chemical factors that influence the performance of mine sulphidic paste backfill. Cem Concr Res 32(7):1133–1144

Fall M, Benzaazoua M (2005) Modeling the effect of sulphate on strength development of paste backfill and binder mixture optimization. Cem Concr Res 35(2):301–314

Hassani FP, Archibald JF (1998) Mine backfill handbook. Canadian Institute of Mining, Metallurgy and Petroleum, Montreal, Quebec, Canada

Landriault DA (2001) Backfill in underground mining. In: Underground mining methods engineering fundamentals and international case studies, Society of Mining Engineers, Littleton, Colorado, USA, Chapter 69, p 601–614

Mitchell RJ, Olsen RS, Smith JD (1982) Model studies on cemented tailings used in mine backfill. Can Geotech J 19(1):14–28

Potvin Y, Thomas E, Fourie AB (2005) Handbook on mine fill. Australian Centre for Geomechanics, Western Australia, Perth, Australia

Yilmaz (2015) Geotechnical characterization of cemented paste backfill. LAP LAMBERT Academic Publishing, Saarbrucken, Deutschland/Germany, ISBN: 978-3-659-60841-4

Yumlu M, Guresci M (2007) Paste backfill bulkhead monitoring: a case study from Inmet's Cayeli Mine, Turkey. In: The 9th Int. symposium on mining with backfill, Montreal, Quebec, p 1–11

Chapter 11
Boulby Mine Backfill System

Barend Jacobus Snyman

1 Introduction

1.1 Preamble

Paterson & Cooke's involvement with the Boulby mine backfill system began with the initial investigation studies in the mid-1990s and continued through detailed design, construction, to commissioning in 2003 with further ad hoc post-commissioning assistance. A number of papers were published at international conferences during this period and this chapter brings together these past publications (Fehrsen et al. 2002; Wilkins et al. 2004; Keen et al. 2007) and provides a holistic overview of the project from the initial study phase to full-scale operations.

1.2 Background

Cleveland Potash Limited (CPL) produces approximately 55% of the United Kingdom's potash (potassium chloride) supply, along with halite (rock salt) as a co-product at their Boulby mine. Boulby mine is located in the North York Moors National Park (Fig. 11.1) and comprises a deep underground mine and surface process plant (Boulby 2015). Shaft sinking operations commenced in 1969, with the first shipment of potash product taking place in 1973 (Fehrsen et al. 2002).

B.J. Snyman (✉)
Paterson & Cooke, Sunrise Park, Sunrise Circle, Ndabeni, 7405 Cape Town, South Africa
e-mail: JacoS@PatersonCooke.com

E. Yilmaz, M. Fall (eds.), *Paste Tailings Management*,
DOI 10.1007/978-3-319-39682-8_11

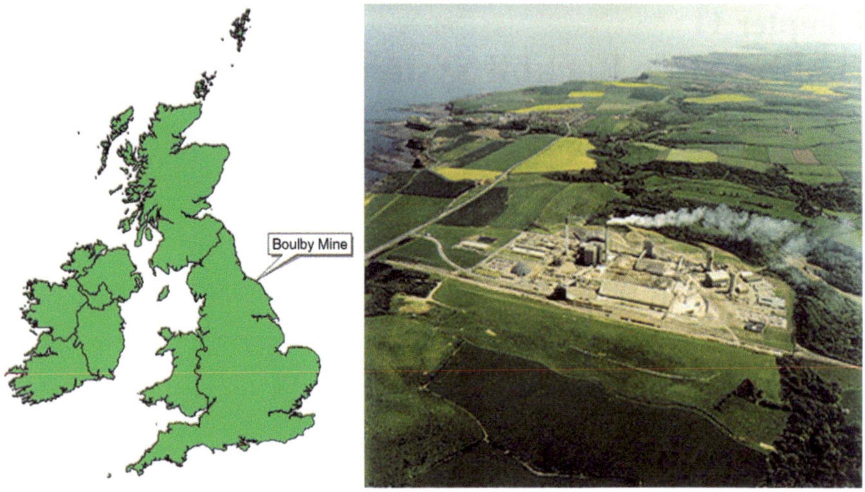

Fig. 11.1 Location map

1.3 Mining

The Boulby potash occurs at depths between 1200 and 1500 m in a seam up to 20 m thick with an average of 7 m. The two shafts (man and rock hoisting) are of 5.5 m diameter and 1150 m deep. In 2001 CPL converted the continuous miners (Fig. 11.2) to a remote-controlled operation for safety to cope with the greater stresses in the ore below 1200 m and to raise output above 3 Mtpa.

The mine uses a two-road-with-stubs stress-relieving room and pillar technique which achieves more mining per shift. The continuous miners discharge to electric shuttle cars, which run to feeder breakers on the main conveyors to the hoisting shaft (1).

Of the 3.0 Mt ore extracted at Boulby mine annually, approximately 1.0 Mt of saleable potash is produced using a conventional floatation process. Two tailings streams are generated as a by-product of the potash processing:

- 1.8 Mtpa (190 tph) centrifuge cake consisting of coarse (+1 mm) salt particles (soluble waste)
- 0.2 Mtpa (30 tph) filter cake comprising fine (<50 μm) montmorillonite clay (insoluble waste), salt and calcium sulphate

During the previous century all process waste was re-pulped with seawater and discharged into the North Sea. Due to the presence of trace quantities of heavy metals (mercury and cadmium) in the insoluble clay, the permitted amount of insoluble waste that CPL can discharge into the sea has been substantially reduced.

Assisted by EU funding, CPL developed an underground backfill system for disposing filter cake mixed with seawater as backfill in the worked-out areas of the mine. The Boulby backfill system is thus unique in that the backfill placed is not

Fig. 11.2 Continuous miners at Boulby mine

used for the usual purposes of support, pillar extraction or reducing ventilation requirements and was specifically implemented for environmental reasons. The backfilling operation started in May 2003 and has operated as required to maintain the discharge into the North Sea to below permitted levels.

2 Backfill Properties and Testing

2.1 Backfill Slurry Properties

When the initial study to investigate using backfill began in 1996, the addition of a binder to the backfill was investigated to allow the backfill to be used for support. The results of the tests showed that high binder addition rates were required and it was decided not to add binder to the backfill due to the cost of the binder, logistical problems of transporting the binder (if added underground) and concerns that the backfill would be mechanically stiffer than the surrounding rock mass leading to possibly localised ground failures.

The advantages of no cement addition to the backfill meant that flushing or pigging was not required following a pour to prevent scaling.

2.2 On-Site Flow Characterisation Tests

On-site flow characterisation tests were initiated as part of the investigation studies with the following objectives (Fehrsen et al. 2002):

- Demonstrate the feasibility of a gravity-fed backfill distribution system.
- Identify the most suitable composition and optimum density of the backfill by analysis of the flow behaviour of the range of slurries considered by CPL.
- Obtain sufficient backfill flow behaviour data to characterise the slurry for the purpose of designing the underground backfill distribution system.
- Investigate whether flushing of the system was required after a system shutdown.

The slurry properties determined during the on-site flow behaviour tests are presented in Table 11.1 and the rheological correlations are graphically presented as Figs. 11.3 and 11.4.

Table 11.1 Slurry backfill properties

Property	Value
Filter cake solids density	2525 kg/m^3 (typical)
Seawater density	1026 kg/m^3
Slurry density	1495–1585 kg/m^3
Slurry concentration by volume	31.3–37.3%

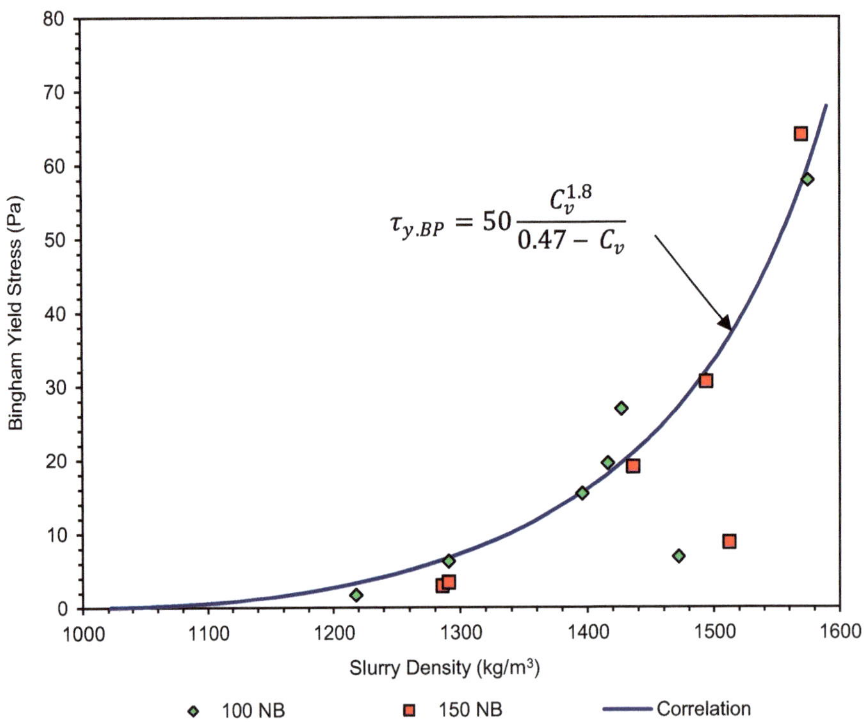

Fig. 11.3 Bingham yield stress vs. slurry density

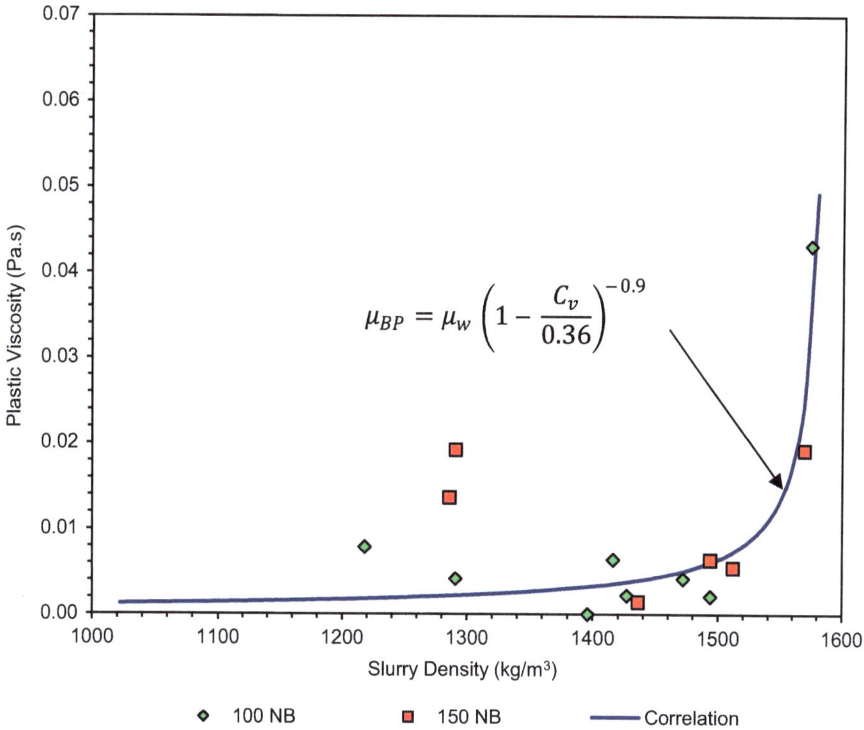

Fig. 11.4 Plastic viscosity vs. slurry density

3 Backfill System Requirements

The function of the backfill system is to re-pulp 200,000 tonnes of filter cake per year with seawater and hydraulically place it underground at as high a solids concentration as possible. The basic requirements for the backfill system are detailed in Table 11.2.

4 Backfill Preparation Plant

4.1 Process Description

The basic backfill preparation plant process flow sheet is illustrated in Fig. 11.5. Filter cake is fed into a batch mixing tank where it is blended with seawater in a high-shear environment to produce the backfill slurry. The ratio of water to filter cake is determined by the required backfill density. After the addition of the last filter cake, the flow is re-circulated through a quality control loop to set the slurry viscosity.

Table 11.2 Backfill system requirements

Property	Value
Backfill material	Filter cake
Carrying fluid	Seawater
Backfill density	Maximum practicable (1400–1600 kg/m^3)
Operational days	330 day/year
Placement	3 times per day
Placement distance	5500–11,040 m
Gravity head available	908–1100 m

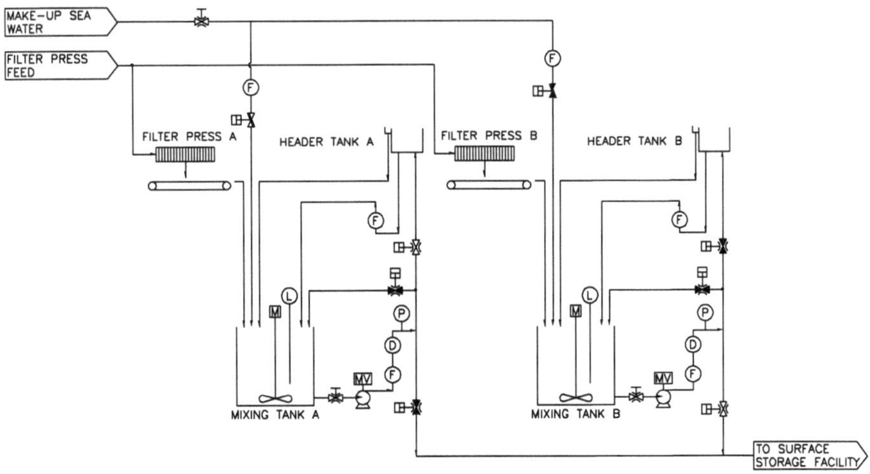

Fig. 11.5 Basic backfill preparation plant flow sheet

The quality control loop maximises the backfill density in order to maintain the design flow rate. This is achieved by changing the viscosity to balance the available head and friction losses for each panel to be backfilled. When the required backfill quality is achieved (Fig. 11.6), the batch is transferred to the surface storage tank which provides a surge capacity between the backfill preparation plant and the distribution system.

5 Backfill Distribution System

5.1 Design Philosophy

The backfill distribution pipelines were selected (200 mm shaft column and 150 mm haulage piping) to ensure that the system operates under gravity without the need for a surface pump station. The haulage piping is designed to operate in turbulent flow to avoid the possible accumulation of particles on the pipe invert because of laminar flow settling; however laminar flow operation is permitted for the shaft column.

Fig. 11.6 Discharge from quality control loop

The system is designed to operate as a full-flow system, i.e. the pipeline friction losses match the available gravity head and the shaft column is fully pressurised. To maintain the design flow rates and turbulent flow in the haulage pipes, the allowable backfill slurry viscosity reduces as the discharge distance from the shaft increases.

The underground roadways at Boulby are in the salt layer and thus it was desirable to avoid flushing the pipeline between pours to minimise the water sent underground. The system is designed to operate without pre- and post-flushing and remains full of slurry unless shut down for an extended period. To achieve this, a valve station is provided at the bottom of the shaft column. This ensures that when the system was shut down, the shaft column remained full, eliminating free-fall conditions on start-up in the large-bore vertical shaft column.

To minimise pressure transients, the flow rate through the valve station is controlled using two banks of energy dissipaters operating in parallel. Figure 11.7 shows the energy dissipater units installed at the valve station. The large dissipater is rated as 200 m slurry head loss at a flow rate of 115 m^3/h and the smaller two dissipaters together are rated for 1100 m at 15 m^3/h.

5.2 *Process Description*

The basic underground distribution system flow sheet is presented in Fig. 11.8. Backfill slurry is stored in the surface storage tank until it is required underground. During placement, the centrifugal charge pump transfers backfill slurry to the shaft column from where the gravity head is used to feed it to the designated placement panel.

The valve station at the bottom of the shaft column, shown in Fig. 11.9, isolates the column when the feed to the column is stopped.

A low-flow high-pressure positive displacement pump is provided to flush the backfill piping in the event of a viscous plug forming that stops the flow. A shaft column drain valve and associated piping are provided to drain the column or to dump the shaft column contents during flushing.

Fig. 11.7 Energy dissipater during installation

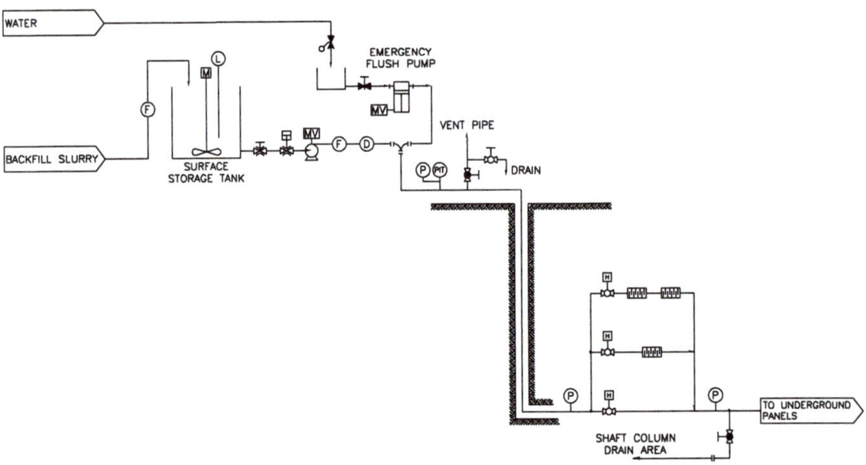

Fig. 11.8 Basic underground distribution system flow sheet

To ensure that the maximum allowable pressures are not exceeded, pressure profiles for slurry delivery to each panel were calculated and the slurry viscosity (and hence backfill density) determined to deliver the maximum backfill density without exceeding the allowable pipeline stresses.

The distribution system is designed to deliver backfill up to 11,040 m from the shaft at a rate of between 200 and 250 m^3/h, depending on which panel backfill is being placed into.

Fig. 11.9 Underground valve station

6 Underground Placement

6.1 Placement Panels

The principal considerations in selecting mine areas suitable for backfilling are the following:

- The placement panels need to be dipping generally away from the point of access in order to form a natural sump.
- The panel should be configured such that whilst undergoing gradual closure the placed backfill would not be squeezed into any other area where future access would be required.
- The physical conditions within the panel must be suitable for re-entry in order to enable installation of the placement pipe work.
- The placement panel conditions should be such that re-entry and observation of the placed fill material would be possible long after filling operations had ceased.

The placed backfill was not expected to have any structural strength even years after placement. Test work by CPL also showed that no significant bleed water was expected. Water within the backfill was expected to be trapped by the high fines content of the backfill and no drying of the backfill is possible due to lack of ventilation in the placement panels. Normally cement added to backfill with high fines content allows for hydration and the consolidation of the backfill. In this case cement would not be added.

Calculations by CPL's rock mechanics department showed that on completion of mining, the pillars between adjacent panels undergo consolidation and lateral deformation. As a result the cross-sectional area of the panel will reduce by 36% within 4 years of the termination of mining activities with a further 26% reduction in the original cross section occurring over the subsequent decade.

Thus the selection of placement panels and the allowed placement volumes had to be determined carefully to prevent the backfill from being "squeezed out" of the placement panels and sterilising future reserves.

7 System Operation and Performance

7.1 Commissioning Experience

Figure 11.10 shows backfill discharging into a placement panel during the commissioning in May 2003.

Apart from the usual start-up and commissioning challenges (fine-tuning cycle times and instrument settings, leaking joints, intermittent communication faults, etc.) no major issues were experienced during the commissioning of the backfill system. The backfill system has proved simple to operate and the system flow rates have been consistent indicating the reliability of the viscosity control.

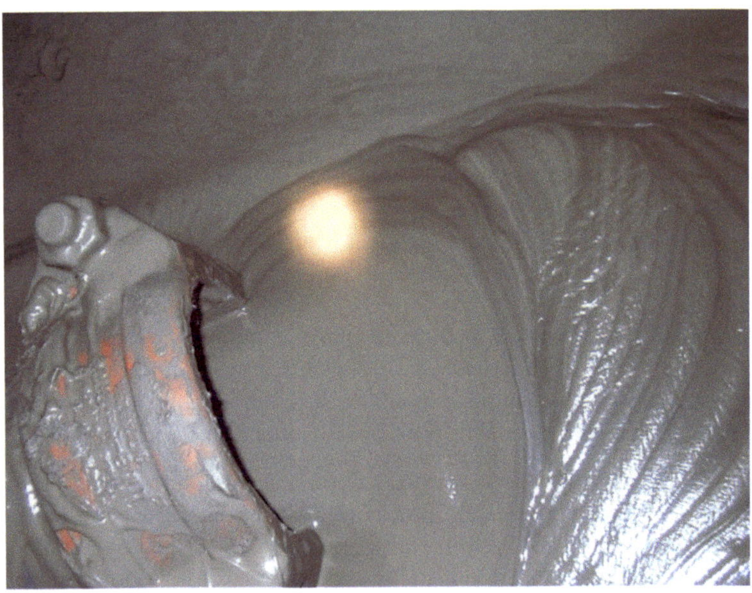

Fig. 11.10 Backfill slurry discharge

7.2 System Utilization

Following the successful commissioning of the backfill plant (Wilkins et al. 2004), the actual backfill tonnage placed underground has been significantly less than the system was designed for as the percentage of insolubles in the mined ore has been less than expected. This has reduced the total effluent discharge into the sea to below permitted levels. This can be seen in Fig. 11.11, where the actual tonnes of effluent decreased as the permitted discharge decreased. Thus the mine has reduced the backfill sent underground due to the limited suitable placement areas underground and is keeping the limited capacity.

7.3 Achieving System Design Criteria

7.3.1 Quality Control Loop Operation

The quality control loop has proved to be the key to being able to control the rheology of each backfill batch and ensuring that the piping system operates in full flow without stalling. During commissioning the rheology of the backfill was slightly reduced to allow the backfill to flow better in the placement panels and ensure that the panels were correctly filled. Thus flow rates achieved in the system are higher than originally designed for.

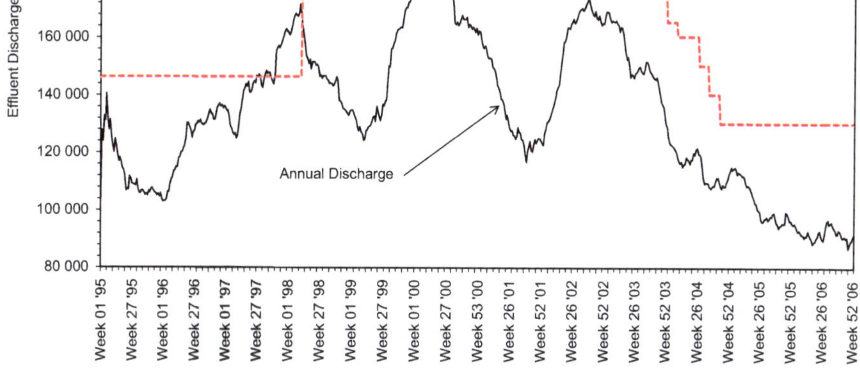

Fig. 11.11 Rolling average effluent discharge values

7.3.2 Shaft Column

It has not proved necessary to flush the shaft column at any time as no significant settling has occurred. During periods when the mine did not need to place backfill, the shaft column was left shut for up to 3 or 4 weeks without any settling problems.

Air entrainment did prove to be problematic however. During 2006, air was being entrained towards the end of the pour into the charge pump on the surface. After shutdown of the system this air was migrating to the top of the shaft column. It was estimated that the air pocket at the top of the shaft column was nearly 70 m. This meant that the system first needed to be recharged before it could be restarted. The air column expanded and contracted significantly with changing temperatures during the day and night. This caused some concern as the drop in pressure was first noticed on the trends which were thought to indicate a leaking valve at the shaft bottom, but fortunately this was not the case.

7.3.3 Valve Station Operation

The successful implementation of the valve station for starting up and shutting down the system while keeping the shaft column full has allowed for the reliable use of a large-bore shaft column over such a long vertical drop. High-speed pressure traces during commissioning confirmed that there are no significant pressure surges in the shaft column on start-up and shutdown of the system. Further traces during the operation of the system showed that pressure build-up in the downstream piping during start-up is slow, confirming that the valve station is continuing to control the system pressures within safe limits.

During commissioning the actuator sleeve on the isolation valve on the second bypass failed (Wilkins et al. 2004). In July 2005 this valve failed following recharging of the shaft column after a flushing scenario. The valve was not closing properly and the high pressure across the valve caused excessive wear of the ball and seat. It is suspected that possibly the positioning of the ball during the repair of the actuator sleeve was affected and eventually resulted in the ball jamming in the partially open position. Figure 11.12 shows the damage caused to the seat.

7.3.4 Filling of Placement Panel

One of the key criteria for panel selection was that gradual closure would not cause the backfill to be squeezed out into any area where future access might be required. During the initial studies calculations were prepared to estimate the closure rates and a monitoring programme was implemented to determine actual closure rates.

Figure 11.13 shows a plan of the placement panel with the location of instrumentation for measuring convergence. Measurements between 2002 and 2007 showed that the vertical average closure was 12.7%, although nearest to the water level (and therefore to the fill) the average was only 6.8%.

Fig. 11.12 Damage to seat of the second bypass valve

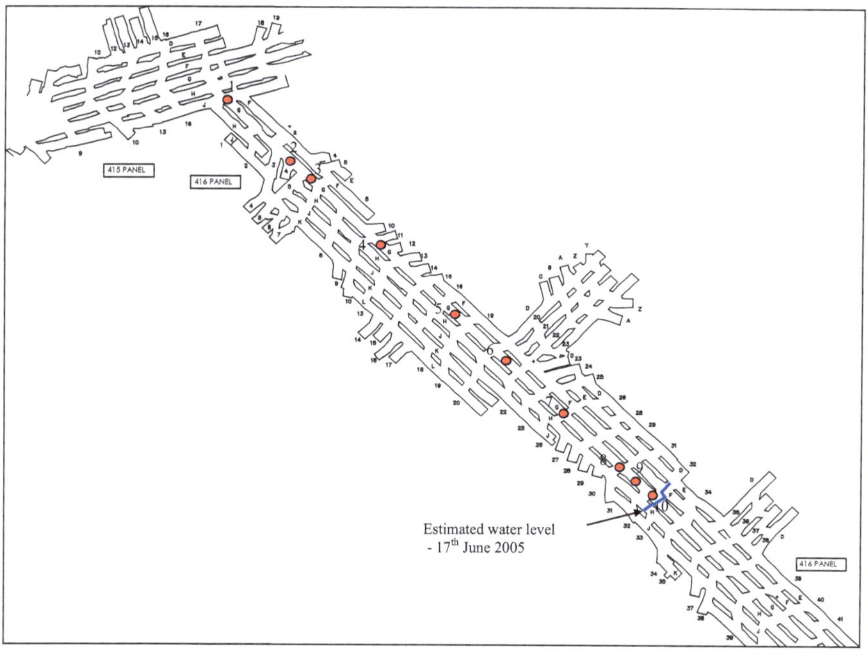

Fig. 11.13 Plan of panel 416 showing instrument positions for measuring convergence

Fig. 11.14 Backfill in panel 416

By making some assumptions on the horizontal convergence, and extrapolating the measurements indicated above, calculations yielded a 25.4% loss of cross-sectional area nearest the fill which was in line with the predicted figure. The access to the panel was becoming limited though as the closure in that area was higher than the average.

There has been more water in the panel than was anticipated. However, considering the slightly lower placement density than originally designed for, more bleed water should be expected. In addition most of the bleed water will run back off the backfill accumulating at the edge, creating the impression of more bleed water. There was also some additional water due to the flushing of a viscous blockage when the shaft had been left for an extended shut beyond 4 weeks, as well as some refilling of the column with water to compensate for the air pockets forming due to the air entrainment.

Figure 11.14 shows the backfill that was placed in panel 416 and shows that the slope of the placed backfill was very flat.

7.4 *Other Operational Experience*

7.4.1 Mixing Tanks

There were many reliability problems with the previous system and the new mixing tanks with their high-shear mixers have proved reliable and efficient at breaking down the clay lumps discharged from the filter cake.

Fig. 11.15 Build-up in backfill tank following draining for inspection

7.4.2 Settling in Storage Tank

During the commissioning it was noted that the agitator in the storage tank did not adequately mix the backfill in the storage tank. This has resulted in gradual build-up as shown in Fig. 11.15 in the tank over the last few years of operation. It has not affected operation significantly, except that the minimum tank level needed to be raised slightly to prevent air entrainment in to the system.

7.4.3 Effect of Filter Upgrade on Quality Control Loop

The plant has upgraded the filter press cloths since commissioning, resulting in a suspected change in the particle size grading of the backfill. Following the change in the filter press cloth, it was found that more water needed to be added to achieve the same backfill rheology.

8 Conclusion

The successful commissioning of the backfill system in May 2003 has provided the potash industry with a proven alternative method of tailings disposal to the conventional options of disposal at sea or placement in a conventional surface tailings depository.

The operation of the Boulby mine backfill system has shown that the design features implemented to achieve the design criteria have been successful. Backfilling for environmental reasons is not common, but it does introduce certain constraints as well as certain design freedoms that allow for some unusual design features.

Acknowledgements The author acknowledges the contributions of the various authors of the previously published papers that are listed at the end of the chapter.

References

Boulby, United Kingdom. http://www.mining-technology.com/projects/boulby. Accessed 14 April 2015

Fehrsen MG, Goosen PE, Wilkins MJ ((2002) On-site backfill slurry characterisation tests at Boulby Mine: a case study. In: Proceeding 15th Int. Conf. on hydrotransport. BHR Group, Banff, pp 161–171

Keen M, Fehrsen M, Cooke R (2007) Boulby mine backfill system: operational experience, Minefill 2007. Montreal Quebec, Canada, May 2007

Wilkins MJ, Gilchrist C, Fehrsen M, Cooke R (2004) Boulby mine backfill system: design, commissioning and operation. Proceedings 16th international conference on hydrotransport. BHR Group, Chile. Also presented at Minefill 2004, Beijing, China, September 2004

Chapter 12
Case Studies on Paste Backfill Plants

Isaac Ahmed

1 Introduction

The inherent characteristics of the mill tailings and backfill strength requirements largely dictate the overall design development of the paste backfill plant (PBP). The former can be understood with a paste-centric laboratory test programme, and the latter is defined by the mining method and stope geometry. As each tailings is unique, the settling behaviour, filterability and ultimately strength properties will govern the equipment selection, sizing and process design. The two case studies presented in this chapter demonstrate the distinct approaches to PBP design based on these key drivers.

1.1 Case Study 1: Barrick Goldstrike Mines Inc. (BGMI)

Commissioned in the first quarter of 2013, the facility at BGMI represents the first PBP in the state of Nevada, USA. The PBP features a cyanide destruction circuit, a cyclone station, high-rate thickener, flocculant system, two vacuum disc filters, continuous mixer, binder system (individual fly ash and normal Portland cement or NPC silos) and a positive-displacement pump. Paste backfill can also be delivered underground via gravity. The PBP has a production capacity of 2700 tonnes/day (dry tailings), with an instantaneous pour rate of 80 m^3/h of wet paste at a 178 mm slump. Considering the average demand from the underground, the PBP has a utilisation rate of 55%.

I. Ahmed (✉)
Paste Engineering & Design, Golder Associates Ltd., Sudbury, ON, Canada
e-mail: iahmed@golder.com

E. Yilmaz, M. Fall (eds.), *Paste Tailings Management*,
DOI 10.1007/978-3-319-39682-8_12

The construction of the PBP took place over a period of about 12 months. The site excavation works, drilling and casing of the boreholes were completed prior to this period. Subsequently, the commissioning and bringing the PBP into operation took another 3 months.

1.2 *Case Study 2: AuRico Gold Inc., Young-Davidson Mine (AuRico)*

With a dry tailings processing capacity of 8000 tonnes/day, this PBP is one of the largest in the world. It consists of a high-rate thickener, flocculant system, two vacuum disc filters, two continuous mixers and a binder system (four pre-blended slag cement silos). Commissioned in the first quarter of 2014, the construction of the plant spanned over 9–10 months in total, plus about 2 months of commissioning.

The PBP has the ability to produce two different paste backfill mix designs simultaneously and transport the paste backfill underground via two operating gravity boreholes. The target paste backfill production rate is 375 tonnes/h (dry tailings) or 230 m^3/h (wet paste backfill at 178 mm slump). Based on the average demand from the underground mine, 230 m^3/h of wet paste at 70% utilisation is required for backfilling.

2 Case Study 1: BGMI Paste Backfill Plant

2.1 *Process Overview*

A portion of the whole-mill tailings from the roaster facility is diverted to the cyanide destruct (CN-D) tank. Prior to placing paste backfill underground the cyanide concentration of the tailings stream had to be reduced. This was a stipulated requirement from the Nevada Division of Environmental Protection (NDEP). BGMI carried out a diffusion test programme to demonstrate that major ions are effectively sequestered in the highly alkaline backfill matrix (Barrick 2013). To decrease the cyanide concentration from 20 ppm to below the allowable 8–10 ppm limit, the slurry is sparged with air and dosed with ammonium bisulphite (NH_4SO_3) solution. Copper sulphate ($CuSO_4$) solution was added as a catalyst for the oxidation reaction. The method is similar to that described by Devuyst et al. (1989) and is a technology that the site was already using to treat the tailings prior to discharge to the tailings storage facility (TSF). After the CN-D step, dilution water is added prior to feeding the cyclones, as the cyclones operate most effectively at a certain feed solids content. The cyclones are designed to remove the ultra-fines portion from the tailings. The cyclone underflow at 48–53 wt% solids reports to a 12 m diameter high-rate thickener, commencing the tailings dewatering sequence. By removing the ultra-fines, the settling rate, filtration rate and backfill strength are all enhanced. This implies that the thickener, disc filter and binder system size can be reduced. Further, the consumption of binder can be optimised to achieve a given target strength. Landriault and Primeau (2013) discussed the merits of the using cyclones in paste backfill systems. In one case, the de-slimed tailings resulted in increasing the filtration rate by approximately 40%. At the same time, to reach the target of 28-day UCS, a 1 wt% decrease in NPC consumption was observed when de-slimed tailings were used.

Thickened tailings at about 60 wt% solids, from the thickener underflow, are pumped to an agitated filter feed tank, before it is fed to the disc filters to produce a high-solids-density filter cake. The filter cake is combined with thickened tailings, water and binder in a continuous mixer, to a target consistency. With a dual-side discharge-style high-intensity mixer, the resulting paste can be diverted to one of the two paste hoppers. One paste hopper discharges backfill underground via a positive displacement, while the other allows the paste to flow by gravity to the underground. Both boreholes are cased with 150 mm diameter carbon steel schedule 80 piping.

2.2 Cyclone System

The cyclone will separate the tailings based on size fraction, with the cyclone underflow (coarse fraction at 48–53 wt% solids) reporting to the thickener feed box and the cyclone overflow (fine fraction at 17–22 wt% solids) reporting to the existing roaster CN-Dtank. A target PSD with 20–25% passing 20 μm and 15–20% passing 10 μm is desired for the cyclone underflow. The cyclone feed contains about 50–60% passing the 20 μm sieve.

The cyclone cluster consists of eight individual cyclones (Fig. 12.1). Spare cyclones and feed ports are provided to allow the site the ability to adjust to changes in the tailings characteristics. Individual cyclones can be closed and opened to maintain a consistent cyclone feed pressure in order to attain the target underflow PSD. Maintaining the design cyclone feed pressure and feed solids content is integral to the effective operation of the cyclones. These key parameters dictate how well the cyclones operate to cut the tailings feed.

2.3 Thickener System

The cycloned underflow tailings are received via a gravity pipeline at the thickener feed box. Diluted flocculant at 0.025 wt% solids is introduced to the thickener feed well to facilitate the settling process at a dosing rate of 20 g of flocculant per tonne tailings. The dewatered tailings are discharged as thickener underflow and are pumped to the agitated filter feed tank via a centrifugal pump. Overflow from the thickener reports to the wastewater tank.

Fig. 12.1 Cyclone cluster located above the paste plant cyanide destruct tank. Tailings slurry is fed through the centre of the cyclone cluster to a distribution pot. The cyclone overflow returns to the existing cyanide destruct tank at the mill, while the cyclone underflow reports to the thickener. The elevation of the cyclone cluster allows for gravity flow of both the overflow and underflow to their respective locations

The speed of the underflow pump will be controlled by the output of the flow and density meters to maintain the target solids density discharge from the thickener and into the filter feed tank. The underflow pump is equipped with a variable speed drive (VSD). Instrumentation is used to support the monitoring of mud bed pressure as an auxiliary means to predict thickened tailings solids content. A bed profile instrument is also used to monitor the mud bed and supernatant interface.

Thickener torque is measured and the rake mechanism is protected by a series of safety mechanisms in order to prevent damage (in addition to the operator alarms). At high torque, the thickener rakes will automatically lift while under rotation until the high-torque condition is removed. At cutout torque, the rakes will automatically stop. The operator may manually lower the rakes (while under rotation) to the normal operating elevation once the conditions causing high torque or cutout torque have been removed. If the thickener needs to be placed on hold, the recirculation loop will be engaged to allow the underflow to be pumped back to the thickener feed box to avoid compaction in the thickener. At the start-up of the process, the thickener underflow is recirculated in the thickener to build up the mud bed.

2.4 Vacuum Disc Filter System

The 60 wt% solids thickened tailings in the filter feed tank are pumped via filter feed pumps to two disc filters. Each disc filter is comprised of ten operating discs that are 3.2 m in diameter. Dedicated filter feed pumps deliver the thickened tailings to individual disc filters to better control the flow requirements and disc filter bypass slurry addition. At discharge, the filter cake is deposited onto a common filter cake conveyor (1050 mm belt).

The thickened tailings enter the eductors at the bottom of the filters (Fig. 12.2). The residual pressure and increased slurry velocity introduced to the disc filter vat serve to agitate the slurry and prevent settling to augment the filtration of the tailings. The filter cake is removed from the disc filter sectors when snap air is applied, expelling the dewatered material. Each vacuum disc filter is coupled with a snap air receiver containing a reservoir of air. This snap air is released according to the disc filter rotational position, blowing the cake off the filter cloth. The duration of the snap air blow is controlled by an adjustable timer.

A barometric leg is employed that equilibrates the applied vacuum with the filtrate water column. As a closed system, the filtrate pipes from the filtrate receiver are connected to a filtrate pot located at the ground floor of the PBP.

Fig. 12.2 Thickened slurry feed to eductor distribution across the length of the disc filter vat. Overflow and drain lines from the disc filter also shown in larger diameter HDPE piping

Fig. 12.3 Illustration of the binder silo bottom cone where the rotary valve discharges to the binder weigh conveyor; picture taken during construction

Taking advantage of the PBP elevation and the overall height of the PBP, the use of a barometric leg eliminates the need for filtrate pumps that are typically installed on filtrate receivers.

2.5 Binder System

The binder system consists of two binder storage silos: one that contains NPC, and the other, fly ash. The storage silos are of equal volume and have a 225-tonne and 140-tonne capacity for NPC and fly ash, respectively. The contents are metered onto individual binder weigh conveyors through a rotary valve. From the weigh conveyors, the NPC and fly ash are deposited onto a single screw conveyor and finally into the continuous mixer. The rotary valve and weigh conveyor work in unison, by speeding up or slowing down to maintain a consistent binder bed depth on the belt (Fig. 12.3). This allows the binder weigh conveyor to operate within the optimal range. The target mass flow rate of NPC required is calculated based on the ratio of dry tailings to NPC. This is derived from the mass flow rate of dry tailings solids calculated from the filter cake weigh conveyor and measured filter cake moisture. The rotary valve and binder weigh conveyor speeds are automated to maintain the target mass flow rate of NPC and fly ash. At Goldstrike, the binder system is designed to deliver NPC and fly ash to the continuous mixer, producing a paste backfill that could contain 1.5–8 wt% NPC and the equivalent fly ash. On average the paste backfill mix is expected to contain 3.5 wt% NPC and 3.5 wt% fly ash.

2.6 Mixing and Paste Backfill Delivery

Filter cake, dilution slurry, dilution water, fly ash and NPC are all introduced into the high-intensity twin-shaft continuous mixer (Fig. 12.4). The working volume of the continuous mixer is approximately 5.35 m^3. As illustrated, the blended fly ash and NPC mixture is introduced through the top of the conveyor discharge chute. Dilution water and slurry are introduced from the back of the conveyor discharge chute near the inlet of the continuous mixer to intercept the filter cake and the binder. The paste backfill consistency or slump is correlated with the mixer power draw. Dilution water is used as a means to fine-tune the paste mixture to achieve the target slump. As a dual-side discharge mixer, paste is received by either a dedicated gravity hopper on one side or a dedicated pump hopper on the other side (Fig. 12.5). Depending on the location of the stope, paste delivery underground may be via gravity or a positive-displacement pump. On the gravity line a vacuum-rated pinch valve is used to modulate the flow of paste backfill to the underground distribution system (UDS).

At BGMI, an admixture dosing system is installed. Diluted admixture can be introduced to the mixer to improve the pumpability and UCS of the paste backfill at

Fig. 12.4 Continuous mixer view from the discharge end illustrating the binder screw conveyor feed, dual-side discharge chutes and planetary drive system of the twin-shaft mixer

Fig. 12.5 Illustration on the *right* depicts the paste hopper and its connection to the piston pump. Illustration on the *left* shows the gravity (foreground) and pumped inclined boreholes. Note that the paste hoppers are outfitted with freshwater line (blue piping) for clean-up

an otherwise reduced water-to-NPC ratio. This allows the operation to optimise binder use for a certain stope location. Weatherwax et al. (2010) describe a series of trials conducted at Barrick Gold Corporation's David Bell and William's mines in Northern Ontario. The studies demonstrated the viability of using such admixtures in an operating facility, improving the rheological characteristics of the paste backfill and the UCS.

Freshwater is used to flush the paste hoppers, piston pump and the UDS. Freshwater is used instead of process water to minimise the quantity of cyanide that could report to the underground. Also, drawing from the freshwater tank, a high-pressure low-volume, triplex-piston diaphragm pump can also be used to flush the underground distribution system in an emergency situation.

3 Case Study 2: AuRico Paste Backfill Plant

3.1 Process Overview

With the PBP located remotely from the mill, tailings are pumped from the mill tailings pump box to the PBP thickener feed box. When paste backfill is not required the tailings is directed to the PBP tailings discharge pump box. From here the tailings slurry is transported via multistage centrifugal pumps to the tailings storage facility (TSF). At AuRico, the PBP is situated between the mill and the tailings TSF and hence the PBP acts as the intermediary for tailings disposal.

Entering the 22 m diameter thickener at about 45 wt% solids, the slurry is thickened to approximately 70 wt% solids. The thickener underflow is then delivered to the filter feed tank. The filter feed tank serves to provide both surge capacity and recirculation loading capacity from the disc filters. From the filter feed tank, the thickened slurry is delivered via dedicated pumps to the two disc filters. Water is further removed to produce a high-solids-density filter cake, around 85 wt% solids. The disc filter discharges filter cake to its dedicated filter cake conveyor, which feeds a continuous mixer. In essence, there are two equipment trains, with each train consisting of a disc filter, filter cake conveyor, continuous mixer and binder system.

The filter cake is combined with water and binder in a continuous mixer, to produce paste backfill of a target consistency or slump, with a particular binder content. With dual-side discharges, each continuous mixer may deliver paste backfill to one of the two paste hoppers. Each paste hopper is in turn connected to gravity boreholes. The resulting paste backfill flows by gravity to the underground via two 200 mm diameter schedule 80 carbon steel-cased boreholes.

3.2 Thickener System

Tailings are pumped from the mill to the 22 m diameter tailings thickener at the PBP (Fig. 12.6). Diluted flocculant is introduced to the thickener feedwell. Flocculant addition is proportional to the solids feed rate which is derived from the mill tailings flowrate and solids density. The targeted flocculant addition rate is 25 g of flocculant per tonne tailings. Reclaim water addition, if necessary, is added to maintain a target feed concentration of 35–45 wt% solids at the thickener feed well. The location of the thickener feed box allows for gravity flow to the thickener feed well. The thickener feedwell has an auto-dilution system, where overflow (or supernatant) is added back to maintain the desired feed density. The resulting discharge from the thickener feed well is intended to be about 15 wt% solids. This further dispersion of solid tailings particles in the feed well augments the tailings agglomeration process to improve the settling characteristics. The thickened tailings exit as thickener underflow at about 70 wt% solids and are pumped to the filter feed tank via a centrifugal underflow pump. Overflow from the thickener reports to the tailings discharge pump box.

When the PBP is shut down, the mill tailings feed to the thickener is diverted directly to the tailings discharge pump box. From here, the tailings are delivered to the TSF.

3.3 Flocculant System

Solid flocculant in bulk bags is unloaded at the flocculant system. The bag is cut, and the solid flocculant is dispensed into a hopper. A screw feeder meters the solid flocculant into a wetted cone where it is mixed with water and fed into the mix tank.

Fig. 12.6 22 m diameter high-rate thickener—bolted tank construction. This thickener receives tailings from the mill and dewaters the slurry to about 70 wt% solids

The flocculant is diluted in the mix tank to a concentration of 0.50 wt%. The flocculant is transferred from the mix tank to the storage tank. The storage tank allows sufficient time for mixed flocculant to be aged for optimal effectiveness. The mixed flocculant is fed to the thickener feed well by dosing pumps. An in-line mixer allows the mixed flocculant to be diluted prior to delivery to the thickener feedwell to a concentration of 0.10 wt%. The flocculant flow rate to the thickener feedwell is controlled such that flocculant addition is maintained at a consistent concentration of 25 g of flocculant per tonne tailings.

3.4 Disc Filter System

The 70 wt% solids thickened tailings in the filter feed tank are pumped via filter feed pumps to two disc filters. Each disc filter carries 15 discs that are 3.8 m in diameter (Fig. 12.7). The filters are fed simultaneously by separate pumps to accommodate the flow requirements. At discharge, the filter cake is deposited onto the filter cake conveyors. The filter feed pump discharge flow rate is controlled by the filter feed pressure controller to maintain minimum operating pressure, allowing for slurry agitation in the disc filter vat. Thickened tailings from the filter feed tank are pumped to the disc filters in excess capacity such that there is a constant overflow. The overflow is brought back to the filter feed tank. The disc filter is an eductor style and operates without a paddle agitator. Automated valving is in place to allow either filter feed pump to feed either disc filter for added operational flexibility.

Sections of the rotating discs which are subjected to vacuum are submerged into the slurry at the bottom of the filter. Solids are attached to the surface of the rotating discs due to the vacuum suction. The disc filter rotational speed is manually set, governed by the cake thickness and filtration rate. The filter cake is removed from the discs when snap air is applied to the cake-laden sectors, expelling the dewatered material. A snap air receiver contains a reservoir of air that is released according to the disc filter rotational position. The filter cake is scraped off the discs as it enters the cake discharge chute. The duration of the snap air blow is controlled by an adjustable timer.

A barometric leg is employed to deliver the filtrate from the filtrate receivers to a common filtrate pot. Fluid from the filtrate pot is pumped to the tailings discharge pump box. The filtrate pot is comprised of two compartments separated by an overflow-style interior baffle. The filtrate discharge pipes enter one side of the filtrate pot and remain fully submerged, creating a closed system with the filtrate receiver. As the filtrate is removed from the disc filter and into the filtrate pot, it overflows to the pump side of filtrate pot.

Under normal operating conditions, both continuous mixers will be discharging simultaneously into two separate boreholes. Should the production of paste backfill and hence consumption of tailings be reduced to a level less than that supplied by the mill, the thickener will continue to operate at full capacity. The excess thickened tailings will overflow from the filter feed tank to the tailings pump box. From there, the excess tailings will be delivered to the TSF. Dilution water from the reclaim water tank can be added to the tailings pump box as necessary to reduce the solids content and enable the tailings discharge pumps to operate seamlessly.

Fig. 12.7 A 3.8 m diameter × 15 disc filter in operation producing filter cake. This is one of the two disc filters in operation. Each disc filter is also fitted with an automatic wash system for cloth clean-up

3.5 Binder System

There are two complete stand-alone binder systems at the PBP: one system for each mixer. There are a total of four binder storage silos for the twin systems, with each silo capable of holding 150 tonnes of pre-blended slag cement (90:10, slag:NPC) and a storage volume of 140 m^3. Each binder system will consist of two binder storage silos. The binder discharge flow rate from each silo will be controlled through a rotary valve. Individual binder weigh conveyors provide an accurate means of measuring the binder weight. From the weigh conveyors, binder from two silos will be deposited into a single screw conveyor, where it will be discharged into the continuous mixer.

The target binder mass flow rate is calculated based on the required ratio of slag cement and the mass flow rate of dry tailings solids. The dry tailings solids are comprised of the filter cake solids plus the tailings solids from the slurry bypass line. The filter cake moisture content data is obtained by periodically analysing the grab samples. The operating speed of the rotary valve and binder weigh conveyor is controlled to maintain the target mass flow rate of binder. The average binder content in the paste backfill is anticipated to be 3.25 wt% and the binder system is designed to deliver from 1 wt% to about 7 wt%.

3.6 Continuous Mixing System

There are two parallel continuous mixing systems in the PBP (Fig. 12.8). Each continuous mixer has a dry mix volume capacity of 10.5 m^3 and an operating capacity of 8.1 m^3. This translates to approximately 4 min of mixing residence time. Tailings filter cake, binder, slurry and dilution water are added to the continuous mixer and brought to the target mix design, including the slump or consistency of the paste. This consistency is correlated to the target strengths for paste backfill. A relationship between the mixer power draw and slump is determined at the commissioning stage and examined periodically during plant operations to determine the validity of this relationship. Subsequently, the mixed tailings are discharged to paste hoppers. Delivery to the underground is by gravity fill. If necessary, provisions for a positive-displacement pump have been made and can be installed in the future to pump paste to locations that are not serviceable by gravity.

Dilution water will be added into the mixer for final slump adjustment. If the actual density is too high, the power draw will be above the targeted set point and more dilution water will be added to adjust the density to its desired consistency, as indicated by the actual power draw. The AuRico PBP is configured to be able to deliver two different types of paste backfill mix designs to the underground simultaneously. This allows the operation the ability of pouring paste backfill to two different locations with different target strengths or slumps. Further, with each continuous mixer being dual-side discharge each mixer can deliver paste backfill to either hopper

Fig. 12.8 An overview of the PBP layout, illustrating the two separate equipment trains: disc filter, conveyor and mixer. The binder screw conveyor can be seen penetrating the building wall and discharging at the conveyor head chute

and hence either borehole. This added level of flexibility also allows the operation to service one train of equipment while running the PBP at half capacity. Paste backfill delivery to a particular borehole can be readily switched over from one continuous mixer to the other. As the mixers must therefore be placed discharge end to discharge end, and incorporating clearance requirements for maintenance, this means that the paste hoppers had to be much larger than typical installations. The paste hoppers are needed to span between both mixer discharge chutes. This presented another challenge on how to monitor the paste hopper level as level instruments cannot scan across such a large surface area. As a result, load cells were used to measure the weight of the paste hopper contents to infer a level.

3.7 Paste Pipeline Flushing

The paste distribution pipeline will require flushing when a production cycle is complete, or when a plant shutdown is required. A dedicated borehole flush pump is used to deliver reclaim water to the paste pipeline in combination with flush air. In the event of a power failure and should the flush water pumps not be operational, a water line, connecting the reclaim water tank to the borehole, can be used to direct flush water to the underground.

3.8 Underground Distribution System

At approximately 40° from vertical, the 250 m long boreholes intersect the 10,130 level underground. The boreholes are located inside the PBP and considerations have been made to allow a raise bore to be set up inside the building to ream out the boreholes if a plug occurs. Inter-level boreholes are established to connect the upper levels of the mine to lower levels (Fig. 12.9). 200 mm diameter carbon steel schedule 80 piping is used for the borehole casing. These are grouted through the entire length of the borehole.

Newly excavated or slashed areas from the main drift are used to accommodate inter-level borehole penetrations and switchover downholes from mainline piping. These switchover downhole areas connect the mainline piping to the lower levels of the mine, effectively allowing the paste backfill a means to be transported to different levels from one area. These switchover locations as well as any connections that divert the paste backfill along a level for example are manually connected via spool pieces of piping and elbows.

There are three different types of horizontal piping used to optimise the cost of the UDS:

1. Main-line piping—200 mm diameter carbon steel schedule 80 piping. This is used to transport paste backfill from one level to another and connected via inter-level boreholes. Main-line piping will see the most use throughout the life of mine.
2. Level piping—200 mm diameter carbon steel schedule 40 piping. This is used to transport paste backfill along the drift at the particular level.

Fig. 12.9 Typical borehole penetration from upper levels of the mine showing low-pressure uphole support arrangement

3. Stope piping—200 mm diameter high-density polyethylene (HDPE) SR9 piping. Stope piping penetrates the barricade and is used to connect the stope to be filled with the level piping run.

The UDS must be flexible as it needs to be configured for each pour. Having a thorough understanding of the friction loss for different mix designs gives the operation a means to predict the extent of paste backfill distribution. This provides further guidance for the PBP to produce a paste backfill that would satisfy the target strength at particular stope location.

4 Tailings Characteristics

As each tailings exhibit unique dewatering characteristics, it is important to understand the conditions by which this occurs. Further, it has been demonstrated that tailings can react with various types of binder with strength gain results that can be significantly different. A laboratory test programme that examines these characteristics and its rheological properties is an important first step. The results of the laboratory test programme are used as input into the PBP design. In many cases, the laboratory test programme will change according to the intermediate results uncovered. For example, if the UCS results are poor and the paste backfill does not achieve the target strengths, it would normally be worthwhile to try to remove some of the ultra-fines (−20 μm fraction), add aggregate, use a different binder or implement some combination of these. Each of these actions will have an impact on the process design. This section presents some of the results from the BGMI and AuRico tailings test work.

4.1 Mineralogy and Specific Gravity

The major, moderate and minor mineralogical components as well as the specific gravity of the tailings are presented in Table 12.1. This information can often be used to identify any potential challenges with dewatering or interferences with different binders that could result in low strength gain. Conversely, in some cases, high strength gain with certain binders can be explained.

Table 12.1 Mineralogy and specific gravity of tailings samples

		BGMI	AuRico
Mineralogy	Major	Quartz	Orthoclase
	Moderate	Calcite, dolomite	Albite, quartz
	Minor	Gypsum, mica, pyrite	Calcite, dolomite
Specific gravity		2.82	2.71

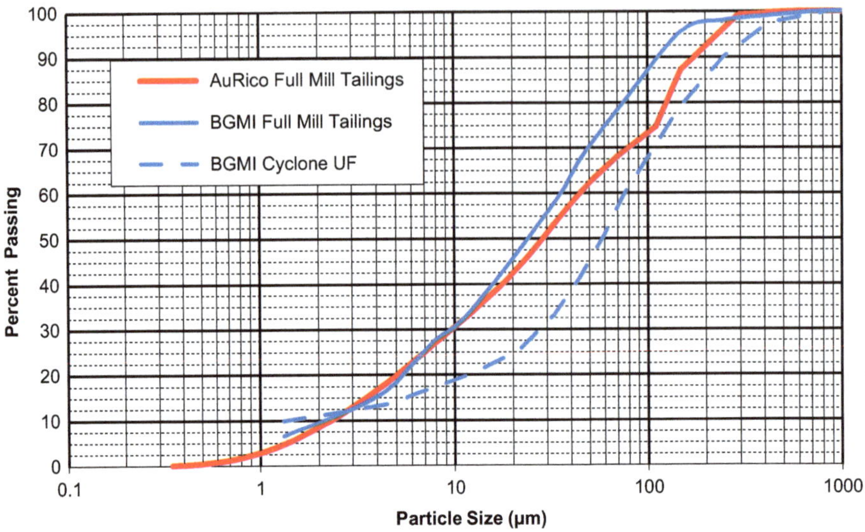

Fig. 12.10 Indicative particle size distribution of the BGMI (full mill and cyclone underflow) and AuRico (full mill) tailings

4.2 Particle Size Distribution

The PSD of the full-mill tailings from BGMI and AuRico is presented in Fig. 12.10. Additionally, the PSD of the cyclone underflow for BGMI, the feed to the PBP, is also illustrated. The PSD of the material provides an initial assessment of the potential for a tailings to be dewatered. Typically, the finer the material, the more difficult it is to thicken and filter. In general, finer tailings also have a tendency to produce a weaker strength paste that would require more binder to achieve the equivalent strength. As depicted, by cycloning the tailings, a coarser tailings fraction is produced. It may even be conjectured that certain deleterious minerals could be selectively separated via this process to further improve the dewatering and strength-making properties of the tailings.

4.3 Settling Flux Rate

The PBP process often begins with the need to dewater the tailings, with thickeners employed as the initial step. Both the BGMI and the AuRico PBPs use thickeners and their respective settling flux rates and underflow densities are presented in Table 12.2. As observed there is a marginal improvement in the settling flux rate of the cyclone UF tailings compared to the full-mill tailings, but a marked increase in underflow density. The sizing of the thickener is directly proportional to the settling flux rate (presented as m^2/tonne/day).

Table 12.2 Settling flux rate for BGMI (full mill and cyclone underflow) and AuRico (full mill) tailings

	BGMI		AuRico
Parameter	Full mill	Cyclone UF	Full mill
Settling flux rate (m^2/tonne/day)	0.034	0.031	0.039
Predicted underflow density (wt% solids)	60	64	70

Table 12.3 Cake loading rate for BGMI (full mill and cyclone underflow) and AuRico (full mill) tailings

	BGMI		AuRico
Parameter	Full mill	Cyclone UF	Full mill
Cake loading rate (tonne/m^2/h)	0.56 (2003)	1.22 (2003) 0.46 (design)	0.65
Filter feed solids content (wt% solids)	56 (2003)	58 (2003) 60 (design)	69
Filter cake solids content (wt% solids)	77 (2003)	79 (2003) 77 (design)	85

4.4 Filter Cake Loading Rate

After the thickening step, the thickened tailings are further dewatered. In the case of BGMI and AuRico, disc filters are employed. The cake loading rate (in tonne/m^2/h) is proportional to the sizing of the disc filters, with a larger cake loading rate requiring a smaller surface area for filtration. For the BGMI tailings two sets of results are shown, one with test work completed in 2003 and the current design values (Table 12.3). For the 2003 samples, the full-mill tailings contained 40% passing the 20 μm sieve while the cyclone UF tailings had 25% passing the 20 μm sieve. When comparing the 2003 BGMI cyclone underflow to that of the BGMI full-mill tailings, it can be seen that the cake loading rate more than doubled. It can be seen that the removal of the ultra-fines has a significant impact on the performance of the vacuum filtration step. However, the cake loading rate is also a function of the feed solids content and the slightly higher value for the BGMI cyclone UF positively influences the overall cake loading rate. As illustrated in Fig. 12.11 with the AuRico tailings, the higher feed solids content yields a higher cake loading rate. Often the sizing of the vacuum disc filters is performed with cake loading rates that are representative of feed solids content below predicted underflow density to allow for some level of conservatism.

The design values for the BGMI tailings were used for the development of the PBP and represent the mining and milling conditions at the time of the design.

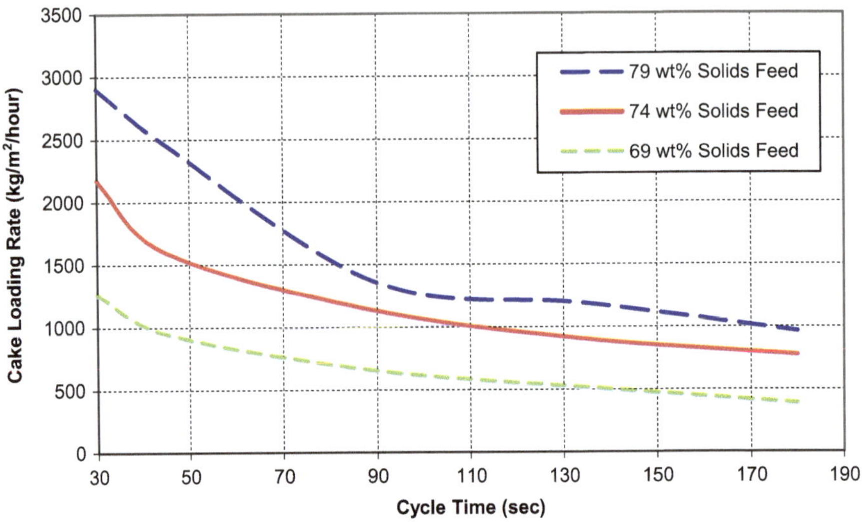

Fig. 12.11 Cake loading rate as a function of cycle time and the effect of feed solids content

4.5 Paste Backfill Unconfined Compressive Strength

For a typical paste backfill, it is well understood that for a given tailings matrix, the UCS of a paste backfill is a function of the water content, binder content and binder type. As indicated earlier in the chapter, the removal of the ultra-fines portion of the tailings can improve the paste backfill UCS. The test results from Fig. 12.12 were derived from earlier stages of the laboratory programme and not necessarily indicative of the current operating conditions. However, the results do serve to illustrate that the cycloned tailings can improve the UCS under similar conditions.

The AuRico tailings performed well with all three binder types tested but are most compatible with the 90:10 slag cement (Fig. 12.13). Both the 50:50 fly ash/NPC and the 90:10 slag cement showed slower UCS gain on the day 3 break. With subsequent breaks it became clear that the 90:10 slag cement outperformed the other binders and with the 14- and 28-day break, the UCS results were more than double that of the 50:50 fly ash/NPC and NPC breaks. Additional stages of testing were completed and demonstrated that the paste backfill with 90:10 slag cement continued to gain up to 40% UCS from the 28- to 56-day break. The UCS was maintained up to the 112-day break, the last set of cylinders cast.

5 Summary

The laboratory results and key design aspects of the BGMI and AuRico PBPs are depicted in this chapter. It can be seen that a laboratory test programme that focuses on the ability of the material to make paste backfill is key and central to the design of PBPs.

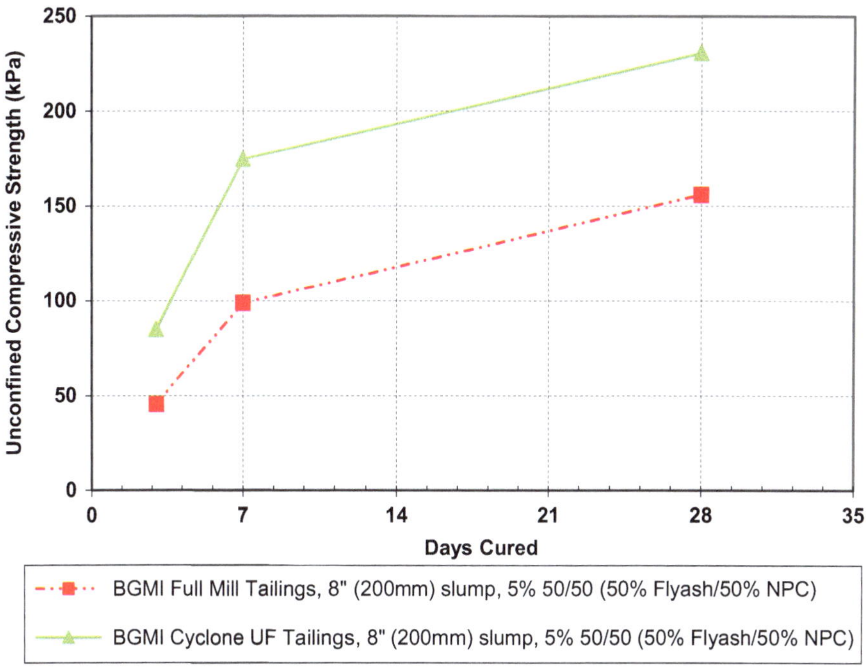

Fig. 12.12 UCS comparison of BGMI full-mill tailings versus cyclone underflow tailings

5.1 BGMI Paste Backfill Plant

With the BGMI PBP, it was clear from earlier testing that dates back to 2003 that the removal of ultra-fines was integral to the UCS development. The UCS test programme focused on binders that were readily available in the area and hence slag cement was not trialled. Due to the need to remove the ultra-fines a cycloning stage had to be added to the overall proÎss design. This improved both the settling flux and the cake loading rates for thickening and vacuum filtration operations, respectively. Although a minor decrease in the settling rate flux rate was observed, there was a marked increase in the predicted underflow density. The small decrease in settling flux rate resulted in a smaller thickener size, which impacts not only the capital cost of the thickener, but also its foundation requirements and the associated costs. Similarly, it is observed that the cake loading rate improves with the removal of the ultra-fines. As the equipment layout of a PBP is heavily dependent on the number and size of disc filters employed, a higher cake loading rate will result in lower vacuum filtration needs. The vacuum pumps represent a noticeable portion of the overall PBP power draw and a reduction in the motor load has a positive effect on the operating costs. While cycloning has been successfully used to achieve higher UCS, it should be borne in mind that the operation must also treat the cyclone overflow. Containing the ultra-fines, this material is more difficult to settle. In the case of BGMI, only a portion of the tailings was used for paste backfill and the cyclone

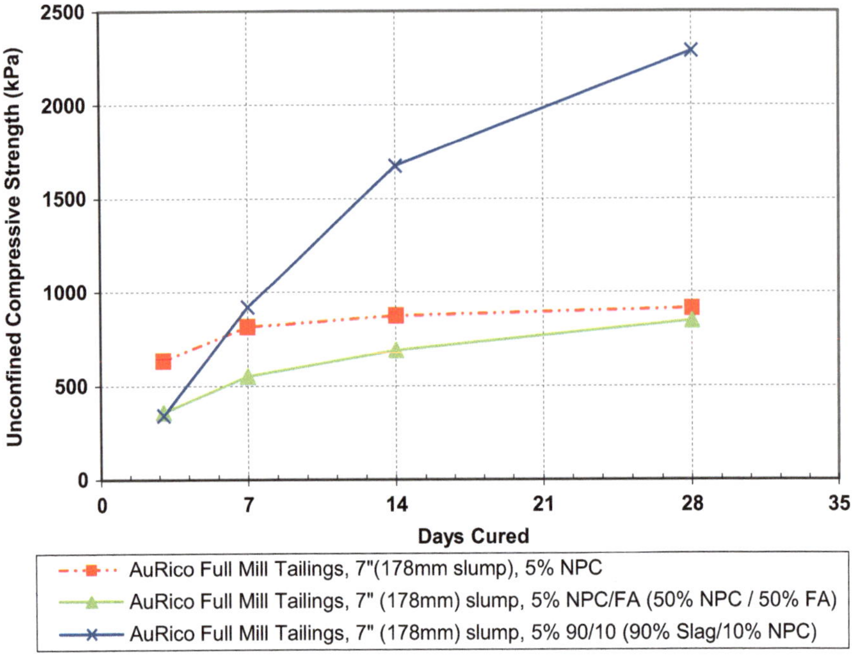

Fig. 12.13 UCS comparison of AuRico full-mill tailings with different binder types

overflow was returned to the tailings disposal system. In other cases, the cyclone overflow may need to report to a clarifier to recover the water and concentrate the fines for disposal.

As the first PBP operating in the state of Nevada, the BGMI PBP carries some important design features that enable it to be constructed and operated in a cost-effective manner. Through extensive testing both on the past backfill side and on the environmental side, with diffusion testing to demonstrate the effectiveness of paste backfill to sequester major ions, the design of the plant was shaped.

5.2 *AuRico Paste Backfill Plant*

The AuRico PBP entails some unique design features. The inherent characteristics of the tailings and its propensity to readily dewater, filter and gain UCS allowed for the construction and operation of a streamlined plant. The full-mill tailings without difficulty dewater to 70 wt% solids. The operators do not necessarily run the thickener to this underflow density; however, it does demonstrate that the tailings consolidate exceptionally well. Consequently, the high thickener underflow density (or filter feed solids content) augments the cake loading rate. These favourable tailings characteristics allowed for a relatively compact, two-vacuum disc filter (albeit 3.8 m diameter) plant design. The tailings reacted well to different types of binder, but most advantageous with the 90:10 slag cement. As a result the average binder consumption is expected to be about 3.25 wt%. With binder costs typically in the range of 50–70% of the overall PBP operating costs, binder consumption is a key consideration in the viability of a plant design.

A continuous PBP design is most economical from a capital perspective. And this is especially pronounced with a high-throughput PBP. A batch plant is more cumbersome requiring more operating equipment, such as mixers and weigh hoppers, but has better product quality control. Slump control is an important process parameter as it is linked to the paste backfill water-to-NPC ratio and hence the UCS. With the AuRico paste backfill, the relationship between solids content and slump is approximately 1.5 wt% solids from 7″ (178 mm) to 10″ (254 mm) slump; that is, at 7″ slump (178 mm), the solids content is 81 wt% and at 10″ slump (254 mm), the solids content is 79.5 wt%. Slump control is therefore relatively tight, with small amounts of water addition resulting in noticeable changes in slump. Nevertheless, with high-intensity mixing and sufficient continuous mixer residence time, the operation is able to control the slump to within ¼″ (6.4 mm). The standard method for slump measurement is accurate to the same ¼″ (6.4 mm) tolerance.

Owing to the high throughput, even during the early stages of the design process, two active 200 mm nominal diameter boreholes were envisioned. The objective was to maintain paste velocity at 1–1.5 m/s and this was possible with the use of two boreholes. The use of larger boreholes and pipelines was considered; however, this would have meant heavier piping sections to manage and possibly special order high-pressure couplings for certain runs in the UDS. As discussed, with the use of two boreholes the PBP has the added flexibility of producing two different mix designs simultaneously. This level of flexibility offsets the additional piping costs.

Acknowledgements The author wishes to thank both BGMI and AuRico for permission to reference their respective paste backfill plants and related data for this chapter. Also, the author would like to extend his gratitude to Ibrahim Karajeh for reviewing this chapter.

References

Barrick Goldstrike Mines Inc., (2013) Cyanide code recertification—summary audit report

Devuyst EA, Conard BR, Vergunst R, Tandi B (1989) Cyanide removal process using sulfur dioxide and air. J Metals 41(12):43–45

Landriault J, Primeau PA (2013) Application of cyclone technology in past backfill plant design. In: Jewell R, Fourie A (eds) Proceedings of the 16th international seminar on paste and thickened tailings, Belo Horizonte, Brazil, pp 17–20

Weatherwax T, Brosko W, Evans R, Champa J (2010) Role of admixtures in the optimisation of paste backfill systems. In: Jewell R, Fourie A (eds) Proceedings of the 13th international seminar on paste and thickened tailings, Toronto, Canada, pp 3–6